Null-Additive Set Functions

Mathematics and Its Applications

Volume 337

Null-Additive Set Functions

by

Endre Pap
Institute of Mathematics,
University of Novi Sad,
Novi Sad, Yugoslavia

ISTER SCIENCE
Bratislava

KLUWER ACADEMIC PUBLISHERS
Dordrecht – Boston – London

A C.I.P. Catalogue record for this book is available from the Library of Congress.

ISBN 978-90-481-4602-4

Published by Kluwer Academic Publishers,
P.O. Box 17, 3300 AA Dordrecht, The Netherlands.
in co-edition with
Ister Science Ltd.,
Staromestská 6, 811 03 Bratislava, Slovakia

Kluwer Academic Publishers incorporates
the publishing programmes of
D. Reidel, Martinus Nijhoff, Dr W. Junk and MTP Press.

Sold and distributed in the Czech Republic and Slovak Republic
by Ister Science Ltd.,
Staromestská 6, 811 03 Bratislava, Solvakia

Sold and distributed in the U.S.A. and Canada
by Kluwer Academic Publishers,
101 Philip Drive, Norwell, MA 02061, U.S.A.

In all other countries, sold and distributed
by Kluwer Academic Publishers Group,
P.O. Box 322, 3300 AH Dordrecht, The Netherlands.

Printed on acid-free paper

Contents

Preface

The present book is the result of about twenty years work by the author on non-additive set functions, mainly from the point of view of pure mathematics. Although non-additive set functions appear very early as a part of classical measure theory, they played an auxiliary role in mathematics. Very recently it turned out that non-additive set functions are important in applications in very different fields, such as, knowledge engineering, artificial intelligence, game theory, statistics, economy, sociology, industry, and generally in problems with uncertainties and vague data which connect them with the theory of fuzzy sets and systems. These applications stimulated the need for developing some theories for special kinds of non-additive set functions, mainly by non-mathematicians such as engineers, economists, system engineers, computer scientists, etc.

In this book I shall present the mathematical theory of a wide class of non-additive set functions (so called null-additive set functions), which includes the classical measure theory, too. This class includes such important set functions as capacities, triangular set functions, some fuzzy measures, submeasures, decomposable measures, possibility measures, distorted probabilities, etc.

On the other side, the theory of non-additive set functions gives feedback on mathematics, for example in the theory of nonlinear differential equations and difference equations, and in the theory of diagonal theorems as a universal method for proving general and fundamental theorems in functional analysis and measure theory.

We shall not present here the wide range of applications of null-additive set functions. We shall only show a few examples of some different applications.

My intention was to prepare a book not only for mathematicians but also for a wide range of scientists and readers who want to apply non-additive set functions.

Some parts of the book were used by me in lectures at the many seminars at the Institute of Mathematics at the University of Novi Sad and at the Mathematical Institute in Belgrade, in Austria at the Johannes Kepler University, in Italy at the University Federico II of Naples and at the University of Basilicata of the Potenza, in Hungaria at the Bolyai Mathematical Institute in Szeged, in Poland at the Institute of Mathematics of the Polish Academy of Sciences in Katowice and in Slovakia at the Mathematical-Physical Faculty of Comenius University in Bratislava.

Acknowledgment

I am grateful to Professors P. de Lucia, E. P. Klement, H. Weber, P. Antosik, A. Dvurečenskij and Dr R. Mesiar for many discussions, criticism, help, and their work on the subject related to this book.

I would like to express my thanks to Marčičev Merima for her patience during the typing and corrections of the manuscript. I am grateful to my wife Darinka and my daughter Danijela for their encouragement and patience.

I also want to thank the Institute of Mathematics in Novi Sad and the Institute of Mathematics in Belgrade for working conditions and financial support.

November, 1994 Endre Pap

List of Symbols

$\mathbf{N}$	natural numbers
$\mathbf{R}$	real numbers
X	nonempty basic set
$\mathbf{T}$	Hausdorff locally compact topological space
$\liminf A_n$	limes inferior of sets
$\limsup A_n$	limes superior of sets
A^c	complement of the set A
χ_A	characteristic function of the set A
$\|A\|$	cardinal number of the set A
$\cap$	union of sets
$\cup$	intersection of sets
Δ	symmetric difference
$\setminus$	set difference
$\mathcal{B}$	Borel σ-algebra
$\mathcal{P}(X)$	power set of X
$\mathcal{D}$	paving, family of subsets of X with $\emptyset \in \mathcal{D}$
$\mathcal{R}$	ring of subsets of X
$\mathcal{A}$	algebra of subsets of X
Σ	σ-algebra of subsets of X
$\mathcal{F}, \mathcal{F}(X)$	family of all finite subset of X
μ	additive (σ-additive) measure
$\vee$	sup, max in real or join in lattice
$\wedge$	min, inf in reals or met in lattice
m, v	general set functions
η	submeasure
Bel	belief measure

Pl	plausibility measure
π	possibility measure
$\mu_{\mathbf{H}}$	Haar measure
μ_A	membership grade function
$\overline{m}$	disjoint variation
$\|m\|$	chain variation
$\\|m\\|_c$	composite variation
m^*	outer set function of m
$m_*, \hat{m}$	inner set function of m (semivariation)
BV	set of all set functions with bounded chain variation
CV	set of all set functions with bounded composite variation
DV	set of all set functions with bounded disjoint variation
CCV	subset of CV with σ-additive representation
$(C)\int$	Choquet integral (asymmetric)
$(\check{S})\int$	Symmetric integral
$(S)\int$	Sugeno (fuzzy) integral
$\int^{\oplus}$	integral with respect to $\oplus$-measure
S	t-conorm
$S_{\lambda}^{\mathbf{S}}$	Sugeno family of t-conorms
$\overline{\lim}$	limes superior
$\underline{\lim}$	limes inferior
$\mathbf{S}$	semigroup
$\oplus$	pseudo-addition; operation in MV-algebra and D-poset
$\otimes$, $\odot$	pseudo-multiplication
$\mathbf{L}$	lattice, D-poset
$\ominus$	difference in D-poset
$\dim_{\mathbf{H}}$	Hausdorff dimension
$\dim_{\mathbf{B}}$	Box dimension
$\mathcal{P}m$	family of pseudometrics
$\overline{A}$	closure of a set A in a topological space
$m \perp v$	set function m is v-singular
$a \perp b$	elements a and b from orthomodular lattice are orthogonal
$a^{\perp}$	complement of an element a from D-poset
a'	orthocomplement of an element a from orthomodular lattice

Chapter 1

Introduction

1.1 Non-Additive Set Functions

In this book we will consider some general classes of set functions which include different types of important non-additive set functions.

Non-additive set functions, as for example outer measures, semi-variations of vector measures, naturally appeared earlier in classical measure theory concerning countable additive set functions or more general finite additive set functions. The pioneer in the theory of non-additive set functions was G. Choquet (from 1953) with his theory of capacities. This theory had influences on many parts of mathematics and different areas of sciences and technique.

Non-additive set functions are extensively used in decision theory, mathematical economy, social choice problems, with early traces by Aumann and Shaplay in their monograph from 1974. Recently, many authors have investigated different kinds of non-additive set functions, as subadditive and superadditive set functions, submeasures, k-triangular set functions, t-conorm and pseudo-addition decomposable measures, null-additive set functions, and many other types of set functions. Although in many results the monotonicity of the observed set functions was supposed, there are also some results concerning some classes of set functions which include non-monotone set functions (for example superadditive set functions, k-triangular set functions).

On the other hand, fuzzy measures such as monotone and continuous set functions were investigated by Sugeno in 1974 with the purpose to evaluate non-additive quantity in systems engineering. This notion of fuzziness is different from the one given by Zadeh. Namely, instead of taking membership grades of a set, we take (in the fuzzy measure approach) the measure that a given unlocated element belongs to a set. There are many different type of fuzzy measures which are used. For example belief, possibility, decomposable measures.

Specially in different branches of mathematics there are many types of non-additive set functions. They appeared in potential theory, harmonic analysis, fractal geometry, functional analysis, the theory of nonlinear differential equations, and in the theory of difference equations and optimizations.

There are many different fields in which the interest in non-additive set functions is developing. In artifical intelligence, belief functions have been applied to model uncertainty. Belief functions, corresponding plausibility measures and other kinds of non-additive set functions are used in statistics. Non-additive expected utility theory has been applied, for example, in multi-stage decision and economics.

We can compare additive set functions (which form the basis for classical measure theory) and non-additive set functions in the following simple way. We have for a fixed set $A \in \Sigma$ and the classical measure $\mu: \Sigma \to [0, +\infty]$ that for every set B from Σ such that $A \cap B = \emptyset$ the difference $\mu(A \cup B) - \mu(B)$ is always equal to the constant $\mu(A)$, i.e., it is independent of B. In contrast, for a non-additive set function m, the difference $m(A \cup B) - m(B)$ depends on B and it can be interpreted as the effect of A joining B.

Many different problems rose when non-additive set functions produced not only many different kinds of non-additive set functions, but also many different methods and tools. So it turns out a unified mathematical theory of non-additive set functions, which would be covered by many important non-additive set functions, independently of their applications, is needed.

The unification of many investigated different kinds of non-additive set functions is achieved by a new wide class, so called null-additive set functions. A set function m defined on a family $\mathcal{D}$ of sets such that it is closed with respect to the union and $\emptyset \in \mathcal{D}$, and with values in $[-\infty, \infty]$ (or more general in a semigroup with a neutral element 0) and such that $m(B) = 0$ implies $m(A \cup B) = m(A)$ whenever $A, B \in \mathcal{D}$, $A \cap B = \emptyset$ is called null-additive set function. These set functions have very nice properties with respect to the usual notions in measure theory and they occur naturally in integration theory.

The origins of null-additive set functions were in the papers of V. N. Aleksjuk (under the name 'quasimeasures') in 1968, I. Dobrakov (under the name 'submeasures') in 1974 and L. Drewnowski in 1978. These set functions were rediscovered and investigated by Wang in 1984 under the name null-additive set functions. In all these papers the monotonicity of set functions was supposed. Many papers have appeared recently on this subject.

1.2 The Concept of the Book

Now about the book. There are certain sections in which some notations occur very frequently. In order to avoid repetitions we adopt the following convention: throughout the book it is assumed that, besides the hypothesis explicitly formulated, the hypothesis stated at the beginning of the section also holds.

Throughout this book $\mathcal{R}$ and Σ always denote a ring and a σ-algebra of subsets of the given set X, respectively. We will always suppose, if otherwise is not stated, that the considered set functions vanish on empty set.

We will use also the usual convention in measure theory that $0 \cdot \infty = 0$. Sometimes we will use some results from classical measure theory, but the definitions and proofs in the book mainly are independent of it and shed new light also on classical results.

Chapter 2 defines and discusses the objects and notions connected with non-additive set functions, taking out the null-additivity of set functions and presents major examples: S-decomposable measures, $\oplus$-decomposable measures, convex and concave monotone set functions, submeasures, distorted measures, k-triangular set functions. Some applications of capacities to potential theory are presented. Maxitive measures are applied in fractal geometry through fractal dimensions.

Chapter 3 is devoted to two kinds of positive set functions which correspond to an arbitrary real valued set function, named 'disjoint variation' and 'chain variation', which kinds of set functions were used in classical measure theory, where they were used for measures. We present the main properties of these two kinds of variations and we examine which properties of set functions are transferred to the variations.

Chapter 4 contains two types of correspondence between arbitrary real valued set functions and additive set functions. The first representation is based on the classical result in the theory of finite cooperative game theory, where a set function is considered as a game, which can be spanned in the base of unanimity games (additive set functions). We present also an extension of the canonical representation to general (infinite) case. Therefore a third type of variation — the composition variation — is introduced and the relation to the two previously introduced variations is discussed. The second representation, whose trace is in the theory of fuzzy sets, is devoted to the representation of positive monotone set functions according to the classical measure.

Chapter 5 consists primarily of discussion of different kinds of autocontinuities (from above, from below, uniform) and connections with the usual continuity from above and from below and null-additivity. Special attention is given to the countable case, where following a recent result the null-additivity for finite case reduces on the autocontinuity. Submeasures and the corresponding Frechet-Nikodym topology are topologically connected with monotone autocontinuous set functions. This correspondence is very useful and makes it possible to transfer some classical results on submeasures to the monotone autocontinuous set functions. We will use this relation in the proof of a general extension theorem for positive monotone autocontinuous and continuous set function. At the end we present a recently introduced notion of the linearly indexed family of submeasures as a generalization of the notion of families of submeasures, which are closely related to the set functions represented by integrals. There is a close relation of this notion with modular spaces and generalized Orlicz spaces. We present some nonlinear singular integral operators as an illustration of applications in the approximation of functions in modular spaces.

In Chapter 6 we discuss Hahn and Jordan decomposition type theorems for so-called signed revised monotone set functions. Lebesgue decomposition type theorems are proved for null-additive and σ-null-additive set functions. The notion of atoms for monotone set functions is introduced, and it turns out that again null-additivity gives good description of them. If the monotone null-additive set function is μ-continuous with respect to a measure μ, then it can be represented by a monotone null-additive set function defined on $\mathcal{P}(\mathbf{N})$, where $\mathbf{N}$ is the set of all natural numbers. For null-additive exhaustive and order continuous set functions the Saks decomposition theorem is obtained. At the end we present a result related to recent investigations of generalizations of domains and ranges of the considered set functions, which are initiated by the need of unified theory of many different applications. We consider a Hewitt-Yosida type theorem for a function which is defined on a lattice and with values in an ordered semigroup. The generalized measures are a very recent lattice generalization of σ-additive measure, probability measures on fuzzy events, possibility measures, fuzzy probability measures, fuzzy-valued fuzzy measures, σ-S-decomposable measure, measures on fuzzy events, $\oplus$-decomposable measures, Stone and W^* algebra-valued positive measures.

Chapter 7 treats two major integrals for monotone set functions: Choquet's integral and Sugeno integral, both very important in many different applications. It turns out that again null-additivity of set functions is very important for nice properties of these integrals. We present two extensions of the Choquet integral: symmetric and asymmetric. In the presentation of the symmetric Choquet integral we use the Šipoš approach and that is the reason why we will call this integral also the Šipoš integral.

Chapter 8 presents a part of the theory of the important subclass of null-additive set functions, the so-called decomposable measures, mainly treating the induced integrals. Three types of integrals are presented briefly: Murofushi-Sugeno, Weber and Maslov-Pap. In a special case the last integral gives the g-integral, which induces a g-calculus. The usefulness of these notions illustrated by applications on nonlinear differential and difference equations.

Chapter 9 is devoted to an important problem in measure theory (mainly for the applications) as it describes the properties of the range of the set function. We present theorems on the Darboux property and Lyapunov's type theorem. We prove a theorem on singleton characterization of an atom of a regular null-additive set function and that inner regularity together with autocontinuity and exhaustivity on compact sets imply regularity of this set function.

Chapter 10 is devoted to some existing but still not quite applicable Radon-Nikodym type theorems for Choquet and Sugeno integrals. The representation of a set function as a linear combination of additive set functions gives the corresponding representation for a Choquet integral. We present a representation theorem for nonlinear autocontinuous from above functional in the form of a Choquet integral.

Chapter 11 considers an important subclass of null-additive set functions, the so-called k-triangular set functions. These generally non-monotone set functions were investigated by many mathematicians. We present a Brooks-Jewett theorem, Dieudonne boundedness type theorem and Brooks-Jewett type theorem on orthomodular lattices which are important in quantum logic. The methods of proofs are based on the so-called diagonal theorems (matrix methods), which are obtained here as consequences of a diagonal theorem for triangular set functions, although they were originally proved independently. There are many applications of diagonal theorems in functional analysis and measure theory. We present only a few of them with special attention to the adjoint theorem (connected with closed graph theorem) and the 'fatness' condition.

We make some very general considerations in Chapter 12. This chapter treats functions defined on D-poset, partially ordered set endowed with a special partial operation (difference), and with values in an arbitrary uniform space without any algebraic operation. D-poset, originating in the theory of BCK-algebras and sets of effects, includes many important cases as MV-algebras, orthomodular lattices, orthoalgebras, fuzzy sets, and sets of effects. A Nikodym boundedness theorem with necessary and sufficient conditions and a Brooks-Jewett type theorem are proved.

1.3 Notes and References

The classical measure theory is presented in many books, among others in [Constantinescu 1984], [Constantinescu and K. Weber 1985], [Dinculeanu 1967], [Dunford and Swartz 1958], [Halmos 1950], [M. M. Rao 1987], [K. P. S. B. Rao and M. B. Rao 1983].

The classical paper on non-additive set functions is [Choquet 1953–1954]. Non-additive set functions were considered in the following books: [Aumann and Shapley 1974], [Shafer 1976], [Denneberg 1994], [Dinculeanu 1967], [Pap 1982], [Z. Wang and Klir 1992]. Early traces of null-additive set functions were in [Aleksjuk 1968], [Aleksjuk 1970], [Dobrakov 1974], [Drewnowski 1978] and [Dobrakov and Farkova 1980]. Null-additivity was introduced in [Z. Wang 1984].

Chapter 2

General Properties of Set Functions

2.1 Basic Definitions

2.1.1 Preliminaries

For the reader's convenience, we repeat some known definitions of collections of subsets, which will be used in this book.

Throughout this book, X always denotes a non-empty set.

Definition 2.1 Let X be a set and $\mathcal{D}$ a collection of subsets of X. Then

(1) $\mathcal{D}$ is a paving if $\emptyset \in \mathcal{D}$.

(2) $\mathcal{D}$ is a lattice of subsets if the following conditions are satisfied
  (i) $A, B \in \mathcal{D}$ implies $A \cup B \in \mathcal{D}$;
  (ii) $A, B \in \mathcal{D}$ implies $A \cap B \in \mathcal{D}$.

(3) $\mathcal{D}$ is a ring on X if the following conditions are satisfied
  (i) $\emptyset \in \mathcal{D}$;
  (ii) $A, B \in \mathcal{D}$ implies $A \cup B \in \mathcal{D}$;
  (iii) $A, B \in \mathcal{D}$ implies $A \setminus B \in \mathcal{D}$.

(4) $\mathcal{D}$ is a algebra on X if $\mathcal{D}$ is ring and $X \in \mathcal{D}$.

(5) $\mathcal{D}$ is a σ-ring on X if the following conditions are satisfied
  (i) $\emptyset \in \mathcal{D}$;
  (ii) $\{A_n\} \subset \mathcal{D}$ implies $\bigcup_{n=1}^{\infty} A_n \in \mathcal{D}$;
  (iii) $A, B \in \mathcal{D}$ implies $A \setminus B \in \mathcal{D}$.

(6) $\mathcal{D}$ is a σ-algebra if $\mathcal{D}$ is a σ-ring and $X \in D$.

Throughout this book $\mathcal{D}$, $\mathcal{R}$, $\mathcal{A}$ and Σ always denote a paving, a ring, an algebra and a σ-algebra of subsets of the given set X, respectively.

$\mathcal{P}(X)$ will denote the power set of X and $\mathcal{F}(X)$ the family of all finite subsets of X.

The algebra $\mathcal{A}(\mathcal{D})$ generated by the family $\mathcal{D}$ is given in the following way

$$\mathcal{A}(\mathcal{D}) = \cap\{\mathcal{A} : \mathcal{A} \subset \mathcal{P}(X) \text{ is an algebra such that } \mathcal{D} \subset \mathcal{A}\}.$$

In an analogous way we also introduce the σ-algebra $\sigma(\mathcal{D})$ generated by the family $\mathcal{D}$.

Definition 2.2 Let $\{E_n\}$ be a sequence from $\mathcal{D}$. Then we define

$$\limsup_n E_n = \bigcap_{n=1}^{\infty} \bigcup_{k=n}^{\infty} E_k$$

and

$$\liminf_n E_n = \bigcup_{n=1}^{\infty} \bigcap_{k=n}^{\infty} E_k,$$

whenever the right parts are meaningful. If

$$\limsup_n E_n = \liminf_n E_n$$

holds we denote this set by $\lim_n E_n$.

We now give a list of basic notions related to set functions.

Definition 2.3 A function $m\colon \mathcal{D} \to [-\infty, +\infty]$, with $m(\emptyset) = 0$, is called an extended real-valued set function.
We say that

(1) m is a (finite) real-valued set function if $-\infty < m(A) < +\infty$ for all $A \in \mathcal{D}$.

(2) m is monotone if $A \subset B$ implies $m(A) \leq m(B)$ for every $A, B \in \mathcal{D}$.

(3) m is normalized if $m(X) = 1$.

(4) m is non-negative if it is finite and $m(A) \geq 0$ for all $A \in \mathcal{D}$.

(5) $m\colon \mathcal{D} \to [0, +\infty]$ is positive.

(6) $m\colon \mathcal{R} \to (-\infty, +\infty)$ is supermodular (convex) if for every $A, B \in \mathcal{R}$ we have

$$m(A \cup B) + m(A \cap B) \geq m(A) + m(B).$$

(7) $m: \mathcal{R} \to (-\infty, +\infty)$ is superadditive if for every $A, B \in \mathcal{R}$ with $A \cap B = \emptyset$ we have

$$m(A \cup B) \geq m(A) + m(B).$$

(8) $m: \mathcal{R} \to (-\infty, +\infty)$ is submodular (concave) if for every $A, B \in \mathcal{R}$ we have

$$m(A \cup B) + m(A \cap B) \leq m(A) + m(B).$$

(9) $m: \mathcal{R} \to (-\infty, +\infty)$ is subadditive if for every $A, B \in \mathcal{R}$ with $A \cap B = \emptyset$ we have

$$m(A \cup B) \leq m(A) + m(B).$$

(10) $m: \mathcal{R} \to [0, +\infty)$ is totally monotone if for every $n \in \mathbf{N}$ and $A_1, \ldots, A_n \in \mathcal{R}$ we have

$$m\Big(\bigcup_{i=1}^{n} A_i\Big) \geq \sum_{\{I: I \subset \{1,\ldots,n\}\}} (-1)^{|I|+1} m\Big(\bigcap_{i \in I} A_i\Big).$$

(11) $m: \mathcal{R} \to (-\infty, +\infty)$ is additive (a signed finitely additive measure) if we have

$$m(A \cup B) = m(A) + m(B)$$

for all $A, B \in \mathcal{R}$ with $A \cap B = \emptyset$.

(12) $m: \mathcal{R} \to (-\infty, +\infty)$ is σ-additive (a signed measure) if we have

$$m\Big(\bigcup_{i=1}^{\infty} E_i\Big) = \sum_{i=1}^{\infty} m(E_i)$$

for all pairwise disjoint sequences $\{E_i\}$ from $\mathcal{R}$, i.e., $E_n \cap E_m = \emptyset$ for $n \neq m$, and $\bigcup_{i=1}^{\infty} E_i \in \mathcal{R}$.

(13) $m: \mathcal{R} \to (-\infty, +\infty)$ is a measure if it is nonnegative and σ-additive. If the value $+\infty$ is allowed, then m is called a positive measure. If a measure m additionally satisfies the condition $m(X) = 1$, then it is called a probability.

(14) $m: \mathcal{R} \to (-\infty, +\infty)$ is continuous from below if for every sequence $\{E_n\}$ such that $E_1 \subset E_2 \subset \ldots$, $E_n \in \Sigma$, $\bigcup_{n=1}^{\infty} E_n \in \mathcal{R}$, we have $m(\bigcup_{n=1}^{\infty} E_n) = \lim_{n\to\infty} m(E_n)$.

(15) $m: \mathcal{R} \to (-\infty, +\infty)$ (or m is positive and monotone) is continuous from above if for every sequence $\{E_n\}$ such that $E_1 \supset E_2 \supset \ldots$, $E_n \in \mathcal{R}$, $\bigcap_{n=1}^{\infty} E_n \in \mathcal{R}$ (and there exists n_0 such that $m(E_{n_0}) < \infty$) we have $m(\bigcap_{n=1}^{\infty} E_n) = \lim_{n\to\infty} m(E_n)$. If m is continuous from above at $\emptyset$, then m is called order continuous.

(16) m is exhaustive (s-bounded) if $\lim_{n\to\infty} m(E_n) = 0$ for every sequence $\{E_n\}$ of pairwise disjoint sets from $\mathcal{D}$.

(17) A positive set function m is called σ-finite if there exists a sequence $\{X_n\}$ such that

$$X_1 \subset X_2 \subset \ldots, \quad \bigcup_{n=1}^{\infty} X_n = X \quad \text{and} \quad m(X_n) < \infty \quad (n \in \mathbf{N}).$$

We will take the space of all set functions as a vector space over $\mathbf{R}$ with pointwise operations.

Remark 2.1 We can take for the range of a set function a subset of a commutative semigroup $\mathbf{S}$ with a neutral element 0, i.e., $m: \mathcal{D} \to \mathbf{S}$. Of course, in this case not all notions in Definition 2.3 are meaningful. Alternatively, we can take for the range a set of complex numbers and in this case we speak about complex valued set functions.

Remark 2.2 Non-negative superadditive set function on an algebra is always monotone. This is not always true for non-negative submodular set function. An additive set function is monotone if and only if it is positive.

Propositon 2.1 *If a finite monotone set function m is defined on a σ-algebra Σ, then m is continuous from above and below if and only if it is continuous, i.e., for every sequence $\{A_n\}$ from Σ such that $\lim_n A_n = A \in \Sigma$ we have $\lim_{n\to\infty} m(A_n) = m(A)$.*

Proof. Let $\{A_n\}$ be a sequence from Σ such that $\lim_n A_n = A \in \Sigma$. Then we have

$$m(A) = m(\limsup_n A_n) = \overline{\lim_{n\to\infty}}\, m\Big(\bigcup_{k=n}^{\infty} A_k\Big) \ge \overline{\lim_n}\, m(A_n)$$

$$\ge \underline{\lim_n}\, m(A_n) \ge \underline{\lim_n}\, m\Big(\bigcap_{k=n}^{\infty} A_k\Big) = m(\liminf_n A_n) = m(A).$$

Hence

$$\lim_{n\to\infty} m(A_n) = m(A). \qquad \square$$

If the domain of m in the preceding proposition is not a σ-ring then this proposition may be not true. Take for example the family $\mathcal{F}(\mathbf{N})$ of all finite subsets of $\mathbf{N}$. Then $\mathcal{F}(\mathbf{N})$ is a ring, which is not a σ-ring. If we define $m(A) = 1$ for $A \in \mathcal{F}(\mathbf{N})$ and $A \neq \emptyset$, and $m(\emptyset) = 0$, then m is not continuous (take $A_n = \{n, n+1, \ldots, 2n\}$, then $\lim_n A_n = \emptyset$ but $1 = \lim_{n\to\infty} m(A_n) \neq 0$), although it is continuous from above and below.

The following example shows that a set function's continuity from above does not imply its exhaustivity.

Example 2.1 Let $X = \mathbf{N} \cup \{\infty\}$ and

$$\mathcal{R} = \{E : E \in \mathcal{F}(\mathbf{N}) \text{ or } E = (\mathbf{N} \setminus A) \cup \{\infty\} \quad \text{if } A \in \mathcal{R} \setminus \mathcal{F}(\mathbf{N})\}.$$

Let a series $\sum_{n=1}^{\infty} x_n$ be convergent but not unconditionally to x_∞. We define m on $\mathcal{R}$ in the following way $m(A) = \sum_{n\in A} x_n$ if $A \in \mathcal{F}(\mathbf{N})$ and $m(A) = x_\infty - \sum_{\mathbf{N}\setminus A} x_n$ if $A \in \mathcal{R} \setminus \mathcal{F}(\mathbf{N})$. Then m is even σ-additive but not exhaustive.

In a special case we have

Propositon 2.2 *If $m, m\colon \Sigma \to [0, +\infty)$, is a monotone set function which is continuous from above at $\emptyset$ (order continuous), then it is exhaustive.*

Proof. Let $\{D_n\}$ be a sequence of pairwise disjoint sets from Σ. Monotonicity of m implies

$$0 \leq m(D_n) \leq m\Big(\bigcup_{i=n}^{\infty} D_i\Big).$$

Letting $n \to \infty$ we obtain by order continuity of m that $\lim_{n\to\infty} m(D_n) = 0$. □

Two set functions which are defined on all subsets of X can be given for a positive monotone set function.

Definition 2.4 Let m be a positive monotone set function defined on a paving $\mathcal{D}$.

Outer set function of m is a function $m^*\colon \mathcal{P}(X) \to [0, +\infty]$ defined by

$$m^*(A) = \inf\{m(B) : A \subset B \in \mathcal{D}\} \qquad (A \in \mathcal{P}(X)),$$

taking $\inf \emptyset = \infty$.

Inner set function (semivariation) of m is a function $m_*: \mathcal{P}(X) \to [0,+\infty]$ (denoted also by $\hat{m}$) defined by

$$m_*(A) = \sup\{m(C) : C \subset A,\ C \in \mathcal{D}\} \qquad (A \in \mathcal{P}(X)).$$

Since m^* and m_* coincide with m on $\mathcal{D}$ we have that m_*, m^* are the smallest, respectively greatest, extensions of m onto the entire $\mathcal{P}(X)$.

Remark 2.3 It is obvious that m^* and m_* are also monotone and that if m is defined on $\mathcal{R}$ and it is positive supermodular (superadditive), m_* is too, and if m is submodular (subadditive), that m^* is too. Moreover, it can be proved that if m is submodular and continuous from below on Σ, then m^* is also continuous from below.

2.1.2 Null-Additive Set Functions

We have now the definition of the main notion in this book.

Definition 2.5 A set function m, $m: \mathcal{D} \to [-\infty, \infty]$, is called null-additive, if we have

$$m(A \cup B) = m(A)$$

whenever $A, B \in \mathcal{D}$, $A \cap B = \emptyset$, and $m(B) = 0$.

It is obvious that for a null-additive set function m we have $m(\emptyset) = 0$ whenever there exists $B \in \mathcal{D}$ such that $m(B) = 0$. We shall always suppose that $m(\emptyset) = 0$, if nothing otherwise is not explicitly stated.

In the next section we present many important examples of null-additive set functions.

Two simple examples of set functions follow.

Example 2.2 Let $m(A) \neq 0$ whenever $A \in \Sigma$, $A \neq \emptyset$. Then m is null-additive.

Example 2.3 Let $X = \{x, y\}$ and define m in the following way:

$$m(X) = 1 \quad \text{and} \quad m(A) = 0 \quad \text{for} \quad A \neq X\,.$$

Then m is not null-additive.

We have some obvious properties of null-additive set functions.

Theorem 2.1 *Let m be a positive monotone set function defined on a ring $\mathcal{R}$. Then the following statements are equivalent:*

(1) *m is null-additive.*

(2) *If $E \in \mathcal{R}$, $F \in \mathcal{R}$, $m(F) = 0$, then $m(E \cup F) = m(E)$.*

(3) *If $E \in \mathcal{R}$, $F \in \mathcal{R}$, $F \subset E$ and $m(F) = 0$, then $m(E \setminus F) = m(E)$.*

(4) *If $E \in \mathcal{R}$, $F \in \mathcal{R}$, $m(F) = 0$, then $m(E \setminus F) = m(F)$.*

(5) *If $E \in \mathcal{R}$, $F \in \mathcal{R}$, $m(F) = 0$, then $m(E \Delta F) = m(F)$.*

Now we have an example of a null-additive non-monotone set function.

Example 2.4 Let $m\colon \Sigma \to [0,1]$ be defined by

$$m(A) = \begin{cases} 2\mu(A), & \text{if } \mu(A) \leq \frac{1}{2}, \\ -2\mu(A) + 2 & \text{if } \mu(A) \geq \frac{1}{2}, \end{cases}$$

where $\mu, \mu\colon \Sigma \to [0,1]$, is a σ-additive measure. Then m is null-additive and not monotone, but it is continuous from above and from below.

We can introduce more general class of set functions.

Definition 2.6 A set function m, $m\colon \mathcal{D} \to [-\infty, \infty]$, is called supernull-additive, if we have

$$m(A \cup B) \geq m(A)$$

whenever $A, B \in \mathcal{D}$, $A \cap B = \emptyset$, and $m(B) = 0$.

The class of all supernull-additive set functions includes all null-additive set functions, all monotone set functions (defined on a ring) and all super-additive set functions.

Theorem 2.2 *Let m be a null-additive, positive, monotone and continuous from above set function defined on Σ. If $A \in \Sigma$, then*

$$m(A \cup B_n) \to m(A)$$

for any decreasing sequence $\{B_n\}$ from Σ for which $m(B_n) \to 0$ and there exists at least one n_0 such that $m(A \cup B_{n_0}) < \infty$ as $m(A) < \infty$.

Proof. Without loss of generality we can suppose $m(A) < \infty$. Taking $B = \bigcap_{n=1}^{\infty} B_n$ we obtain $m(B) = \lim_{n\to\infty} m(B_n) = 0$. By $A \cup B_n \searrow A \cup B$ and continuity from above and null-additivity of m we obtain

$$\lim_{n\to\infty} m(A \cup B) = m(A). \qquad \square$$

Theorem 2.3 *Let m be a null-additive, positive, monotone and continuous set function on Σ. If $A \in \Sigma$, then we have*

$$m(A \setminus B_n) \to m(A)$$

for any decreasing sequence $\{B_n\}$ from $\mathcal{R}$ for which $\lim_{n\to\infty} m(B_n) = 0$.

Proof. Since we have

$$m\Big(\bigcup_{n=1}^{\infty} B_n\Big) = \lim_{n\to\infty} m(B_n) = 0,$$

and

$$A \setminus B_n \nearrow A \setminus \Big(\bigcup_{n=1}^{\infty} B_n\Big),$$

by (4) in Theorem 2.1 we obtain

$$\lim_{n\to\infty} m\Big(A \setminus \Big(\bigcup_{n=1}^{\infty} B_n\Big)\Big) = m(A). \qquad \square$$

We introduce the following generalization of the σ-additiveness.

Definition 2.7 Let $\mathcal{R}$ be a σ-ring. A set function $m\colon \mathcal{R} \to [-\infty, \infty]$ with $m(\emptyset) = 0$ is σ-null-additive if for every sequence $\{B_i\}$ of pairwise disjoint sets from $\mathcal{R}$ such that $A \cap B_i = \emptyset$ and $m(B_i) = 0$ we have

$$m\Big(A \cup \bigcup_{i=1}^{\infty} B_i\Big) = m(A).$$

Propositon 2.3 *Let $\mathcal{R}$ be a σ-ring and let m be a function $m\colon \mathcal{R} \to [-\infty, \infty]$. A function m is σ-null-additive if and only if m is null-additive and $m(B_i) = 0$ $(i \in \mathbf{N})$ implies $m(\bigcup_{i=1}^{\infty} B_i) = 0$ for a sequence $\{B_i\}$ of pairwise disjoint sets from $\mathcal{R}$.*

Proof. We have for σ-null-additive set function m

$$m\Big(\bigcup_{i=1}^{\infty} B_i\Big) = m\Big(B_1 \cup \bigcup_{i=2}^{\infty} B_i\Big) = m(B_1) = 0\,, \quad \text{where } B_1 \cap \bigcup_{i=2}^{\infty} B_i = \emptyset\,.$$

If $m(B) = 0$ for $B \in \mathcal{R}$, then taking $B_1 = B$ and $B_i = \emptyset$ for $i \geq 2$ we have by Definition 2.7

$$m(A \cup B) = m\Big(A \cup \bigcup_{i=1}^{\infty} B_i\Big) = m(A)$$

for any $A \in \mathcal{R}$, i.e., m is null-additive. The inverse statement is obvious.□

Propositon 2.4 *Let $\mathcal{R}$ be a σ-ring. If m is null-additive and continuous from below, then it is σ-null-additive.*

Proof. Let $\{B_i\}$ be a sequence of pairwise disjoint sets from $\mathcal{R}$ such that $A \cap B_i = \emptyset$ and $m(B_i) = 0$ $(i \in \mathbf{N})$. Then, since

$$A \cup \bigcup_{i=1}^{n} B_i \nearrow A \cup \bigcup_{i=1}^{\infty} B_i,$$

we have by continuity of m

$$\lim_{n\to\infty} m\Big(A \cup \bigcup_{i=1}^{n} B_i\Big) = m\Big(A \cup \bigcup_{i=1}^{\infty} B_i\Big)\,.$$

But the left part of the preceding equality, by null-additivity of m is the limit of a stationary sequence with all members equal to $m(A)$. Hence the limit is also equal to $m(A)$ and so we have

$$m(A) = m\Big(A \cup \bigcup_{i=1}^{\infty} B_i\Big),$$

i.e., m is σ-null-additive. □

2.2 Important Classes of Set Functions

2.2.1 Triangular Conorm Decomposable Measures

We introduce an operation which generalizes the usual addition on the interval $[0,1]$.

Definition 2.8 A triangular conorm S (t-conorm briefly) is a function $S\colon [0,1]^2 \to [0,1]$ such that

(s_1) $S(x,y) \leq S(x,z)$ for $y \leq z$ (monotonicity);

(s_2) $S(x,y) = S(y,x)$ (commutativity);

(s_3) $S(x,S(y,z)) = S(S(x,y),z)$ (associativity);

(s_4) $S(x,0) = x$ (boundary condition).

Example 2.5 The following are the most important t-conorms

$$S_{\mathbf{M}}(x,y) = \max(x,y)\,, \qquad S_{\mathbf{P}}(x,y) = x + y - xy\,,$$

the bounded sum is given by

$$S_{\mathbf{L}}(x,y) = \min(1, x+y)\,,$$

and a non-continuous t-conorm

$$S_{\mathbf{W}}(x,y) = \begin{cases} \max(x,y) & \text{if } \min(x,y) = 0, \\ 1 & \text{otherwise.} \end{cases}$$

There are many other important t-conorms. Two families of t-conorms are given in the next example.

Example 2.6 The family of Sugeno t-conorms is defined for $\lambda > -1$ by

$$S^{\mathbf{S}}_{\lambda}(x,y) = \min(x + y + \lambda xy, 1)\,.$$

The family of Frank t-conorms is defined for $s \in [0,\infty]$ by

$$S^{\mathbf{F}}_{s}(x,y) = \begin{cases} \max(x,y) & \text{if } s = 0, \\ x + y - x \cdot y & \text{if } s = 1, \\ \min(x+y, 1) & \text{if } s = \infty, \\ 1 - \log_s\left(1 + \frac{(s^{1-x}-1)(s^{1-y}-1)}{s-1}\right) & \text{if } s \in (0,\infty) \setminus \{1\}\,. \end{cases}$$

We introduce now a generalization of the notion of finitely additive set functions.

Definition 2.9 An S-decomposable measure m, $m\colon \mathcal{D} \to [0,1]$, with respect to a t-conorm S is such that $m(\emptyset) = 0$ and

$$m(A \cup B) = m(A) S m(B)$$

whenever $A, B \in \mathcal{D}$, $A \cup B \in \mathcal{D}$ and $A \cap B = \emptyset$.

We remark that an S-decomposable measure on a ring $\mathcal{R}$ is always a monotone null-additive set function.

We introduce finite S-sum

$$S_{i=1}^{n} a_i = (S_{i=1}^{n-1} a_i) S a_n \quad \text{for} \quad n > 1$$

and an infinite S-sum

$$S_{i=1}^{\infty} a_i = \lim_{n\to\infty} S_{i=1}^{n} a_i,$$

whenever $\{a_i\}$ is a sequence from $[0,1]$ (the limit on the right side always exists since it is the limit of a bounded non-decreasing sequence of reals).

Definition 2.10 A set function m, $m\colon \mathcal{D} \to [0,1]$ is σ-S-decomposable if it satisfies $m(\emptyset) = 0$ and

$$m\left(\bigcup_{i=1}^{\infty} A_i\right) = S_{i=1}^{\infty} m(A_i)$$

for every sequence $\{A_i\}$ from $\mathcal{D}$ of pairwise disjoint sets such that $\cup_{i=1}^{\infty} A_i \in \mathcal{D}$.

We remark that any σ-S-decomposable measure is always σ-null-additive.

Theorem 2.4 *A set function m, $m\colon \Sigma \to [0,1]$, is a σ-S-decomposable measure if and only if m is S-decomposable and continuous from below.*

Proof. Suppose that m is a σ-S-decomposable measure. If $\{A_i\}$ is any increasing sequence from Σ, then we can define a disjoint sequence $\{B_n\}$ in the following way $B_1 = A_1$ and $B_n = A_n \setminus A_{n-1}$. Hence

$$m\left(\bigcup_{i=1}^{\infty} A_i\right) = m\left(\bigcup_{i=1}^{\infty} B_i\right) = S_{i=1}^{\infty} m(B_i) = \lim_{n\to\infty} m\left(\bigcup_{i=1}^{n} B_i\right) = \lim_{n\to\infty} m\left(\bigcup_{i=1}^{n} A_i\right).$$

Conversely, if m is a continuous from below S-decomposable measure and $\{B_i\}$ is a sequence from Σ of pairwise disjoint sets, then introducing an increasing sequence $\{A_i\}$ in the following way $A_n = \bigcup_{i=1}^n B_i$ we obtain

$$\lim_{n\to\infty} m\Big(\bigcup_{i=1}^n B_i\Big) = \lim_{n\to\infty} m(A_n) = \lim_{n\to\infty} S_{i=1}^n m(B_i) = S_{i=1}^\infty m(B_i). \qquad \square$$

A continuous t-conorm S is called Archimedean if $S(x,x) > x$ for all $x \in (0,1)$. A continuous t-conorm S is strict if $S(x,y) < S(x,z)$ whenever $x \in (0,1)$ and $y < z$. Each strict t-conorm S is Archimedean. Non-strict continuous Archimedean t-conorms are called nilpotent.

We will need the following two representation theorems for t-conorms, which we will not prove here.

Theorem 2.5 *S is a continuous Archimedean t-conorm if and only if there is a continuous strictly increasing function $g\colon [0,1] \to [0,\infty]$ such that $g(0) = 0$ and*

$$S(a,b) = g^{(-1)}(g(a) + g(b)),$$

where $g^{(-1)}$ is the pseudoinverse of g, i.e.,

$$g^{(-1)}(x) = g^{-1}(\min(x, g(1))).$$

S is strict if and only if $g(1) = +\infty$, i.e., g is bijective and $g^{(-1)} = g^{-1}$.

The function g is called an additive generator of S and it is unique up to a positive multiplicative constant.

Theorem 2.6 *S is a continuous t-conorm if and only if there is a unique countable family $\{(\alpha_k, \beta_k)\}_{k\in K}$ of pairwise disjoint open subintervals of the interval $[0,1]$ and a unique family of continuous Archimedean t-conorms $\{S_k\}_{k\in K}$ such that S is the ordinal sum of $\{(\alpha_k, \beta_k, T_k)\}_{k\in K}$, i.e.,*

$$S(x,y) = \begin{cases} \alpha_k + (\beta_k - \alpha_k) S_k\left(\frac{x-\alpha_k}{\beta_k-\alpha_k}, \frac{y-\alpha_k}{\beta_k-\alpha_k}\right) & \text{if } x,y \in [\alpha_k, \beta_k], \\ \max(x,y) & \text{otherwise.} \end{cases}$$

The preceding considerations can be transferred on the interval $[0, M]$ (where also $M = \infty$ is allowed) instead of the interval $[0,1]$.

The representation of an Archimedean t-conorm S on $[0, M]$ by an additive generator g yields the following classification for S-decomposable measures $m: \Sigma \to [0, M]$. m is of type (A) if $g \circ m$ is a σ-additive measure on Σ, i.e.,

$$(g \circ m)\Big(\bigcup_{n\in\mathbf{N}} A_n\Big) = \sum_{n\in\mathbf{N}} (g \circ m)(A_m);$$

m if of type (P) if $g \circ m$ is a finite pseudo-σ-additive measure in the sense that there exists a sequence $\{A_n\}$ of disjoint sets in Σ with

$$(g \circ m)\Big(\bigcup_{n\in\mathbf{N}} A_n\Big) = g(M) < \sum_{n\in\mathbf{N}} (g \circ m)(A_n).$$

An S-decomposable measure is of type (A) if S is a strict t-conorm (in this case the measure $g \circ m$ is infinite) or if S is non-strict with finite σ-additive $g \circ m$.

The following result allows us to construct S-decomposable measures by a σ-additive measure and an Archimedean t-conorm.

Propositon 2.5 *Let μ be a σ-additive measure on Σ and g some additive generator of an Archimedean t-conorm S on $[0, M]$. Then $m = g^{(-1)} \circ \mu$ is an S-decomposable measure on Σ. If $g(M) > \mu(X)$, then m is of type (A). If, additionally, there exists $E \in \Sigma$ such that $0 < \mu(E) < \mu(X)$, then m is of type*

(A) if and only if $g(M) \geq \mu(X)$,

(P) if and only if $g(M) < \mu(X)$.

Proof. We can easily check that $m = g^{(-1)} \circ \mu$ is a σ-S-decomposable measure. This means that for every sequence $\{E_n\}$ of pairwise disjoint sets from Σ we have

$$g \circ m\Big(\bigcup_{i=1}^{\infty} E_i\Big) = \min\left\{\sum_{i=1}^{\infty} g \circ m(E_i), g(M)\right\}.$$

Hence $g \circ m$ is σ-additive if and only if

$$\min\left\{\sum_{i=1}^{\infty} g \circ m(E_i), g(M)\right\} = \sum_{i=1}^{\infty} g \circ m(E_i),$$

which is possible if and only if $\sum_{i=1}^{\infty} g \circ m(E_i) \leq g(M)$, which is equivalent with

$$\sum_{i=1}^{\infty} \min\{\mu(E_i), g(M)\} \leq g(M).$$

Therefore, for $E \in \Sigma$ such that $0 < \mu(E) < \mu(X)$, we obtain

$$\min\{\mu(E), g(M)\} + \min\{\mu(A^c), g(M)\} \leq g(M).$$

Hence $\mu(X) \leq g(M)$. □

As an application of Proposition 2.5 we obtain the following relationship between additive and decomposable measures.

Example 2.7 Let X be a discrete (finite or countable infinite) set, $\Sigma = \mathcal{P}(X)$, and the counting measure μ given by $\mu(A) = |A|$, the cardinality of A. Proposition 2.5 enables us to construct $S_\lambda^{\mathbf{S}}$-decomposable measures $m_{\lambda,\alpha}$ for $\lambda > -1$, $\lambda \neq 0$ and $\alpha > 0$ in the following form

$$m_{\lambda,\alpha}(A) = \begin{cases} \frac{1}{\lambda}[\exp(\frac{\lambda}{\alpha}|A|) - 1] & \text{if } |A| \leq \frac{\lambda}{\alpha}\ln(1+\lambda), \\ 1 & \text{if } |A| > \frac{\alpha}{\lambda}\ln(1+\lambda). \end{cases}$$

If X is a finite set, then the measure $m_{\lambda,\alpha}$ is of type (A) if $\alpha = \lambda|X|/\ln(1+\lambda)$, and of type (P) if $\alpha < \lambda|X|/\ln(1+\lambda)$. For infinite X, the measure $m_{\lambda,\alpha}$ is of type (A) if and only if $S_\lambda^{\mathbf{S}}$ is strict, i.e., if $\lambda = -1$.

Propositon 2.6 *(i) The infinite σ-additive measures are exactly the $+$-decomposable measures of type (A), $+$ being the usual addition on $[0,\infty]$.*

(ii) The finite σ-additive measures μ are exactly the $S_{\mathbf{L}}$-decomposable measures of type (A), $S_{\mathbf{L}}$ being the bounded sum on $[0,\mu(X)]$.

(iii) The $S_{\mathbf{L}}$-decomposable measures of type (P) ($S_{\mathbf{L}}$ being the bounded sum on $[0,M]$ with $M < \infty$) are not σ-additive.

Example 2.8 For $X = \mathbf{N}$, $M \in \mathbf{N}$, define the $S_{\mathbf{L}}$-decomposable measure m by $m(A) = \min(|A|, M)$. If we choose $M = 3$, $A = \{1,2,3,4\}$, $B = \{3,4,5,6\}$, then we obtain $m(A\cap B)+m(A\cup B) = 5$, but $m(A)+m(B) = 6$. The role of the fixed natural number M can be interpreted as the largest number one is able to count.

Example 2.9 $S_\lambda^{\mathbf{S}}$-measure (introduced a special case in Example 2.7) is a decomposable measure in the following sense: Let p be a probability measure on a measurable space (X,Σ) and $S = S_\lambda^{\mathbf{S}}$, $\lambda > -1$, the non-strict Archimedean t-conorm on $[0,1]$ with normed generator g_λ. Then m_λ defined by $m_\lambda(A) = g_\lambda^{-1}(p(A)) = [(1+\lambda)^{p(A)} - 1]/\lambda$ is a σ-$S_\lambda^{\mathbf{S}}$-decomposable measure of type (A) with $m_\lambda(X) = 1$. On the other hand, each σ-$S_\lambda^{\mathbf{S}}$-decomposable measure m_λ of type (A) induces a probability p. The classical case corresponds to the case $\lambda = 0$: $S_0^S = S_L$ on $[0,1]$, $g_0 = \mathrm{id}_{[0,\infty]}$ by continuous extension, and $m_0 = p$.

This means that the σ-S-decomposable measures of type (A) may be viewed as transformations of σ-additive measures, the transformation being given by the additive generator g (and its pseudo-inverse $g^{(-1)}$). The S-decomposable measures of type (P), however, are non-additive in nature: they never can be transformed into an additive measure. The information carried by a measure of type (P) is incomplete or, more precisely, it gives information about the 'additive' part only. This generalization not only produces additive situations, as for the type (A), but is a true extension, as for type (P).

The essence of this generalization can be formulated in terms of a special valuation.

Propositon 2.7 *Let* $m: \Sigma \to [0, M]$ *be a mapping and* S_L *the bounded sum on* $[0, M]$. *Then*

$$m(A \cap B) + m(A \cup B) = m(A) + m(B)$$

implies

$$m(A \cap B) \;\; S_L \;\; m(A \cup B) = m(A) \;\; S_L \;\; m(B).$$

The converse is not true in general.

2.2.2 Pseudo-Addition Decomposable Measures

Let $\oplus$ be a pseudo-addition on $[0, +\infty]$, i.e., a binary operation on $[0, \infty]$ such that

(P1) $x \oplus 0 = 0 \oplus x = x$ (neutral element);

(P2) $(x \oplus y) \oplus z = x \oplus (y \oplus z)$ (associativity);

(P3) $x \leq x', y \leq y' \Rightarrow x \oplus y \leq x' \oplus y'$ (monotonicity in both components);

(P4) $x_n \to x,\ y_n \to y \Rightarrow x_n \oplus y_n \to x \oplus y$ (continuity).

Pseudo-addition $\oplus$ can be represented by a family of one place functions (and vice-versa),

$$\oplus \approx \{\langle(\alpha_k, \beta_k), g_k\rangle : k \in K\},$$

where $\{(\alpha_k, \beta_k)_{k \in K}\}$ are pairwise disjoint intervals contained in $(0, \infty)$, $g_k: [\alpha_k, \beta_k] \to [0, +\infty]$ is (for any $k \in K$) a continuous strictly increasing function with $g_k(\alpha_k) = 0$. Then

$$x \oplus y = \begin{cases} g_k^*(g_k(x) + g_k(y)) & (x, y) \in [\alpha_k, \beta_k]^2, \\ \max(x, y) & \text{otherwise.} \end{cases}$$

Here g_k^* is the pseudo-inverse of g_k defined by

$$g_k^*(x) = g_k^{-1}(\min(x, g_k(\beta_k))) .$$

Note that if $g_k(\beta_k) = +\infty$, then $g_k^* = g_k^{-1}$. Further, functions g_k are determined uniquely by $\oplus$ up to a positive multiplicative constant. For example, the ordinary addition $+$ has the representation

$$\{\langle(0, +\infty), i\rangle\}, \quad \text{where } i(x) = x, \ \forall x \in [0, +\infty].$$

If $\oplus = \max$, than its representation is $\emptyset$.

Let (X, Σ) be a measurable space. A set function $m\colon \Sigma \to [0, +\infty]$ is said to be a pseudo-additive measure (with respect to $\oplus$) if and only if m satisfies the following conditions:

(1) $m(\emptyset) = 0$;

(2) $A, B \in \Sigma$, $A \cap B = \emptyset \Rightarrow m(A \cup B) = m(A) \oplus m(B)$;

(3) $\{A_n\} \subset \Sigma$, $A_n \uparrow A \Rightarrow m(A_n) \uparrow m(A)$.

m will be called briefly a $\oplus$-measure.

We can consider also $\oplus$-decomposable measure $m\colon \Sigma \to [a, b]$. Namely, let $[a, b]$ be a closed (in some cases semiclosed) real interval. The operation $\oplus$ (pseudo-addition) is a function $\oplus\colon [a, b] \times [a, b] \to [a, b]$ which is commutative, nondecreasing, associative and has a zero element, denoted by $\mathbf{0}$. A set function $m\colon \Sigma \to [a, b]$ is a $\oplus$-decomposable measure if there hold $m(\emptyset) = \mathbf{0}$ and

$$m(A \cup B) = m(A) \oplus m(B)$$

whenever $A, B \in \Sigma$ and $A \cap B = \emptyset$. Then m is null-additive.

2.2.3 The Maxitive Measure

A special decomposable measure is very important in many applications.

Definition 2.11 max-decomposable measure m, $m\colon \mathcal{D} \to [0, +\infty]$, i.e., m satisfies the condition

$$m(A \cup B) = m(A) \vee m(B)$$

whenever $A, B \in \mathcal{D}$ and $A \cup B \in \mathcal{D}$, is called a maxitive measure.

m is completely maxitive if for any family $\{E_i\}_{i\in I}$ from $\mathcal{D}$ with $\bigcup_{i\in I} E_i \in \mathcal{D}$ we have

$$m\Big(\bigcup_{i\in I} E_i\Big) = \sup_{i\in I} m(E_i)\,.$$

If I is countable, then m is σ-maxitive.

We can prove this analogously to Theorem 2.4.

Theorem 2.7 *Every completely maxitive measure m such that there exists $E \in \mathcal{D}$ with $m(E) < \infty$ is continuous from below.*

It is interesting that completely maxitive measure may not be continuous from above (in spite of the classical σ-additive measure which is always continuous from above).

Example 2.10 A set function $m\colon \mathcal{P}(\mathbf{R}) \to [0,1]$ defined by

$$m(A) = \begin{cases} 1 & \text{if } A \neq \emptyset, \\ 0 & \text{if } A = \emptyset \end{cases}$$

for $A \in \mathcal{P}(\mathbf{R})$ is a completely maxitive measure but it is not continuous from above. Thus, taking $E_n = (0, \frac{1}{n})$ $(n \in \mathbf{N})$, we obtain a decreasing sequence $\{E_n\}$ with $E_n \searrow \emptyset$, but

$$1 = \lim_{n\to\infty} m(E_n) \neq m(\emptyset) = 0\,.$$

Example 2.11 Every function $f\colon X \to [0, +\infty)$ determines a completely maxitive measure m in the following way

$$m(A) = \sup_{x\in A} f(x) \qquad (A \in \mathcal{P}(X))\,.$$

Conversely, for any maxitive measure defined on $\mathcal{P}(X)$ a corresponding function f can be defined on X (density function) by

$$f(x) = m(\{x\}) \qquad (x \in X).$$

Definition 2.12 A completely maxitive measure $\pi\colon \mathcal{P}(X) \to [0,1]$ with $m(X) = 1$ is called a possibility measure.

A set function $m\colon \mathcal{P}(X) \to [0,1]$ is called necessity measure if for every family $\{E_i\}_{i\in I}$ from $\mathcal{P}(X)$ we have

$$m(\bigcap_{i\in I} E_i) = \inf_{i\in I} m(E_i)$$

and $m(\emptyset) = 0$.

It is easy to see that for every possibility measure π we can obtain a corresponding necessity measure by

$$m(A) = 1 - \pi(A^c) \qquad (A \in \mathcal{P}(X)).$$

Possibility theory is based on possibility measures and necessity measures and can be considered in terms of fuzzy sets or as a restriction of the Dempster-Shafer theory (see Section 2.2.7 Belief Measures and Plausibility Measures) to bodies of evidence that are nested.

Since the usual (crisp) set A can be completely characterized by its characteristic function χ_A we can describe the fuzzy set A (without sharp boundaries) in a similar way by its membership grade function μ_A, $\mu_A\colon X \to [0,1]$. The union and intersection of two fuzzy sets A and B is defined by

$$\mu_{A\cup B}(x) = \mu_A(x) \vee \mu_B(x) \qquad (x \in X)$$

and

$$\mu_{A\cap B}(x) = \mu_A(x) \wedge \mu_B(x) \qquad (x \in X),$$

respectively.

We link a fuzzy set A defined on X and a variable α in X with a possibility distribution function $f_{\alpha,A}\colon X \to [0,1]$ defined by $f_{\alpha,A}(x) = \mu_A(x)$ $(x \in X)$ and interpret $f_{\alpha,A}(x)$ as the degree of possibility that the value of α is x. Now, the corresponding possibility measure is given by

$$\pi_{\alpha,A}(E) = \sup_{x\in E} f_{\alpha,A}(x) \qquad (E \in \mathcal{P}(X)).$$

We can interpret $\pi_{\alpha,A}(E)$ as the degree of possibility that the value of α belongs to the crisp set E.

2.2.4 Fractals, Hausdorff Measure and Dimension

We will present some special maxitive measures in this section. In the last decade fractals have become very popular as a model of a wide variety of physical and technical phenomena. This popularity is growing, owing mainly to beautiful computer graphics and animation.

Classical mathematics concentrated on sufficiently smooth sets and functions although some irregular (non-smooth) sets give a much better picture of many natural events, for example in dynamical systems, economics, biology, etc. These irregular sets, called fractals, still do not have a unique definition. It is more convenient to make a list of characteristic properties of fractals. These properties are:

(a) A fractal has a fine structure (detail on arbitrary small scales);

(b) A fractal is too irregular to be described by traditional geometry (both locally and globally);

(c) A fractal has some form of self-similarity;

(d) Usually the fractal dimension of a fractal defined in some way is greater than its topological dimension;

(e) In most cases of interest, a fractal is defined in a very simple way, for example recursively.

Some well known examples of fractals are:

(1) **The middle third Cantor set**, which is obtained as a subset of the interval $[0, 1]$ deleting its middle third and continuing always this procedure on the remaining subintervals;

(2) **von Koch curve**, which is obtained removing the middle third of the unit segment and replacing it by the other two sides of the equilateral triangle based on the removed segment and continuing always this procedure on the remaining segments;

(3) **Sierpinski gasket**, which is obtained starting from an equilateral triangle and at each stage removing from each equilateral triangle the inverted equilateral triangles;

(4) **Cantor dust**, which is obtained starting from a square and at each stage each remaining square is divided into 16 smaller equal squares of which four are kept and the rest discarded.

One of the main characteristics of the fractal is the property (d), i.e., that the fractal dimension is different from the topological dimension. Unfortunately, this fractal dimension is not uniquely defined. There are many different definitions of the fractal dimension, depending on the examined problem.

One of the oldest and most important one is the Hausdorff dimension, defined for arbitrary subsets of $\mathbf{R}^n$, where $\mathbf{R}$ is the set of all real numbers.

Thus, for a subset F of $\mathbf{R}^n$, a non-negative number s and any $\delta > 0$ we introduce first the set function

$$\mathcal{H}^s_\delta(F) = \inf\left\{\sum_{i=1}^{\infty} |U_i|^s : \{U_i\} \quad \text{is a } \delta\text{-cover of } F\right\}.$$

Then

$$\mathcal{H}^s(F) = \lim_{\delta \to 0} \mathcal{H}^s_\delta(F)$$

is called the s-dimensional Hausdorff measure of F.

Now the Hausdorff dimension of the set F is given by

$$\dim_{\mathbf{H}} F = \inf\{s : \mathcal{H}^s(F) = 0\} = \sup\{s : \mathcal{H}^s(F) = \infty\}.$$

The Hausdorff dimension is a σ-maxitive measure on $\mathcal{P}(\mathbf{R}^n)$ (in the theory of fractals this property is usually called countable stability).

Another type of dimension used also in the theory of fractals is the box-counting measure (Kolmogorov entropy). We have

Definition 2.13 The lower and upper box-counting dimensions of a subset F of $\mathbf{R}^n$ are given by

$$\underline{\dim}_{\mathbf{B}} F = \underline{\lim}_{\delta \to 0} \frac{\log N_\delta(F)}{-\log \delta}$$

$$\overline{\dim}_{\mathbf{B}} F = \overline{\lim}_{\delta \to 0} \frac{\log N_\delta(F)}{-\log \delta}$$

and the box-counting dimension of F by

$$\dim_{\mathbf{B}} F = \lim_{\delta \to 0} \frac{\log N_\delta(F)}{-\log \delta}$$

(if this limit exists), where $N_\delta(F)$ is any of the following:

(i) the smallest number of closed balls of radius δ that cover F;

(ii) the smallest number of cubes of side δ that cover F;
(iii) the number of δ-mesh cubes that intersect F;
(iv) the smallest number of sets of diameter at most δ that cover F;
(v) the largest number of disjoint balls of radius δ with centers in F.

The upper box-counting dimension $\overline{\dim}_{\mathbf{B}}$ is finitely maxitive, but $\underline{\dim}_{\mathbf{B}}$ is not. The box-counting dimension generally is not a σ-maxitive measure. Take, for example, for F the set of all rational numbers between 0 and 1. The box-counting dimension of each rational number as one-point is zero, but the box-counting dimension of F is 1, since F is a dense subset of the interval $[0, 1]$.

2.2.5 Fuzzy Measures

Definition 2.14 A fuzzy measure m is a non-negative extended real-valued set function m defined on a σ-algebra Σ and with the properties:
(FM_1) $m(\emptyset) = 0$,
(FM_2) $E \subset F$ implies $m(E) \leq m(F)$,
(FM_3) $E_1 \subset E_2 \subset \ldots$, $E_n \in \Sigma$ implies $m(\cup_{n=1}^{\infty} E_n) = \lim_{n\to\infty} m(E_n)$,
(FM_4) If $E_1 \supset E_2 \supset \ldots$, $E_n \in \Sigma$ and there exists n_0 such that $m(E_{n_0}) < \infty$, then $m(\cap_{n=1}^{\infty} E_n) = \lim_{n\to\infty} m(E_n)$.

Fuzzy measures are often considered only with the properties (FM_1) and (FM_2). In this case we shall say that the fuzzy measure is considered in the general sense.

Fuzzy measures, together with different types of corresponding integrals (Choquet, Sugeno, etc.) turn out to be very useful in applications in image processing, information fusion from diverse sources, and synthetic evaluation of objects in terms of multiple factors.

2.2.6 Submeasures

A submeasure η is a positive, subadditive, monotone set function which vanishes at the empty set. Submeasures play an important role in many investigations on non-additive set functions (we will show some of these applications later in this book). It is null-additive.

Theorem 2.8 *(i) Every order continuous submeasure η on a ring $\mathcal{R}$ is σ-subadditive;*

(ii) If a submeasure η on a ring $\mathcal{R}$ is exhaustive and σ-subadditive then it is order continuous.

(iii) If $\mathcal{R}$ is a σ-ring then a submeasure η on $\mathcal{R}$ is order continuous if and only if it is exhaustive and σ-subadditive.

Proof.

(i) Let $\{E_n\}$ be a sequence of pairwise disjoint sets from $\mathcal{R}$ and such that $\bigcup_{i=1}^{\infty} E_i \in \mathcal{R}$. Then we have

$$\eta(\bigcup_{i=1}^{\infty} E_i) \leq \sum_{i=1}^{n} \eta(E_i) + m(\bigcup_{i=n+1}^{\infty} E_i).$$

Taking $n \to \infty$, we obtain

$$\eta(\bigcup_{i=1}^{\infty} E_i) \leq \sum_{i=1}^{\infty} \eta(E_i).$$

(ii) If $\{A_n\}$ is from $\mathcal{R}$ such that $A_n \searrow \emptyset$, by exhaustivity of η there exists an increasing sequence $\{n_k\}$ of natural numbers such that

$$\eta(A_{n_k} \setminus A_{n_{k+1}}) < 2^{-k} \qquad (k \in \mathbf{N}).$$

Therefore for $n > n_k$

$$A_n \subset \bigcup_{i=k}^{\infty} (A_{n_i} \setminus A_{n_{i+1}})$$

and by σ-subadditivity of η

$$\eta(A_n) < \sum_{i=k}^{\infty} 2^{-i}.$$

Letting $k \to \infty$ (and so also $n \to \infty$) we obtain $\eta(A_n) \searrow 0$.

(iii) Consequence of (i) and (ii). □

Definition 2.15 A submeasure η on a ring $\mathcal{R}$ is called complete ($\mathcal{R}$ is complete with respect to η) of for each Cauchy sequence $\{A_n\}$ from $\mathcal{R}$, i.e., such that

$$\eta(A_n \Delta A_m) \to 0$$

as $n, m \to \infty$, there exists $A \in \mathcal{R}$ such that

$$\eta(A \Delta A_n) \to 0 \quad \text{as} \quad n \to \infty.$$

Propositon 2.8 *A submeasure η on a ring $\mathcal{R}$ is exhaustive if and only if every monotone sequence $\{A_n\}$ from $\mathcal{R}$ is Cauchy with respect to η.*

Proof. Let η be exhaustive and suppose that a nondecreasing sequence $\{A_n\}$ is not Cauchy. Hence there exist $\epsilon > 0$ and a strictly increasing sequence $\{n_k\}$ of natural numbers such that

$$\eta(A_{n_{k+1}} \Delta A_{n_k}) > \epsilon \qquad (k \in \mathbf{N}),$$

but this contradicts the exhaustivity of η. The converse statement can be proved in an analogous way. □

We point out at the end of this section a classical (unsolved) problem related to submeasures. We will say that a submeasure is non-trivial if $\eta(X) > 0$ and that a submeasure η_1 dominates a submeasure η_2 if $\eta_1(A) \geq \eta_2(A)$ $(A \in \mathcal{A})$. A submeasure is pathological if it is nontrivial, but dominates no nontrivial measure.

Maharam's submeasure problem.

Does no algebra exist which carries a non-trivial pathological exhaustive submeasure?

This problem is equivalent with **Control Measure Problem:**

Does there exist for every σ-additive vector measure ν on a σ-algebra Σ a probability p such that for $A \in \Sigma$, $p(A) = 0$ implies $\nu(A) = 0$?

There are examples of non-exhaustive pathological submeasures.

2.2.7 Belief Measure and Plausibility Measure

Definition 2.16 A set function $m: \mathcal{P}(X) \to [0, 1]$ is a basic probability assignment if it satisfies the conditions

(i) $m(\emptyset) = 0$;

(ii) $\sum_{A \in \mathcal{P}(X)} m(A) = 1$.

Definition 2.17 Let m be a basic probability assignment on $\mathcal{P}(X)$. The belief measure Bel : $\mathcal{P}(X) \to [0, 1]$ induced by m is defined by

$$\text{Bel}\,(A) = \sum_{B \subset A} m(B) \qquad (A \in \mathcal{P}(X)).$$

Belief measure satisfies the following conditions $\mathrm{Bel}(\emptyset) = 0$, $\mathrm{Bel}(X) = 1$ and

$$\mathrm{Bel}(A_1 \cup A_2 \cup \cdots \cup A_n) \geq \sum_{i=1}^{n} \mathrm{Bel}(A_i) - \sum_{i<j} \mathrm{Bel}(A_i \cap A_j) + \cdots + (-1)^{n+1} \mathrm{Bel}(A_1 \cap A_2 \cap \cdots \cap A_n)$$

for $A_1, \ldots, A_n \in \mathcal{P}(X)$.

Hence a belief measure is monotone and superadditive. It can be proved that belief measure is continuous from above.

Definition 2.18 Let m be a basic probability assignment on $\mathcal{P}(X)$. Plausibility measure $\mathrm{Pl} : \mathcal{P}(X) \to [0,1]$ induced by m is defined by

$$\mathrm{Pl}(A) = \sum_{B \cap A \neq \emptyset} m(B) \qquad (A \in \mathcal{P}(X)).$$

If Bel and Pl are induced by the same basic probability assignment then

$$\mathrm{Bel}(A) = 1 - \mathrm{Pl}(A^c) \quad \text{and} \quad \mathrm{Bel}(A) \leq \mathrm{Pl}(A) \quad (A \in \mathcal{P}(X)).$$

This implies that the plausibility measure satisfies the following condition

$$\mathrm{Pl}(A_1 \cap A_2 \cap \cdots \cap A_n) \leq \sum_{i=1}^{n} \mathrm{Pl}(A_i) - \sum_{i<j} \mathrm{Pl}(A_i \cup A_j) + \cdots + (-1)^{n+1} \mathrm{Pl}(A_1 \cup A_2 \cup \cdots \cup A_n)$$

for $A_1, \ldots, A_n \in \mathcal{P}(X)$.

Hence the plausibility measure is monotone and subadditive and so it is a submeasure. Moreover it is continuous from below.

Example 2.12 Let X be a countable set. A Sugeno σ-$S_\lambda^{\mathbf{S}}$-decomposable measure m_λ defined on $\mathcal{P}(X)$ and with values in $[0,1]$ is a belief measure for $\lambda > 0$ and it is a plausibility measure for $\lambda < 0$. So, if we define for $\lambda > 0$ the basic probability assignment by

$$m(A) = \begin{cases} \lambda^{|A|-1} \prod_{x_i \in A} m_\lambda(\{x_i\}) & \text{if } A \neq \emptyset, \\ 0 & \text{if } A = \emptyset \end{cases}$$

for $A \in \mathcal{P}(X)$, then using that m_λ satisfies the σ-λ-rule, i.e.,

$$m_\lambda(\bigcup_{n=1}^{\infty} A_n) = \begin{cases} \frac{1}{\lambda}(\prod_{n=1}^{\infty}(1+\lambda m_\lambda(A_n)) & \text{if } \lambda \neq 0, \\ \sum_{n=1}^{\infty} m_\lambda(A_n) & \text{if } \lambda = 0, \end{cases}$$

for a disjoint sequence $\{A_n\}$.

It is easy to show

$$m_\lambda(A) = \sum_{B \subset A} m(B) = 1\,.$$

For $\lambda < 0$, we have for $\lambda^* = \frac{-\lambda}{\lambda+1} > 0$

$$m_\lambda(A) = 1 - m_{\lambda^*}(A) \qquad (A \in \mathcal{P}(X))\,.$$

Since m_{λ^*} is a plausibility measure it follows that m_λ is a belief measure.

The Dempster-Shafer theory is based on plausibility and belief measures. Namely, roughly speaking, probabilities between the corresponding lower probability (belief measure) and upper probability (plausibility measure) are considered.

2.2.8 k-Triangular Set Functions

Definition 2.19 A set function $m\colon \Sigma \to [0,\infty)$ is said to be k-triangular for $k \geq 1$ if $m(\emptyset) = 0$ and

$$m(A) - km(B) \leq m(A \cup B) \leq m(A) + km(B),$$

whenever $A, B \in \Sigma, A \cap B = \emptyset$.

We remark that even for the classical signed measure μ, the set function $|\mu(\cdot)|$ is not monotone, but it is 1-triangular.

Propositon 2.9 *Let $\mathcal{R}$ be a σ-ring. A monotone and σ-k-subadditive set function $m\colon \mathcal{R} \to [0,\infty]$ with $m(\emptyset) = 0$ is σ-null-additive.*

Proof. Follows by Proposition 2.3 and the inequality

$$m(B_1) \le m(\bigcup_{i=1}^{\infty} B_i) \le m(B_1) + k\sum_{i=2}^{\infty} m(B_i),$$

where $\{B_i\}$ is a sequence of pairwise disjoint sets from $\mathcal{R}$, such that $m(B_i) = 0$ $(i \in \mathbf{N})$. □

More generally, for a uniform semigroup $\mathbf{S}$ (with a neutral element 0) valued set function v, $v\colon \Sigma \to \mathbf{S}$ is said to be k-triangular if the preceding inequality for $m_i(A) = f_i(v(A))$ $(i \in I)$ holds, where $\{f_i\}_{i\in I}$ is the family of triangular functionals, i.e., such that

$$|f_i(x) - f_i(y)| \le f_i(x+y) \le f_i(x) + f_i(y) \quad (x, y \in \mathbf{S}) \quad \text{and} \quad f_i(0) = 0,$$

defined by

$$d_i(x, 0) = f_i(x) \qquad (x \in \mathbf{S}),$$

where $\{d_i\}_{i\in I}$ is the family of pseudometrics which generates the uniformity in $\mathbf{S}$.

It is obvious that a k-triangular set function m is always null-additive, although it may not be monotone . Chapter 11 is devoted to these set functions.

2.2.9 Distorted Measures

If μ is a signed measure on a σ-algebra Σ and f a function $f\colon \mathbf{R} \to \mathbf{R}$ with $f(0) = 0$, then a set function m defined by

$$m(A) = f(\mu(A)) \qquad (A \in \Sigma) \tag{2.1}$$

is called a distorted measure. If μ is a probability measure, i.e., $\mu(\emptyset) = 0$ and $\mu(X) = 1$, and f is an increasing function with $f(0) = 0$ and $f(1) = 1$, then m defined by (2.1) is called a distorted probability. It is always monotone and null-additive (for strictly increasing f).

If f is a convex (concave) function, then $m = f \circ \mu$ is supermodular (submodular) set function.

We repeat the classical notion of nonatomic measure.

Definition 2.20 A signed measure μ is nonatomic if for every $A \in \Sigma$ such that $\mu(A) \neq 0$ there exists $B \subset A$ such that $\mu(A) \neq \mu(B) \neq 0$.

We denote by NA the set of all nonatomic signed measures and by NA^+ its subspace of all non-negative measures.

Then the space $p\mathrm{NA}$ generated by the powers of measures from NA^+ is of great interest in game theory and mathematical economies. It is interesting that $m \in p\mathrm{NA}$ if and only if $m = f \circ \mu$ for absolutely continuous functions f.

2.2.10 Capacities

We will present in this subsection some results without proofs.

Let k be a positive continuous function on $\mathbf{R}^n \backslash \{0\}$ having at the origin the limit $+\infty$ (such a function will be called a kernel; we will always define it at the origin by the equality $k(0) = \infty$ and assume that it is locally summable). If μ is a nonnegative measure with compact support (we are using the relation between continuous functionals and measures), then the potential of μ with respect to the kernel k is the function

$$u_\mu(t) = \int_{\mathbf{R}^n} k(t-s)\, d\mu(s),$$

and the corresponding energy of μ (with respect to the same kernel) is

$$I(\mu) = \iint_{\mathbf{R}^n \times \mathbf{R}^n} k(t-s)\, d\mu(s)\, d\mu(t).$$

The capacity of a compact set C, $C \subset \mathbf{R}^n$, is the number $\mathrm{cap}_k C$,

$$\mathrm{cap}_k(C) = \sup\{\|\mu\|^2 : \mathrm{supp}\ \mu \subset C,\ I(\mu) \leq 1\}. \tag{2.2}$$

The most important case is when $k(t) = |t|^{-\alpha} = k_\alpha(t)$, $0 \leq \alpha < n$, where $|\cdot|$ is the usual Euclidean norm. The kernels k_α are called the Riesz kernels. In this case the notions just introduced are studied in an important branch of analysis known as potential theory. Here we give the connections with harmonic analysis, in particular with exceptional sets. First let us mention that the natural 'exceptional' sets are now the sets of capacity zero.

Theorem 2.9 *There exists a unique measure μ for which the least upper bound in (2.2) is assumed with the kernel $k = k_\alpha$.*

This measure μ is called the equilibrium distribution on C.

Theorem 2.10 *Let μ be the equilibrium distribution on C. Then the potential u_μ satisfies the inequality $u_\mu \leq 1$ everywhere on the support of μ, and $u_\mu \geq 1$ holds everywhere on C except on a set on which each measure of finite energy vanishes.*

It follows from the maximum principle, formulated below, that if $\alpha \geq n-2$ one has $u_\mu \leq 1$ everywhere.

Theorem 2.11 (maximum principle). *If the inequality $u_\mu \leq M$ holds in the support of μ and if $n-2 \leq \alpha < n$, then $u_\mu \leq M$ everywhere.*

Theorem 2.12 (crude maximum principle). *If the same assumption on v one has, in the case $0 < \alpha < n-2$, everywhere $u_\mu \leq 2^\alpha M$.*

Capacity is a non-additive set function but it is subadditive on compact sets:

$$\mathrm{cap}_\alpha(C_1 \cup C_2) \leq \mathrm{cap}_\alpha(C_1) + \mathrm{cap}_\alpha(C_2)\,.$$

So cap_α is null-additive on compact sets. If $n-2 \leq \alpha \leq n$ one has stronger inequality ('convexity of capacity').

$$\mathrm{cap}_\alpha(C_1 \cup C_2) + \mathrm{cap}_\alpha(C_1 \cap C_2) \leq \mathrm{cap}_\alpha(C_1) + \mathrm{cap}_\alpha(C_2)\,.$$

A capacity can be extended to the system of all Borel (or even analytic) sets in the following way. The inner capacity of an arbitrary set E is the number $\mathrm{cap}_* E = \sup\{\mathrm{cap}\, S : S \subset E,\, S \text{ compact}\}$. The outer capacity of E is defined as $\mathrm{cap}^* E = \inf\{\mathrm{cap}_* U : U \supset E,\, U \text{ open}\}$. Finally, E is called measurable with respect to capacity, if $\mathrm{cap}^* E = \mathrm{cap}_* E$ (then $\mathrm{cap}^* E$ is called the capacity of E and is denoted simply cap E). One can prove that all analytic sets are measurable with respect to capacity.

The notion 'quasi-everywhere' means 'everywhere except for a set of capacity zero'.

If α grows then the set of measures with finite energy with respect to k_α shrinks. From this it follows that for a fixed E the function $\alpha \mapsto \mathrm{cap}_{k_\alpha} E$ is decreasing. The capacitory dimension of a (closed) set E is defined to be the number $\inf\{\alpha : \mathrm{cap}_{k_\alpha} E = 0\}$.

It turns out that it is possible to describe capacitory dimension in other terms connected with Hausdorff measures (see Section 2.2.4).

Theorem 2.13 (Frostman). *The capacitory dimension and the Hausdorff dimension of a compact set coincide.*

2.2.11 Approximately Additive Set Functions

Let $\mathcal{A}$ be an algebra.

Definition 2.21 A set function $m\colon \mathcal{A} \to \mathbf{R}$ is called Δ-approximately additive if $m(\emptyset) = 0$ and

$$|m(A \cup B) - m(A) - m(B)| \leq \Delta \qquad (A, B \in \mathcal{A}).$$

The following close relation with additive set functions was proved.

Theorem 2.14 *There exists a universal constant $K < 45$ with the property that if $m\colon \mathcal{A} \to \mathbf{R}$ is Δ-approximately additive there is an additive set function $m\colon \mathcal{A} \to \mathbf{R}$ such that*

$$|m(A) - \mu(A)| \leq K \cdot \Delta \qquad (A \in \mathcal{A}).$$

The existing proof of Theorem 2.14 is based on many different tools, such as concentrators, Hall's Marriage Lemma, and covering index linear programming. So it is natural to ask whether a direct measure theoretical proof exists?

For the universal constant it is known that $\frac{1}{2} \leq K < 45$, but the best constant K is unknown. We can introduce in an analogous way the following generalization.

Definition 2.22 A set function $m\colon \mathcal{A} \to \mathbf{R}$ is called Δ-approximately null-additive if $m(\emptyset) = 0$ and

$$|m(A \cup B) - m(A)| \leq \Delta \qquad (A, B \in \mathcal{A})$$

whenever $m(B) = 0$.

It is then natural to ask whether a universal constant C and a null-additive set function m' exist such that

$$|m'(A) - m(A)| \leq C \cdot \Delta \qquad (A \in \mathcal{A}).$$

2.3 Notes and References

The basic facts on non-additive set functions can be found in [Aumann and Shapley 1974], [Denneberg 1994], [Pap 1982], [Z. Wang and Klir 1992].

The representations of t-conorms and pseudo-additions in Subsections 2.2.1 and 2.2.2 are results of Aczel, Mostert and Shields contained in [Ling 1965].

More details about decomposable measures can be found in [Dubois and Prade 1980], [Dubois and Prade 1982], [Ichihashi, Tanaka and Asai 1988], [Klement and S. Weber 1991], [Kruse 1982], [Pap 1990a], [Pap 1990b], [Pap 1990c], [Pap 1991a], [Pap 1991b], [Pap 1992], [Sugeno and Murofushi 1987], [Z. Wang and Klir 1992], [H. Weber 1984].

Maxitive measures appear especially in [Dubois and Prade 1988], [Shilkret 1971] and fractals theory for example in [Falconer 1990].

Fuzzy measures can be found in many papers at the end in the references, we only mention here the books [Dubois and Prade 1980] and [Z. Wang and Klir 1992].

Submeasures appears in many papers such as [Brook 1984], [Drewnowski 1972a], [Drewnowski 1972b], [Drewnowski 1978].

Belief and plausibility appears also in many papers and books, we take here only few of them [Shapley 1965], [Dubois and Prade 1980], [Dubois and Prade 1986], [Z. Wang and Klir 1992].

k-triangular set functions appears in [Agafanova and Klimkin 1974], [Antosik and Saeki 1992], [Drewnowski 1978], [Guariglia 1990a], [Guariglia 1990b], [Guariglia 1991], [Guselnikov 1975], [Pap 1982], [Pap 1986], [Pap 1987].

Distorted probability and measures appear in [Aumann and Shapley 1974], [Chateauneuf 1987], [Z. Wang and Klir 1992].

The detailed discussion of the material in the applications of capacity theory can be found in [Choquet 1953–1954], [Landkopf 1966], [Hayman and Kennedy 1976], [Berg and Forst 1975] and [O'Brien and Vervaat 1994].

Approximately additive set functions appear in [Kalton and Roberts 1983], and Maharam's problem was stated in [Maharam 1947] (more details can be found in [Kalton 1989]).

Chapter 3

Variations

In this chapter we will match to every set function with some special positive set functions with some additional properties.

We suppose in this chapter that a real valued set function $m\colon \mathcal{D} \to (-\infty, +\infty)$ (or with values in $[0, +\infty]$) satisfies the condition $m(\emptyset) = 0$.

We shall introduce two kinds of variations of a set function m.

3.1 Disjoint Variation

Definition 3.1 For an arbitrary but fixed subset A of X and a set function m we define the disjoint variation $\overline{m}$ by

$$\overline{m}(A) = \sup_{I} \sum_{i \in I} |m(D_i)|,$$

where the supremum is taken over all finite families $\{D_i\}_{i \in I}$ of pairwise disjoint sets of $\mathcal{D}$ such that $D_i \subset A$ $(i \in I)$.

Propositon 3.1 *If $A \in \mathcal{A}$, then in the previous definition we can take the supremum for all finite families $\{D_i\}_{i \in I}$ of disjoint sets such that $\bigcup_{i \in I} D_i = A$.*

Proof. If $D_1, D_2, \ldots, D_n$ are disjoint sets of $\mathcal{A}$ such that $D_i \subset A$ $(i = 1, 2, \ldots, n)$, then introducing one set more D_{n+1} by

$$D_{n+1} = A \setminus \bigcup_{i=1}^{n} D_i,$$

we obtain a finite family $D_1, D_2, \ldots, D_n, D_{n+1}$ of disjoint sets such that $\bigcup_{i=1}^{n+1} D_i = A$ and

$$\sum_{i=1}^{n} |m(D_i)| \leq \sum_{i=1}^{n+1} |m(D_i)| .$$

Hence

$$\begin{aligned}\overline{m}(A) &= \sup\left\{\sum_{i=1}^{n} |m(D_i)| : \{D_i\}_{i=1}^{n} \subset \Sigma\right\} \\ &\leq \sup\left\{\sum_{i=1}^{n+1} |m(D_i)| : \{D_i\}_{i=1}^{n+1} \subset \Sigma,\ \bigcup_{i=1}^{n+1} D_i = A\right\}.\end{aligned}$$

Since the opposite inequality is obvious, we obtain the desired equality. □

The main properties of disjoint variation are summarized in the following proposition.

Propositon 3.2 *Let m be a set function defined on $\mathcal{D}$ with values in $(-\infty, +\infty)$ (or $[0, +\infty]$), with $m(\emptyset) = 0$. Then the disjoint variation $\overline{m}$ of m has the following properties:*

(i) For every $A \subset X$ we have

$$0 \leq \overline{m}(A) \leq +\infty;$$

(ii) $\overline{m}(\emptyset) = 0$;

(iii) $|m(A)| \leq \overline{m}(A) \quad (A \in \mathcal{D})$;

(iv) $\overline{m}$ is monotone, i.e., if $B \subset A$, then $\overline{m}(B) \leq \overline{m}(A)$;

(v) For every $A \subset X$ we have

$$\overline{m}(A) \geq \sup\{|m(B)| : B \subset A,\ B \in \mathcal{D}\};$$

(vi) $\overline{m}(A) = 0$ if and only if $m(B) = 0$ for every subset B of A from $\mathcal{D}$.

Proof. (i) Follows by the definition of $\overline{m}$.

(ii) The only family is $I = \{\emptyset\}$, and so we have

$$\overline{m}(\emptyset) = \sup \sum |m(\emptyset)| = 0 .$$

(iii) Since one of the families I of disjoint sets of $\mathcal{D}$ contained in A is the family $I = \{A\}$ we obtain by the definition of $\overline{m}$ the desired inequality.

(iv) Since for any family $\{D_i\}_{i\in I}$ of disjoint sets of $\mathcal{D}$ with property $D_i \subset B$ $(i \in I)$, by $B \subset A$, we have $D_i \subset A$, too, we obtain

$$\sum_{i\in I} |m(D_i)| \leq \overline{m}(A),$$

and so $\overline{m}(B) \leq \overline{m}(A)$.

(v) If B is an arbitrary subset of A which belongs to $\mathcal{D}$ we have by the properties (iv) and (iii)

$$\overline{m}(A) \geq \overline{m}(B) \geq |m(B)|.$$

(vi) Let us suppose that $\overline{m}(A) = 0$ for an arbitrary set $A \subset X$. Then by (v) we obtain

$$\overline{m}(B) = 0 \quad \text{for each} \quad B \subset A, \quad B \in \mathcal{D}.$$

Suppose now that $\overline{m}(B) = 0$ for each $B \subset A$, $B \in \mathcal{D}$. Then for each finite family $\{D_i\}_{i\in I}$ of disjoint sets such that $D_i \subset A$ $(i \in I)$ we obtain

$$\sum_{i\in I} |m(D_i)| = 0.$$

This implies $\overline{m}(A) = 0$. □

Theorem 3.1 *Let m be a set function defined on Σ with values in $(-\infty, +\infty)$ (or $[0, +\infty]$), with $m(\emptyset) = 0$. Then $\overline{m}$ is superadditive, i.e.,*

$$\sum_{i\in I} \overline{m}(E_i) \leq \overline{m}\Big(\bigcup_{i\in I} E_i\Big)$$

for each family $\{E_i\}_{i\in I}$ of disjoint sets of X. Moreover, $\overline{m}$ is the smallest of all the positive set functions v defined on $\mathcal{P}(X)$ such that they are monotone, superadditive and satisfy the inequality

$$|m(A)| \leq v(A) \qquad (A \in \Sigma).$$

Proof. Let us suppose that E_1 and E_2 are two disjoint subsets of X. If we take an arbitrary but fixed real number r such that

$$\overline{m}(E_1) + \overline{m}(E_2) > r, \tag{3.1}$$

then there exist two real numbers r_1 and r_2 such that $r = r_1 + r_2$ and

$$\overline{m}(E_1) > r_1 \quad \text{and} \quad \overline{m}(E_2) > r_2 .$$

Then there exists a finite family $\{D_i^1\}_{1 \leq i \leq n}$ of disjoint sets from Σ with $D_i^1 \subset E_1$ $(i = 1, 2, \ldots, n)$ such that

$$\sum_{i=1}^{n} |m(D_i^1)| > r_1, \tag{3.2}$$

and a finite family $\{D_i^2\}_{1 \leq j \leq k}$ of disjoint sets from Σ, with $D_j^2 \subset E_2$ $(j = 1, 2, \ldots, k)$, such that

$$\sum_{j=1}^{k} |m(D_j^2)| > r_2 . \tag{3.3}$$

The family of sets $\{D_1^1, D_2^1, \ldots, D_n^1, D_1^2, D_2^2, \ldots, D_k^2\}$ consists of disjoint sets from Σ which are contained in $E_1 \cup E_2$ and by the definition of $\overline{m}(E_1 \cup E_2)$ we obtain

$$\overline{m}(E_1 \cup E_2) \geq \sum_{i=1}^{n} |m(D_i^1)| + \sum_{j=1}^{k} |m(D_j^2)| .$$

Hence by the inequalities (3.2) and (3.3)

$$\overline{m}(E_1 \cup E_2) > r_1 + r_2 = r . \tag{3.4}$$

Since the number r was arbitrary such that the inequality (3.1) is satisfied, we conclude that

$$\overline{m}(E_1) + \overline{m}(E_2) \leq \overline{m}(E_1 \cup E_2)$$

(since the opposite inequality is impossible by (3.4)).

We extend the preceding inequality by induction for every finite family $\{E_i\}_{i \in J}$ of disjoint subsets of X, i.e., we have

$$\sum_{i \in J} \overline{m}(E_i) \leq \overline{m}\Big(\bigcup_{i \in J} E_i\Big).$$

For an arbitrary family $\{E_i\}_{i\in I}$ of disjoint subsets of X we obtain by the preceding step that for each finite subset J of I

$$\overline{m}\Big(\bigcup_{i\in I} E_i\Big) \geq \overline{m}\Big(\bigcup_{i\in J} E_i\Big) \geq \sum_{i\in J} \overline{m}(E_i).$$

Hence

$$\overline{m}\Big(\bigcup_{i\in I} E_i\Big) \geq \sum_{i\in I} \overline{m}(E_i).$$

The second part of the theorem follows from the properties of any positive, monotone, superadditive set function v taking an arbitrary finite family $\{D_i\}_{i\in I}$ of disjoint sets from Σ contained in A, i.e.,

$$v(A) \geq v\Big(\bigcup_{i\in I} D_i\Big) \geq \sum_{i\in I} v(D_i) \geq \sum_{i\in I} |m(D_i)| .$$

Hence $v(A) \geq \overline{m}(A)$. □

Theorem 3.2 *Let m be a positive set function which is finitely (or countable) subadditive on Σ and $m(\emptyset) = 0$. Then its disjoint variation $\overline{m}$ is finitely (or countable) additive on class $\tau(\Sigma)$ of all subset $A \subset X$ which satisfy that $A \cap E \in \Sigma$ for every $A \in \Sigma$, and so also on Σ.*

Proof. Let $\{E_i\}_{i\in I}$ be a family of disjoint sets of $\tau(\Sigma)$, where I is finite if m is finitely subadditive, and countable if m is countable subadditive.

We take an arbitrary finite family $\{D_j\}_{1\leq j\leq n}$ of disjoint sets from Σ such that $D_j \subset \bigcup_{i\in I} E_i$ $(j = 1, 2, \ldots, n)$. Then $\{E_i \cap D_j\}_{1\leq j\leq n}$ for each $i \in I$ is the family of disjoint sets from Σ such that $E_i \cap D_j \subset A_i$ for $j = 1, 2, \ldots, n$. Since the family $\{E_i \cap D_j\}_{i\in I}$ for each $j = 1, 2, \ldots, n$ consists of disjoint sets from Σ such that

$$\bigcup_{i\in I}(E_i \cap D_j) = D_j \quad \text{for} \quad j = 1, 2, \ldots, n .$$

We have

$$\sum_{j=1}^{n} |m(D_j)| = \sum_{j=1}^{n} \Big| m\Big(\bigcup_{i\in I}(E_i \cap D_j)\Big)\Big| \leq \sum_{j=1}^{n} \Big|\sum_{i\in I} m(E_i \cap D_j)\Big|$$

$$\leq \sum_{j=1}^{n} \sum_{i\in I} |m(E_i \cap D_j)| = \sum_{i\in I} \sum_{j=1}^{n} |m(E_i \cap D_j)| \leq \sum_{i\in I} \overline{m}(E_i) .$$

Since $\{D_j\}_{1\le j\le n}$ was an arbitrary finite family of disjoint sets we obtain by the last inequality

$$\sum_{i\in I}\overline{m}(E_i)\ge\overline{m}\Big(\bigcup_{i\in I}E_i\Big).$$

Since by Theorem 3.1 $\overline{m}$ is superadditive we obtain the conclusion of Theorem 3.2. □

Corollary 3.1 *If* $m\colon\Sigma\to(-\infty,+\infty)$ *is finitely (or countable) additive then* $\overline{m}$ *is finitely (countable) additive.*

Proof. If m is finitely (or countable) additive, then by Theorem 3.2 it follows that $\overline{m}$ is finitely (countable respectively) additive, too. □

Definition 3.2 We say that $m\colon\Sigma\to(-\infty,+\infty)$ with $m(\emptyset)=0$, is with bounded disjoint variation if $\overline{m}(X)<+\infty$.

The set of all set functions with bounded disjoint variation is denoted by DV.

Theorem 3.3 *If* m *is signed additive set function then* m *is countable additive if and only if* $\overline{m}$ *is countable additive.*

Proof. By Corollary 3.1 it follows that if m is countable additive then $\overline{m}$ is countable additive too.

For the converse statement we use the well known fact that each signed measure on a σ-algebra is always of the finite disjoint variation.

By countable additivity of $\overline{m}$ we obtain

$$\lim_{n\to\infty}\overline{m}\Big(\bigcup_{i=n+1}^{\infty}E_i\Big)=\lim_{n\to\infty}\Big(\overline{m}(E)-\sum_{i=1}^{n}\overline{m}(E_i)\Big)=0 \tag{3.5}$$

for a sequence $\{E_i\}$ of disjoint sets from Σ such that $E=\bigcup_{i=1}^{\infty}E_i\in\Sigma$.

On the other hand, for each $n\in\mathbf{N}$ we have

$$\Big|m(E)-\sum_{i=1}^{n}m(E_i)\Big|=\Big|m(E)-m\Big(\bigcup_{i=1}^{n}E_i\Big)\Big|$$
$$=\Big|m\Big(\bigcup_{i=n+1}^{\infty}E_i\Big)\Big|\le\overline{m}\Big(\bigcup_{i=n+1}^{\infty}E_i\Big).$$

Hence by (3.5)

$$m\Big(\bigcup_{i=1}^{\infty} E_i\Big) = \sum_{i=1}^{\infty} m(E_i). \qquad \square$$

Propositon 3.3 *Let m_1 and m_2 be two real valued or with values in $[0,+\infty]$, set functions, with $m_i(\emptyset) = 0 \quad (i = 1,2)$, and c a real number. Then*

$$\overline{m_1 + m_2} \leq \overline{m}_1 + \overline{m}_2$$

and

$$\overline{cm_1} = |c|\,\overline{m}_1 .$$

Proof. Proof goes directly by the Definition 3.1. $\square$

The next theorem gives two sufficient conditions which ensure that the disjoint variation of a set function is completely described by its restriction to a special subclass of Σ.

Theorem 3.4 *Let m be a set function defined on Σ and with values in $(-\infty,+\infty)$ (or $[0,+\infty]$) such that $m(\emptyset) = 0$. If $m_{\mathcal{S}}$ is the restriction of m to a subclass $\mathcal{S} \subset \Sigma$ with $\emptyset \in \mathcal{S}$, then we have*

$$\overline{m}_{\mathcal{S}}(A) = \overline{m}(A) \quad \textit{for every subset } A \textit{ of } X\,,$$

if one of the following two conditions is satisfied

(i) For every $A \in \Sigma$ we have

$$|m(A)| \leq \sup\{|m(B)| : B \subset A,\ \ B \in \mathcal{S}\};$$

(ii) m is finitely (countable) additive and every set of Σ is the union of a finite (respectively countable) family of disjoint sets of $\mathcal{S}$.

Proof. For every subset A of X we have

$$\begin{aligned}
\overline{m}_{\mathcal{S}}(A) &= \sup_I \Big\{ \sum_{i\in I} |m(D_i)| : D_i \subset A,\ D_i \in \mathcal{S},\ D_i \cap D_j = \emptyset \text{ for } i \neq j \Big\} \\
&\leq \sup_I \Big\{ \sum_{i\in I} |m(D_i)| : D_i \subset A,\ D_i \in \Sigma,\ D_i \cap D_j = \emptyset \text{ for } i \neq j \Big\} \\
&= \overline{m}(A) .
\end{aligned}$$

It remains to prove that $\overline{m}_{\mathcal{S}} \geq \overline{m}$.

Case (i). For $A \subset X$ and an arbitrary but finite family $\{D_i\}_{1 \leq i \leq n}$ of disjoint sets from Σ contained in A choose for $\epsilon > 0$ a set D_i' such that $D_i' \subset D_i$, $D_i' \in \mathcal{S}$, and

$$|m(D_i)| \leq |m(D_i')| + \frac{\epsilon}{2^i}.$$

Hence

$$\sum_{i=1}^{n} |m(D_i)| \leq \sum_{i=1}^{n} |m(D_i')| + \epsilon \leq \overline{m}_{\mathcal{S}}(A) + \epsilon.$$

This implies $\overline{m}(A) \leq \overline{m}_{\mathcal{S}}(A) + \epsilon$. Since $\epsilon > 0$ was arbitrary we have $\overline{m} \leq \overline{m}_{\mathcal{S}}$.

Case (ii). Suppose that $A \in \Sigma$ and $\{E_i\}_{i \in I}$ is a family of disjoint sets from Σ such that $\bigcup_{i \in I} E_i = A$. We take I is finite if m is finitely additive and countable if m is countable additive. We have

$$|m(A)| = \left|\sum_{i \in I} m(E_i)\right| \leq \sum_{i \in I} |m(E_i)| = \sum_{i \in I} |m_{\mathcal{S}}(E_i)| \leq \overline{m}_{\mathcal{S}}(A).$$

By Theorem 3.1 $\overline{m}$ is the smallest of all monotone and superadditive set functions v defined on all subsets of X such that

$$|m(A)| \leq v(A) \qquad (A \in \Sigma).$$

Hence $\overline{m} \leq \overline{m}_{\mathcal{S}}$. □

Propositon 3.4 *If a set function m is positive and superaddivite, then for every set $A \subset X$ we have*

$$\overline{m}(A) = \sup\{m(B) : B \subset A,\ B \in \Sigma\} = \sup\{\overline{m}(A \cap B) : B \in \Sigma\}.$$

Moreover, if m is also positive and monotone, then

$$\overline{m}(A) = m(A) \qquad (A \in \Sigma).$$

Proof. Let $\{D_i\}_{1 \le i \le n}$ be a finite family of disjoint sets from Σ, contained in A. Let us denote $B = \bigcup_{i=1}^{n} D_i$. Then $B \subset A$, $B \in \Sigma$ and we have by positivity and superadditivity of m

$$\sum_{i=1}^{n} |m(D_i)| = \sum_{i=1}^{n} m(D_i) \le m\Big(\bigcup_{i=1}^{n} D_i\Big)$$
$$= m(B) \le \sup\{m(B) : B \subset A,\ B \in \Sigma\}.$$

Therefore

$$\overline{m}(A) \le \sup\{m(B) : B \subset A,\ B \in \Sigma\} \le \sup\{m(A \cap B) : B \in \Sigma\} \le \overline{m}(A).$$

The second part of the theorem follows from the first part and by monotonicity of m

$$m(A) \le \overline{m}(A) = \sup\{m(B) : B \subset A,\ B \in \Sigma\} \le m(A) \qquad \square$$

Can we obtain something new if we take the disjoint variation of the disjoint variation of a set function? The next theorem answers negatively to this question.

Theorem 3.5 *If m is a set function with finite variation, then*

$$\overline{\overline{m}} = \overline{m}.$$

Proof. Let $A \subset X$. Since for each finite family $\{D_i\}_{1 \le i \le n}$ of disjoint sets from Σ such that $D_i \subset A$, we have

$$\sum_{i=1}^{n} |m(D_i)| \le \sum_{i=1}^{n} \overline{m}(D_i) \le \overline{\overline{m}}(A),$$

using $|m(D_i)| \le \overline{m}(D_i)$ by (iii) from Proposition 3.2 and by the definition of disjoint variation. Hence

$$\overline{m}(A) \le \overline{\overline{m}}(A).$$

Conversely, by monotonicity of $\overline{m}$ and Proposition 3.4 (since $\overline{m}$ is positive and superadditive) we obtain

$$\overline{m}(A) \ge \sup\{\overline{m}(B) : B \subset A,\ B \in \Sigma\} = \overline{\overline{m}}(A).$$

Hence $\overline{\overline{m}}(A) = \overline{m}(A)$ for each $A \subset X$. $\square$

The relation of the disjoint variation with the notion of null-additivity is given in the next theorem.

Theorem 3.6 *If a set function m, $m(\emptyset) = 0$, is null-additive, then its disjoint variation $\overline{m}$ is also null-additive.*

Proof. Suppose that m is null-additive. Let $B \in \mathcal{D}$ be such that $\overline{m}(B) = 0$. Then by $|m(B)| \leq \overline{m}(B)$ it follows $m(B) = 0$. For arbitrary $A \in \mathcal{D}$ such that $A \cup B \in \mathcal{D}$ and $A \cap B = \emptyset$ we have

$$\begin{aligned}\overline{m}(A \cup B) &= \sup\Big\{\sum_{i=1}^{n} |m(D_i)| : \{D_i\} \text{ disjoint}\Big\} \\ &= \sup\Big\{\sum_{i=1}^{n} |m((D_i \cap A) \cup (D_i \cap B))| : \{D_i\} \text{ disjoint}\Big\} \\ &= \sup\Big\{\sum_{i=1}^{n} |m(D_i \cap A)| : \{D_i\} \text{ disjoint }\Big\} = \overline{m}(A),\end{aligned}$$

where we have used that $\overline{m}(D_i \cap B) = 0, \; i = 1, \ldots, n$, holds by the monotonicity of $\overline{m}$, which implies $m(D_i \cap B) = 0$ and hence by null-additivity of m

$$m((D_i \cap A) \cup (D_i \cap B)) = m(D_i \cap A). \qquad \square$$

3.2 The Chain Variation

The second type of variations of set functions is given in the next definition.

Definition 3.3 The chain variation of a real valued set function m with $m(\emptyset) = 0$ on the set $A \in \mathcal{D}$ is given by

$$\begin{aligned}|m|(A) = \sup\Big\{&\sum_{i=1}^{n} |m(A_i) - m(A_{i-1})| : \\ &\emptyset = A_0 \subset A_1 \subset \cdots \subset A_n = A, \; A_i \in \mathcal{D}, \; i = 1, \ldots, n\Big\}\end{aligned}$$

We remark that the supremum in the previous definition is taken over all chains between $\emptyset$ and A.

We have an analogous proposition to Proposition 3.2.

Propositon 3.5 *Let m be a set function with $m(\emptyset) = 0$, then the chain variation of m has the following properties:*

(i) $0 \leq |m|(A) \leq +\infty \quad (A \in \mathcal{D})$;

(ii) $|m|(\emptyset) = 0$;

(iii) $|m(A)| \leq |m|(A) \quad (A \in \mathcal{D})$;

(iv) $|m|$ *is monotone, i.e.,*

$$|m|(B) \leq |m|(A) \quad \textit{for } B \subset A, \ A, B \in \mathcal{D};$$

(v) $\sup\{m(B) : B \subset A, \ B \in \mathcal{D}\} \leq |m|(A) \quad (A \in \mathcal{D})$;

(vi) $|m|(A) = 0$ *if and only if* $m(B) = 0$ *for every set* $B \subset A, \ B \in \mathcal{D}$.

Proof. (i) Follows by Definition 3.3.

(ii) Every chain connecting $\emptyset$ with $\emptyset$ consists only of an empty set, therefore

$$|m|(\emptyset) = \sup\Big\{\sum |m(A_i) - m(A_{i-1})| : \emptyset = A_0 \subset A_n = \emptyset\Big\} = 0\,.$$

(iii) One of the chains which connect $\emptyset$ and the set A consists only of $\emptyset$ and A, i.e., $\emptyset = A_0 \subset A_n = A$, therefore

$$|m(A)| = \sup\{|m(A) - m(\emptyset)| : \emptyset = A_0 \subset A_n = A\} \leq |m|(A)\,.$$

(iv) Some of chains from $\emptyset$ to A contain chains from $\emptyset$ to B, (exhaust all of them), therefore

$$\begin{aligned} |m|(B) = \sup\Big\{\sum_{i=1}^{n} |m(A_i) - m(A_{i-1})| &: \emptyset = A_0 \subset A_1 \subset \cdots \subset A_n = B, \\ &A_i \in \mathcal{D}, \ i = 1, \ldots, n\Big\} \\ \leq \sup\Big\{\sum_{i=1}^{s} |m(A_i) - m(A_{i-1})| &: \emptyset \subset A_0 \subset A_1 \subset \cdots \subset A_s = A, \\ &A_i \in \mathcal{D}, \ i = 1, \ldots, s\Big\} \\ = |m|(A)\,. \end{aligned}$$

(v) By properties (iii) and (iv) we have for every $B \subset A, \ B \in \mathcal{D}$,

$$|m(B)| \leq |m|(B) \leq |m|(A)\,.$$

(vi) Suppose $|m|(A) = 0$. Then for every set $B \subset A$, $B \in \mathcal{D}$, we have by (v) $m(B) = 0$.

Conversely, if $m(B) = 0$ for every set $B \subset A$, $B \in \mathcal{D}$, then for any chain $\emptyset = A_0 \subset A_1 \subset \cdots \subset A_n = A$ we have $m(A_i) = 0$, $i = 1, 2, \ldots, n$, and so

$$\sum_{i=1}^{n} |m(A_i) - m(A_{i-1})| = 0\,.$$

Hence $|m|(A) = 0$. □

Although disjoint variation and chain variation have many common properties, they are different, as we have to remark that the disjoint variation is defined for every subset A of X and for complex valued set functions, but chain variation is not.

The following theorem shows that even on Σ they are generally different.

Theorem 3.7 *If the set function m with $m(\emptyset) = 0$ is positive and monotone, then*

$$|m|(A) = m(A) \qquad (A \in \mathcal{D}),$$

but there is a set function m such that $\overline{m} \neq m$ on $\mathcal{D}$. If additionally m is superadditive then

$$|m|(A) = \overline{m}(A) = m(A) \qquad (A \in \mathcal{D})\,.$$

Proof. Since for any chain $\emptyset = A_0 \subset A_1 \subset \cdots \subset A_n = A$ in $A \in \mathcal{D}$

$$\sum_{i=1}^{n} |m(A_i) - m(A_{i-1})| = m(A)$$

by monotonicity and positivity of m, we obtain $|m|(A) = m(A)$.

On the other hand taking for m a submeasure which is not additive we obtain by Theorem 3.1 that $\overline{m}$ is additive and so different from m.

The second part of the theorem follows from Proposition 3.4.. □

Consequently, for positive additive set functions and for measures the two variations coincide.

The two kinds of variations coincide also for signed measures.

Theorem 3.8 *If m is an additive set function (real or complex charge) defined on $\mathcal{R}$, then $|m| = \overline{m}$.*

Proof. Take an arbitrary chain $\emptyset = A_0 \subset A_1 \subset \cdots \subset A_n = A$. Then

$$\sum_{i=1}^{n} |m(A_i) - m(A_{i-1})| = \sum_{i=1}^{n} |m(D_i)|,$$

where

$$D_i = A_i \setminus A_{i-1}, \qquad i = 1, 2, \ldots, n,$$

are disjoint. Hence

$$\begin{aligned} |m|(A) &\le \sup\Big\{ \sum_{i=1}^{n} |m(D_i)| : D_1, \ldots, D_n \text{ disjoint}, \\ &\qquad D_i = A_i \setminus A_{i-1},\ i = 1, \ldots, n \Big\} \\ &\le \overline{m}(A). \end{aligned}$$

On the other hand, let $\{D_i\}_{i \in I}$ be a finite family of disjoint sets such that $\bigcup_{i=1}^{n} D_i = A$ where $I = \{1, 2, \ldots, n\}$. Taking $A_i = \bigcup_{k=0}^{n} D_k$ $(i = 1, \ldots, n)$ and $D_0 = \emptyset$, we obtain for this special chain

$$\sum_{i=1}^{n} |m(A_i) - m(A_{i-1})| = \sum_{i=1}^{n} \Big| m\Big(\bigcup_{r=0}^{i} D_r\Big) - m\Big(\bigcup_{r=0}^{i-1} D_r\Big)\Big| = \sum_{i=1}^{n} |m(D_i)|.$$

Hence $|m|(A) \ge \overline{m}(A)$. □

3.3 The Space BV

Definition 3.4 A real-valued set function m with $m(\emptyset) = 0$ is of bounded chain variation if $|m|(X) < +\infty$.

We denote the family of set functions which vanish on empty sets and with bounded chain variation on (X, Σ) by BV. Since for $m, v \in BV$ and $a \in \mathbf{R}$ we have

1. $|m|(X) = 0$ if and only if $m = 0$;
2. $|am|(X) = |a| \cdot |m|(X)$;
3. $|m + v|(X) \le |m|(X) + |v|(X)$,

the functional

$$\|m\| = |m|(X)$$

is a norm on the vector space BV.

Moreover we have

Theorem 3.9 *The normed space* $(BV, \|\ \|)$ *is a Banach space.*

Proof. Take a Cauchy sequence $\{m_n\}$ from BV with respect to the norm $\|\ \| = |\cdot|(X)$. Then we can find an index $n_0 \in \mathbf{N}$ such that

$$\|m_n - m_{n_0}\| \le 1 \quad \text{for } n \ge n_0 . \tag{3.6}$$

For each $A \in \Sigma$ we consider a sequence $\{m_n(A)\}$ of real numbers. Let $m(A) = \lim_{n\to\infty} m_n(A)$. We shall prove that $m \in BV$.

Then for an arbitrary but fixed chain $\emptyset = A_0 \subset A_1 \subset \cdots \subset A_k = X$ and every index $n \ge n_0$ we have by (3.6)

$$\begin{aligned}
&\sum_{i=1}^{k} |m_n(A_i) - m_n(A_{i-1})| - \|m_{n_0}\| \\
&\qquad \le \sum_{i=1}^{k} |m_n(A_i) - m_n(A_{i-1})| - \sum_{i=1}^{k} |m_{n_0}(A_i) - m_{n_0}(A_{i-1})| \\
&\qquad \le \sum_{i=1}^{k} |(m_n - m_{n_0})(A_i) - (m_n - m_{n_0})(A_{i-1})| \\
&\qquad \le \|m_n - m_{n_0}\| \le 1 .
\end{aligned}$$

Hence

$$\sum_{i=1}^{k} |m(A_i) - m(A_{i-1})| \le 1 + \|m_{n_0}\| .$$

Since the considered chain was arbitrary, we have

$$\|m\| \le 1 + \|m_{n_0}\| , \text{ i.e., } m \in BV .$$

It follows that $\|m_n - m\| \to 0$ as $n \to \infty$. □

The following theorem gives an important characterization of elements of BV.

Theorem 3.10 *A set function* m, *with* $m(\emptyset) = 0$, *belongs to* BV *if and only if it can be represented as a difference of two monotone set functions* λ_1 *and* λ_2 *vanishing on an empty set.*

If $m \in BV$ *then*

$$\|m\| = \inf\{v_1(X) + v_2(X)\} , \tag{3.7}$$

where the infimum is taken over all monotone set functions v_1 *and* v_2 *such that* $m = v_1 - v_2$.

Proof. Suppose that $m \in BV$. We define

$$m^+(A) = \sup\Big\{\sum_{i=1}^{n} \max\{m(A_i) - m(A_{i-1}), 0\} :$$
$$\emptyset = A_0 \subset A_1 \subset \cdots \subset A_n = A,\ A_i \in \Sigma,\ i = 1, \ldots, n\Big\},$$

which is called the upper (positive) variation of m. m^+ is monotone. Namely, if $A \supset B$, $A, B \in \Sigma$, then

$$m^+(A) \geq \max\{m(A) - m(B), 0\} + m^+(B) \geq m^+(B).$$

We define $m^- = m^+ - m$. Then m^- is monotone, too. Namely, we have

$$m^+(A) \geq \max\{m(A) - m(B), 0\} + m^+(B) \geq m(A) - m(B) + m^+(B).$$

Hence

$$m^-(A) = m^+(A) - m(A) \geq m^+(B) - m(B) = m^-(B).$$

We remark, that by the definition of m^- it follows

$$m^-(A) = \sup\Big\{\sum_{i=1}^{n} \max\{-(m(A_i) - m(A_{i-1})), 0\}$$
$$\emptyset = A_0 \subset A_1 \subset \cdots \subset A_n = A,\ A_i \in \Sigma,\ i = 1, \ldots, n\Big\}$$

and it is usually called the lower variation of m.

Conversely, let $m = v_1 - v_2$ where v_1 and v_2 are monotone set functions. Then for any chain $\emptyset = A_0 \subset A_1 \subset \cdots \subset A_n = X$ we have

$$\sum_{i=1}^{n} |m(A_i) - m(A_{i-1})|$$
$$\leq \sum_{i=1}^{n} |v_1(A_i) - v_1(A_{i-1})| + \sum_{i=1}^{n} |v_2(A_i) - v_2(A_{i-1})|$$
$$\leq v_1(X) + v_2(X).$$

Hence

$$\|m\| \leq \inf\{v_1(X) + v_2(X)\},$$

where the infimum is taken over all monotone set functions v_1 and v_2 such that $m = v_1 - v_2$, which implies $m \in BV$.

By the preceding inequality, to prove (3.7) we need the opposite inequality

$$\|m\| \geq \inf\{v_1(X) + v_2(X)\}. \tag{3.8}$$

We take $v_1 = m^+$ and $v_2 = m^-$. For an arbitrary but fixed $\epsilon > 0$ we choose a chain

$$\emptyset = A_0 \subset A_1 \subset \cdots \subset A_n = X$$

such that

$$\sum_{i=1}^{n} \max\{m(A_i) - m(A_{i-1}), 0\} \geq m^+(X) - \epsilon.$$

Let $\mathcal{S}_1$ be a subchain which consists of chain pairs $\{A_{i-1}, A_i\}$ for which

$$\sum_{\mathcal{S}_1} |m(A_i) - m(A_{i-1})| \geq m^+(X) - \epsilon \tag{3.9}$$

holds, where $\sum_{\mathcal{S}_1}$ denotes the summation over subchain $\mathcal{S}_1$.

We denote by $\mathcal{S}_2$ the subchain which consists of pairs $\{A_{i-1}, A_i\}$ of the rest with respect to $\mathcal{S}_1$. Then we have

$$\begin{aligned}\sum_{\mathcal{S}_1} |m(A_i) - m(A_{i-1})| - \sum_{\mathcal{S}_2} |m(A_i) - m(A_{i-1})| &= \sum_{i=1}^{n} (m(A_i) - m(A_{i-1})) \\ &= m(X).\end{aligned}$$

Hence

$$\begin{aligned}\sum_{\mathcal{S}_2} |m(A_i) - m(A_{i-1})| &= \sum_{\mathcal{S}_1} |m(A_i) - m(A_{i-1})| - m(X) \\ &\geq m^+(X) - m(X) - \epsilon \\ &= m^-(X) - \epsilon.\end{aligned}$$

Using the preceding inequality and (3.9) we obtain

$$\begin{aligned}\sum_{i=1}^{n} |m(A_i) - m(A_{i-1})| &= \sum_{\mathcal{S}_1} |m(A_i) - m(A_{i-1})| + \sum_{\mathcal{S}_2} |m(A_i) - m(A_{i-1})| \\ &\geq m^+(X) + m^-(X) - 2\epsilon \\ &\geq \inf\{v_1(X) + v_2(X)\} - 2\epsilon.\end{aligned}$$

Since $\epsilon > 0$ was arbitrary we obtain the desired inequality (3.9). □

Corollary 3.2 *For $m \in BV$ we have*

$$\|m\| = m^+(X) + m^-(X) = \|m^+\| + \|m^-\|.$$

Propositon 3.6 *For every $m_1, m_2 \in BV$ we have*

$$\|m_1 \cdot m_2\| \leq \|m_1\| \cdot \|m_2\|.$$

For monotone m_1 and m_2 the equality holds in the preceding inequality.

Proof. The second part of the proposition is obvious.

By Theorem 3.10 we have the following representations of m_1, $m_2 \in BV$

$$m_1 = m_1^+ - m_1^- \quad \text{and} \quad m_2 = m_2^+ - m_2^-,$$

where m_i^+, m_i^-, $(i = 1,2)$ are monotone set functions defined in the proof of Theorem 3.10.

By the second part of this Proposition and Corollary 3.2 we obtain

$$\begin{aligned} \|m_1 \cdot m_2\| &= \|m_1^+ - m_1^-)(m_2^+ - m_2^-)\| \\ &\leq \|m_1^+ m_2^+\| + \|m_1^+ m_2^-\| + \|m_1^- m_2^+\| + \|m_1^- m_2^-\| \\ &= (\|m_1^+\| + \|m_1^-\|)(\|m_2^+\| + \|m_2^-\|) \\ &= \|m_1\| \cdot \|m_2\|. \end{aligned}$$

□

We denote by FA the set of all bounded finitely additive set functions and by FCA its subset of countable additive set functions (measures). Let NA denote the space of all non-atomic measures.

It is easy to check that all spaces: NA^+, NA, CA, FA are closed subspaces of BV with respect to the norm $\|m\| = |m|(X)$.

There are null-additive set functions which do not belong to BV.

Example 3.1 Let $X = [-1, 1]$ and Σ be a family of all Borel subsets of $[-1, 1]$. Taking a measure

$$\mu(A) = \int_A \operatorname{sgn} x \, dx \qquad (A \in \Sigma),$$

we have that

$$m(A) = \sqrt{|\mu(A)|} \qquad (A \in \Sigma)$$

is a null-additive set function. But $m \notin BV$. If we take a special chain

$$\emptyset = A_0 \subset A_1 \subset \cdots \subset A_{2n},$$

with $A_{2i-1} = [-\frac{i-1}{n}, \frac{i}{n}]$ and $A_{2i} = [-\frac{i}{n}, \frac{i}{n}]$ for $i = 1, 2, \ldots, n$. Then

$$\|m\| \geq \sum_{i=1}^{2n} |m(A_i) - m(A_{i-1})| = 2\sqrt{n} \to \infty$$

as $n \to \infty$. Hence $m \notin BV$.

There are also set functions which belong to BV, but which are not null-additive.

3.4 Notes and References

Disjoint and chain variations were used very often in classical measure theory, where they coincide: [Constantinescu and K. Weber 1985], [Dinculeanu 1967], [Diestel and Uhl 1977], [Dunford and Swartz 1958], [M. M. Rao 1987], [K. P. S. B. Rao and M. B. Rao 1983]. Some properties of the variations collected in this section are prepared using results in [Aumann and Shapley 1974], [Constantinescu and K. Weber 1985], [Dinculeanu 1967], [Dunford and Swartz 1958], [M. M. Rao 1987], [K. P. S. B. Rao and M. B. Rao 1983].

Chapter 4

Additive Representations of Set Functions

In this chapter we will match an arbitrary real-valued set function to an additive set function.

4.1 Finite Case

Let $\mathcal{A}$ be an algebra of subsets of given set X. If $\mathcal{A}$ is finite we will suppose without loss of generality that X is also finite and that $\mathcal{A} = \mathcal{P}(X)$.

Lemma 4.1 *If A is a finite set, then we have*

(i) $\sum_{B \subset A} (-1)^{|B|} = 0$, *where $|B|$ denotes the cardinality of the set B.*

(ii) For an arbitrary but fixed subset B of A such that $B \neq A$, we have

$$\sum_{C, B \subset C \subset A} (-1)^{|C|} = 0\,,$$

where $\sum_{C,p(C)}$ denotes summation over all C which satisfies the property $p(C)$: $B \subset C \subset A$.

Proof. (i) If A has n different elements, then there are $\binom{n}{i}$ subsets with i elements, $i = 0, 1, \ldots, n$. Therefore,

$$\sum_{B \subset A} (-1)^{|B|} = \sum_{i=0}^{n} (-1)^i \binom{n}{i} = (1-1)^n = 0\,.$$

(ii) We obtain by (i)

$$\sum_{C,B\subset C\subset A}(-1)^{|C|}=\sum_{D,D\subset A\setminus B}(-1)^{|B\cup D|}=\sum_{D,D\subset A\setminus B}(-1)^{|B|}\cdot(-1)^{|D|}$$
$$=(-1)^{|B|}\cdot\sum_{D\subset A\setminus B}(-1)^{|D|}=0\,.$$

□

Theorem 4.1 *Let m be a real-valued set function defined on $\mathcal{A}=\mathcal{P}(X)$, for $\mathcal{A}$ finite, with $m(\emptyset)=0$. Then there exists a unique family $\{a_A\}_{A\in\mathcal{A}\setminus\{\emptyset\}}$ of real numbers such that*

$$m=\sum_{A\in\mathcal{A}\setminus\{\emptyset\}}a_Am_A\,, \tag{4.1}$$

where

$$m_A(E)=\begin{cases}1, & \text{for } E\supset A\\ 0, & \text{otherwise}\end{cases}$$

(elementary belief function).

The elements of the family $\{a_A\}_{A\in\mathcal{A}\setminus\{\emptyset\}}$ have the following form

$$\begin{aligned}a_A&=\sum_{B\subset A}(-1)^{|A|-|B|}m(B)\\&=m(A)-\sum_{I\subset\{1,2,\ldots,n\},I\neq\emptyset}(-1)^{|I|+1}m\Big(\bigcap_{i\in I}A_i\Big),\end{aligned}\tag{4.2}$$

where $A_i=A\setminus\{x_i\}$ for $A=\{x_1,x_2,\ldots,x_n\}$.

Proof. Taking the family $\{a_A\}_{A\in\mathcal{A}\setminus\{\emptyset\}}$ given by (4.2) we have for $E\in\mathcal{P}(X)$

$$\begin{aligned}\sum_{A\in\mathcal{A}\setminus\{\emptyset\}}a_Am_A(E)&=\sum_{A\subset E}a_Am_A(E)=\sum_{A\subset E}\sum_{B,B\subset A}(-1)^{|A|-|B|}m(B)\\&=\sum_{B,B\subset A}((-1)^{|B|}m(B)\sum_{A,B\subset A\subset E}(-1)^{|A|})\\&=(-1)^{|E|}m(E)(-1)^{|E|}=m(E),\end{aligned}$$

where we have used (ii) from Lemma 4.1.

We will prove now the uniqueness of the representation (4.1) of m. If for m hold

$$m(E) = \sum_{A\in\Sigma\setminus\{\emptyset\}} b_A m_A(E) = \sum_{A\subset E} b_A m_A(E) \qquad (E \in \mathcal{A}),$$

then by Lemma 4.1 for every $A \in \mathcal{A} \setminus \{\emptyset\}$ we have

$$\begin{aligned} b_A &= (-1)^{|A|} b_A (-1)^{|A|} = (-1)^{|A|} \sum_{C\subset A} \Big(b_C m_C(A) \cdot \sum_{C\subset B\subset A} (-1)^{|B|} \Big) \\ &= (-1)^{|A|} \sum_{B\subset A} \Big((-1)^{|B|} \sum_{C\subset B} b_C m_C(B) \Big) = (-1)^{|A|} \sum_{B\subset A} (-1)^{|B|} m(B) \\ &= \sum_{B\subset A} (-1)^{|A|-|B|} m(B) = a_A . \end{aligned}$$

□

As a consequence (which we will not prove here) we have

Theorem 4.2 *Let m be a real-valued set function defined on $\mathcal{P}(X)$, X finite, with $m(\emptyset) = 0$. m is totally monotone if and only if $a_A \geq 0$ ($A \in \mathcal{P}(X) \setminus \{\emptyset\}$), where $\{a_A\}_{A\in\mathcal{A}\setminus\{\emptyset\}}$ is from Theorem 4.1.*

Moreover, m is additive if and only if $a_A = 0$ for all A with $|A| \geq 2$.

Definition 4.1 If $\mathcal{A}$ is finite and m a real-valued set function on $\mathcal{A}$ we define the composition norm of m by

$$\|m\|_c = \sum_{A\in\mathcal{A}\setminus\{\emptyset\}} |a_A| ,$$

where $\{a_A\}_{A\in\mathcal{A}\setminus\{\emptyset\}}$ is from Theorem 4.1.

Theorem 4.3 *If $\mathcal{A}$ is finite and m a real-valued set function on $\mathcal{A}$, then there exist unique totally monotone set functions m^+ and m^- such that*

$$m = m^+ - m^- \quad \text{and} \quad \|m\|_c = \|m^+\|_c + \|m^-\|_c .$$

Moreover, $\|m\|_c = m(X)$ if and only if m is totally monotone.

Proof. Using the representation by Theorem 4.1

$$m = \sum_{A \in \mathcal{A} \setminus \{0\}} a_A m_A ,$$

we have

$$m^+ = \sum_{A \in \mathcal{A}^+} a_A m_A \quad \text{and} \quad m^- = \sum_{A \in \mathcal{A} \setminus \mathcal{A}^+ \setminus \{\emptyset\}} (-a_A) m_A ,$$

where

$$\mathcal{A}^+ = \{A \in \mathcal{A} \setminus \{\emptyset\} : a_A \geq 0\} .$$

Hence $m = m^+ - m^-$, $\|m\|_c = \|m^+\|_c + \|m^-\|_c$ and by Theorem 4.2 we have that m^+ and m^- are totally monotone. □

4.2 General Case

The obtained representation in Theorem 4.1 for the finite case in the previous section we can interpret in the following way. We consider the measurable space $(\mathcal{A} \setminus \{\emptyset\}, \mathcal{P}(\mathcal{A} \setminus \{\emptyset\}))$ and for each real-valued set function m we add a corresponding a signed measure μ_m by

$$\mu_m(C) = \sum_{A \in C} a_A \qquad (C \in \mathcal{P}(\mathcal{A} \setminus \{\emptyset\})) .$$

Then we can interpret the corresponding representation of m as an integral, i.e.,

$$m = \sum_{A \in \mathcal{A} \setminus \{\emptyset\}} a_A m_A = \int_{\mathcal{A} \setminus \{\emptyset\}} a_A \, d\mu_m(A) .$$

This means, that if we want the general infinite case, we have to construct an algebra on $\mathcal{A} \setminus \{\emptyset\}$ on which we will have an analogous representation.

First for every $A \in \mathcal{A}$ we define $\tilde{A} \subseteq \mathcal{A} \setminus \{\emptyset\}$ by

$$\tilde{A} = \{S \in \mathcal{A} \setminus \{\emptyset\} : S \subset A\} .$$

Let $\mathcal{E} = \{\tilde{A} : A \in \mathcal{A} \setminus \{\emptyset\}\}$. Let $\mathcal{A}(\mathcal{E})$ and $\sigma(\mathcal{E})$ be algebra and σ-algebra, respectively, generated by $\mathcal{E}$.

We extend the composition norm (variation) to the general situation in the following way

$$\|m\|_c = \sup\{\|m|_{\mathcal{A}_0}\|_c : \ \mathcal{A}_0 \text{ if a finite sub-algebra of } \mathcal{A}\}.$$

It is easy to check that $\|\cdot\|_c$ is a norm on the set of all real-valued set functions defined on $\mathcal{A}$ and vanishing at the empty set.

If m is additive, then $\|m\|_c = \overline{m}$.

Theorem 4.4 *If m is a real-valued set function defined on algebra $\mathcal{A}$ of subset of X, with $m(\emptyset) = 0$, then there exists a unique signed finitely additive set function μ_m on $(\mathcal{A} \setminus \{\emptyset\}, \mathcal{A}(\mathcal{E}))$ such that*

$$m = \int_{\mathcal{A}\setminus\{\emptyset\}} m_A \, d\mu_m(A). \tag{4.3}$$

Moreover $\|m\|_c = \overline{\mu}_m$ and the map $m \mapsto \mu_m$ is linear and continuous.

Conversely, every additive set function μ on $\mathcal{A}(\mathcal{E})$ by (4.3) defines a real-valued set function m with $m(\emptyset) = 0$.

Furthermore, m is totally monotone if and only if μ_m is non-negative.

Proof. We introduce a basic element as a subset of $\mathcal{A} \setminus \{\emptyset\}$ as

$$\tilde{A} \setminus \bigcup_{i=1}^{n} \tilde{B}_i,$$

where B, $B_i \in \mathcal{A} \setminus \{\emptyset\}$, $i = 1, \ldots, n$ and $\emptyset \neq B_i \subset B$, $B_i \neq B$, $B_i \not\subset B_j$ $(i \neq j)$. Every element of the algebra $\mathcal{A}(\mathcal{E})$ generated by $\mathcal{E} = \{\tilde{A} : A \in \mathcal{A} \setminus \{\emptyset\}\}$ can be represented in its disjunctive normal form

$$\dot{\bigcup_{i=1}^{n}}\Big(\Big(\bigcap_{j=1}^{r_i} \tilde{A}_{ij}\Big) \cap \Big(\bigcap_{j=1}^{s_i} \tilde{B}_{ij}\Big)^c\Big)$$

where $\dot{\bigcup}_{i=1}^{n}$ is disjoint union. Without loss of generality we can take $r_i \geq 1$, for $r_i = 0$ we choose $A_{i1} = X$. Since $A = \bigcap_\alpha A_\alpha$ implies $\tilde{A} = \bigcap_\alpha \tilde{A}_\alpha$ we obtain that every element of $\mathcal{A}(\mathcal{E})$ can be represented as the union of finitely many disjoint basic elements.

We define μ_m on basic elements by

$$\mu_m\Big(\tilde{B} \setminus \bigcup_{i=1}^{n} \tilde{B}_i\Big) = m(B) - \Delta_m(\{B_i\}),$$

where

$$\Delta_m(\{B_i\}_{i=1}^n) = \sum_{I\subset\{1,\ldots,n\},I\neq\emptyset} (-1)^{|I|+1} m\Big(\bigcap_{i\in I} B_i\Big).$$

If $\mathcal{A}_0$ is a finite sub-algebra of $\mathcal{A}$ and $\mathcal{A}(\mathcal{E}_0)$ be the sub-algebra of $\mathcal{A}(\mathcal{E})$ generated by

$$\mathcal{E}_0 = \{\tilde{A} : A \in \mathcal{A}_0\},$$

then by Theorem 4.1

$$\mu_m\Big(\tilde{B}\setminus\bigcup_{i=1}^n \tilde{B}_i\Big) = \sum_{\{A\in\mathcal{A}_0 : A\subset B, A\not\subset B_i, i\leq n\}} a_A \quad (A, B_i \in \mathcal{A}_0,\ i = 1,\ldots,n). \tag{4.4}$$

Since every two elements of $\mathcal{A}(\mathcal{E})$ belong to $\mathcal{A}(\mathcal{E}_0)$ for some finite sub-algebra $\mathcal{A}_0$ we obtain that μ_m is well defined and additive on $\mathcal{A}(\mathcal{E})$.

The equality (4.3) follows by

$$\int_{\mathcal{A}\setminus\{\emptyset\}} m_A(E)\, d\mu_m(A) = \mu_m(\{A : m_A(E) = 1\})$$
$$= \mu_m(\{A : A \subset E\}) = \mu_m(\tilde{E}) = m(E)$$

for every $E \in \mathcal{A}$.

We will prove the uniqeness of μ_m. Namely, let μ be an additive set function satisfying (4.3). Then $\mu(\tilde{A}) = \mu_m(\tilde{A}) = m(A)$ $(A \in \mathcal{A})$.

We have to prove the equality $\mu = \mu_m$ only on basic elements. But this follows by additivity of μ

$$\mu\Big(\tilde{B}\setminus\bigcup_{i=1}^n \tilde{B}_i\Big) = \mu(\tilde{B}) - \mu\Big(\bigcup_{i=1}^n \tilde{B}_i\Big)$$
$$= \mu(\tilde{B}) - \sum_{I\subset\{1,\ldots,n\},I\neq\emptyset} (-1)^{|I|+1}\mu\Big(\bigcap_{i\in I}\tilde{B}_i\Big)$$
$$= \mu(\tilde{B}) - \Delta_m(\{B_i\}_{i=1}^n) = \mu_m\Big(\tilde{B}\setminus\bigcup_{i=1}^n \tilde{B}_i\Big).$$

The equality $\|m\|_c = \overline{\mu}_m$ follows by the equality $\|m|_{\mathcal{A}_0}\|_c = \overline{\mu}_m|_{\mathcal{A}(\mathcal{E}_0)}$ for every finite sub-algebra $\mathcal{A}_0 \subset \mathcal{A}$.

It is obvious that the map $m \mapsto \mu_m$ is linear. Hence it is continuous.

If specially m is totally monotone, then on basic elements μ_m has in (4.4) all coefficients a_A non-negative.

Conversely, if μ_m is nonnegative then m is totally monotone, since for $E_i \in \mathcal{A}$, $i = 1, \ldots, n$ we have

$$m\Big(\bigcup_{i=1}^{n} E_i\Big) - \Delta_m(\{E_i\}_{i=1}^{n}) = \mu_m\Big(\widetilde{\bigcup_{i=1}^{n} E_i \setminus \bigcup_{i=1}^{n} \tilde{E}_i}\Big) \geq 0$$

(the latter brackets contain a basic element). □

4.3 The Space CV

Let us denote by CV the space of all real-valued set functions vanishing on an empty set with bounded composition norm.

Then the following can be proved in an analogous way as for the space BV.

Theorem 4.5 *CV with respect to composition norm is a Banach space.*

We have seen in Theorem 3.10 that each member of the space BV is a difference of two monotone set functions. We can prove more for the members of the space CV.

Theorem 4.6 *If $m \in CV$, then there are totally monotone set functions v_1 and v_2 such that $m = v_1 - v_2$.*

Moreover, there exist unique totally monotone set functions m^+ and m^- such that $m = m^+ - m^-$ and $\|m\|_c = \|m^+\|_c + \|m^-\|_c$.

Proof. Since for totally monotone set functions m we have $\|m\|_c = m(X) < \infty$, we obtain that the difference of every two totally monotone set functions always belongs to the space CV.

Conversely if $m \in CV$, then $\overline{\mu}_m < \infty$. Therefore by the classical Jordan decomposition theorem (see also Chapter 6) there exist finitely additive measures μ^+ and μ^- such that $\mu_m = \mu^+ - \mu^-$ and $\overline{\mu}_m = \overline{\mu}^+ + \overline{\mu}^-$.

Introducing totally monotone set functions m^+ and m^- by (Theorem 4.4)

$$m^{\pm} = \int_{\mathcal{A}\setminus\{\emptyset\}} m_A \, d\mu^{\pm}(A),$$

we obtain the desired unique representation $m = m^+ - m^-$. □

The relation between the spaces CV, DV and BV is given in the next theorem.

Theorem 4.7 *We have $CV \subset BV \cap DV$, where the inclusion is strict.*

Proof. We will show that the composition norm dominates both the disjoint and chain norm variations.

First, compare the composition norm with the norm induced by the chain variation. For that purpose we can write the composition norm in the following equivalent form

$$\|m\|_c = \sup\Big\{ \sum_{i=1}^{n} |m(A_i) - \Delta_m(\{B_i^j\}_{j=1}^{k_i})| : \\ \tilde{A}_i \setminus \bigcup_{j=1}^{k_i} \{\tilde{B}_i^j\} \text{ is a partition of } \tilde{X} \Big\}. \tag{4.5}$$

A special partition of $\tilde{X} = \Sigma \setminus \{\emptyset\}$ into finitely many basic elements for $k_i = 1$, which is of the form $\{(\tilde{A}_i \setminus \tilde{A}_{i-1})\}$ gives

$$|m(A_i) - \Delta_m(\{B_i^j\}_{j=1}^{k_i})| = |m(A_i) - m(A_{i-1})|\,.$$

Hence the supremum over sums of such expressions gives us the chain variation norm and $\|m\| \leq \|m\|_c$.

If we take another special partition of the form

$$\Big\{\Big(\tilde{X} \setminus \bigcup_{j=1}^{k} \tilde{D}^j\Big),\ \tilde{D}^1, \ldots, \tilde{D}^k\Big\}$$

for a partition $\{\tilde{D}^j\}_{j=1}^k$ of the set X, then

$$|m(X) - \Delta_m(\{D^j\}_{j=1}^k)| = \Big|m(X) - \sum_{j=1}^{k} m(D^j)\Big|\,.$$

The supremum over sums of such expressions is greater or equal than $\overline{m}$. Hence $\overline{m} \leq \|m\|_c$.

Finally, we will give an example which shows that $CV \neq BV \cap DV$. We take $X = \mathbf{N}$ and $\Sigma = \mathcal{P}(\mathbf{N})$. If we define m by

$$m(A) = \begin{cases} 1, & \text{if } |A^c| < 1 \\ 0, & \text{otherwise,} \end{cases}$$

then $\|m\| = \overline{m} = 1$, but $\|m\|_c$ is unbounded. To check the last statement it is enough to take a special partition of $\tilde{X}$ which contains the basic element

$$\left(\tilde{X} \setminus \bigcup_{i=1}^{k} \tilde{A}_i\right) \quad \text{for } A_i = X \setminus \{i\} \quad (1 \le i \le k).$$

Hence

$$\|m\|_c \ge |m(X) - \Delta_m(\{A_i\}_{i=1}^k)| = (k-1) \to \infty \quad \text{as} \quad k \to \infty. \qquad \square$$

4.4 Interpreters

For a positive monotone set function m defined on a σ-algebra $\mathcal{X}$ of subsets of X we will link a (classical) measure μ defined on a σ-algebra $\mathcal{Y}$ of subsets of a set Y. First some definitions.

Definition 4.2 A mapping $H : \mathcal{X} \to \mathcal{Y}$ is called an interpreter (for measurable sets) if H satisfies

(i) $H(\emptyset) = \emptyset$ and $H(X) = Y$;

(ii) $H(A) \subset H(B)$, whenever $A \subset B$.

A triple $(Y, \mathcal{Y}, H)$ is called a frame of $(X, \mathcal{X})$ (for representation) if H is an interpreter from $\mathcal{X}$ to $\mathcal{Y}$.

We remark that if $H : \mathcal{X} \to \mathcal{Y}$ is an interpreter and μ is classical measure on $\mathcal{Y}$, then $\mu \circ H$ is monotone on $\mathcal{X}$. Therefore we can define the notion of representation of a set functions:

Definition 4.3 Let m be a positive monotone set function defined on $\mathcal{X}$. A quadruplet $(Y, \mathcal{Y}, \mu, H)$ is called a representation of m (or of $(X, \mathcal{X}, m)$) if H is an interpreter from $\mathcal{X}$ to $\mathcal{Y}$, μ is a classical measure on $(Y, \mathcal{Y})$, and $m = \mu \circ H$.

Theorem 4.8 *Every positive monotone set function m has its representation.*

Proof. Let Y be the open interval $(0, m(X))$ in the real line, $\mathcal{Y}$ the class of all Borel subsets of Y, and μ a Lebesgue measure on $\mathcal{Y}$. We define an interpreter $H: \mathcal{X} \to \mathcal{Y}$ by $H(A) = (0, m(A))$ for all $A \in \mathcal{X}$. Then $(Y, \mathcal{Y}, \mu, H)$ is a representation of m. □

Generally m has infinitely many representations.

Now we define the notion of equivalence between representations. If $(Y, \mathcal{Y}, H)$ is a frame of $(X, \mathcal{X})$, we denote by $\mathcal{Y}$ the σ-algebra generated by $\{H(A) : A \in \mathcal{X}\}$.

If $(Y, \mathcal{Y}, m)$ is a classical measure space, we denote by $(\mathcal{Y}, \mu)$ the measure algebra associated with $(Y, \mathcal{Y}, \mu)$.

Definition 4.4 Let $R_1 = (Y_1, \mathcal{Y}_1, \mu_1, H_1)$ and $R_2 = (Y_2, \mathcal{Y}_2, \mu_2, H_2)$ be representations of $(X, \mathcal{X}, m)$. R_1 and R_2 are said to be equivalent if there exists an isomorphism i from $(\mathcal{Y}_1, \mu_1)$ to $(\mathcal{Y}_2, \mu_2)$ such that $i \circ H_1 = H_2$.

We will show that, for every measurable space $(X, \mathcal{X})$, there exists a frame $(S_X, \mathcal{S}_X, H_X)$ such that, for every monotone set function m on $\mathcal{X}$ and for every representation R of m there exists a measure μ_R on $\mathcal{S}_X$ such that $(S_X, \mathcal{S}_X, \mu_R, H_X)$ is a representation of m which is equivalent to R.

Definition 4.5 A semi-filter in a measurable space $(X, \mathcal{X})$ is a nonempty subclass θ of $\mathcal{X}$ with the properties:

(i) $\emptyset \notin \theta$,

(ii) if $A \in \theta$ and $A \subset B \in \mathcal{X}$, then $B \in \theta$.

Let S_X be the set of all semi-filters in $(X, \mathcal{X})$ and define a mapping $H_X : \mathcal{X} \to \mathcal{P}(S_X)$ by

$$H_X(A) = \{\theta \in S_X : A \in \theta\} \qquad (A \in \mathcal{X}).$$

We denote by $\mathcal{S}_X$ the σ-algebra generated by $\{H_X(A) : A \in \mathcal{X}\}$. The triplet $(S_X, \mathcal{S}_X, H_X)$ is called the universal frame of $(X, \mathcal{X})$ for representation.

S_X is non-empty since the classes $\{A \in \mathcal{X} : A \neq \emptyset\}$ and $\{X\}$ belong to S_X. H_X is an interpreter from $\mathcal{X}$ to $\mathcal{S}_X$.

Theorem 4.9 *For every representation $R = (Y, \mathcal{Y}, \mu, H)$ of $(X, \mathcal{X}, m)$ there exists a measure μ_R on $\mathcal{S}_X$ such that $(S_X, \mathcal{S}_X, \mu_R, H_X)$ is a representation of m which is equivalent to R.*

Proof. We introduce a mapping $\alpha: Y \to S_X$ by

$$\alpha(y) = \{A \in \mathcal{X} : y \in H(A)\} \qquad (y \in Y).$$

Hence $H = \alpha^{-1} \circ H_X$, which implies that α is a measurable mapping. Let μ_R be the image measure $\mu \circ \alpha^{-1}$. Then obviously $R_X = (S_X, \mathcal{S}_X, \mu_R, H_X)$ is a representation of μ. Since $\mathcal{S}_X$ and $\mathcal{Y}$ are generated by $\{H_X(A) : A \in \mathcal{X}\}$ and $\{H(A) : A \in \mathcal{X}\}$, respectively, and since $H = \alpha^{-1} \circ H_X$ and $\mu_R = m \circ \alpha^{-1}$, it follows that $i = \alpha^{-1}$ is an isomorphism from $(\mathcal{S}_X, \mu_R)$ to $(\mathcal{Y}, \mu)$. Therefore, R_X is equivalent to R. □

The next result follows from Theorems 4.8 and 4.9.

Corollary 4.1 *For every positive monotone set function m on $(X, \mathcal{X})$, there exists a classical measure μ on $\mathcal{S}_X$ such that $(S_X, \mathcal{S}_X, \mu, H_X)$ is a representation of m.*

Remark 4.1 If a monotone set function m is continuous from below (respectively above) then the interpreter H in Theorems 4.8 and 4.9 can be chosen such that H is continuous from below (respectively above), i.e., for every sequence $\{A_n\}$ from $\mathcal{X}$ such that $A_n \nearrow A \in \mathcal{X}$ (respectively $A_n \searrow A \in \mathcal{X}$), we have

$$\lim_{n\to\infty} H(A_n) = H(\lim_{n\to\infty} A_n).$$

Remark 4.2 We can give a practical interpretation of the preceding results. Since the monotone set function (fuzzy measure) m on $\mathcal{X}$ measures certain attributes of $\mathcal{X}$, we assume that the subsets of $\mathcal{X}$ interact with each other and we distinguish $\mathcal{X}$ from '$\mathcal{X}$ in the attribute' which is $\mathcal{Y}$. This means that $H : \mathcal{X} \to \mathcal{Y}$ do not have to be a Boolean algebra isomorphism (operation $\bigcup$ and $\bigcap$ are not always preserved).

If we suppose that $H(A) \cup H(B) \subset H(A \cup B)$, with strict inclusion, than we can interpret that $A \cup B$ can have features which neither A nor B have (which we can call cooperative action of A and B). Further, if for $A \cap B = \emptyset$ occurs $H(A) \cap H(B) \neq \emptyset$ then we can say that $H(A) \cap H(B)$ represents the overlap of A and B in attributes.

The Definition 4.2 says that H is an order 0–1 preserving homomorphism where 0 and 1 are the least and greatest elements, respectively.

4.5 Notes and References

The notion of 'composition variation' and the representations in Theorems 4.1 (reinterpreted the 'known result') and 4.4 were introduced in [Gilboa and Schmeidler 1992a] and [Gilboa and Schmeidler 1992b].

The notion of 'interpreter' was initiated in [Höhle 1984] and in the final form was given in [Murofushi and Sugeno 1991a].

Chapter 5

Autocontinuity of Set Functions

In this chapter we will introduce a kind of continuity condition, so called autocontinuity, which will imply null-additiveness. We will obtain a close relation of autocontinuous set functions with submeasures and some corresponding topologies. This enables us to transfer some results from submeasures to autocontinuous set functions. We will use this relation to prove a general extension theorem for monotone autocontinuous set functions.

At the end we will present an application of families of submeasures to approximation theory.

5.1 Autocontinuities

5.1.1 Basic Properties

Definition 5.1 A set function $m, m: \mathcal{R} \to [-\infty, +\infty]$, is called autocontinuous from above (resp. from below) if for every $\epsilon > 0$ and every $A \in \mathcal{R}$, there exists $\delta = \delta(A, \epsilon) > 0$ such that

$$m(A) - \epsilon \leq m(A \cup B) \leq m(A) + \epsilon$$
$$(\text{resp. } \; m(A) - \epsilon \leq m(A \setminus B) \leq m(A) + \epsilon)$$

whenever $B \in \mathcal{R}$, $A \cap B = \emptyset$ (resp. $B \subset A$) and $|m(B)| < \delta$ holds.

m is autocontinuous if it is autocontinuous from above and from below.

It is obvious that each set function which is autocontinuous from above (resp. from below) is null-additive.

Theorem 5.1 *If $m, m\colon \Sigma \to [0, +\infty)$, is a monotone set function which is autocontinuous from below and continuous from above, then it is autocontinuous from above.*

Proof. Suppose that the theorem is not true. Then there exist $\epsilon > 0$, a set E from Σ and a sequence $\{F_n\}$ from Σ such that

$$m(F_n) < \frac{1}{n} \quad \text{and} \quad m(E \cup F_n) > m(E) + \epsilon \quad (n \in \mathbf{N}).$$

Therefore, since m is autocontinuous from below there exists a subsequence $\{F_{n_k}\}$ of $\{F_n\}$ such that

$$m\Big(E \cup \Big(F_{n_s} \setminus \bigcup_{i=s+1}^{k+1} F_{n_i}\Big)\Big) > m(E) + \epsilon \tag{5.1}$$

for $s = 1, 2, \ldots, k$. Since m is a monotone set function we have

$$m\Big(E \cup \Big(\bigcup_{i=s}^{\infty}\Big(F_{n_s} \setminus \bigcup_{i=s+1}^{\infty} F_{n_i}\Big)\Big)\Big) \geq m\Big(E \cup \Big(F_{n_s} \setminus \bigcup_{i=s+1}^{\infty} F_{n_i}\Big)\Big). \tag{5.2}$$

By the continuity from above of m and (5.1) it follows

$$m\Big(E \cup \Big(F_{n_s} \setminus \bigcup_{i=s+1}^{\infty} F_{n_i}\Big)\Big) > m(E) + \epsilon .$$

Hence by (5.2) we obtain

$$m\Big(E \cup \Big(\bigcup_{i=s}^{\infty}\Big(F_{n_s} \setminus \bigcup_{i=s+1}^{\infty} F_{n_i}\Big)\Big)\Big) > m(E) + \epsilon .$$

Since m is continuous from above taking $s \to \infty$ in the last inequality we obtain the contradiction

$$m(E) \geq m(E) + \epsilon . \qquad \square$$

Theorem 5.2 *If* $m, m\colon \Sigma \to [0,+\infty)$, *is a monotone set function which is autocontinuous from above and continuous from below, then it is autocontinuous from below.*

Proof. Suppose that the theorem is not true. Then there exist $\epsilon > 0$, E from Σ and a sequence $\{F_n\}$ from Σ such that

$$m(F_n) < \frac{1}{n} \quad \text{and} \quad m(E) > m(E \setminus F_n) + a\epsilon \quad (n \in \mathbf{N}). \tag{5.3}$$

Since m is autocontinuous from above there exists a subsequence $\{F_{n_k}\}$ of $\{F_n\}$ such that

$$m\Big(\bigcup_{i=s}^{k} F_{n_i}\Big) < \frac{1}{s} \quad \text{for} \quad s = 1, 2, \ldots, k\,.$$

The continuity from below implies

$$m\Big(\bigcup_{i=s}^{\infty} F_{n_i}\Big) \leq \frac{1}{s} \qquad (s \in \mathbf{N})\,. \tag{5.4}$$

Since m is a monotone set function we have

$$m\Big(\bigcap_{s=1}^{\infty}\bigcup_{i=s}^{\infty} F_{n_i}\Big) \leq m\Big(\bigcup_{i=s}^{\infty} F_{n_i}\Big)\,.$$

Hence by (5.4)

$$m\Big(\bigcap_{s=1}^{\infty}\bigcup_{i=s}^{\infty} F_{n_i}\Big) = 0\,. \tag{5.5}$$

By the continuity from below of m we have

$$\lim_{s\to\infty} m\Big(E \setminus \bigcup_{i=s}^{\infty} F_{n_i}\Big) = m\Big(E \setminus \bigcap_{s=1}^{\infty}\bigcup_{i=s}^{\infty} F_{n_i}\Big)\,.$$

Hence by the null-additivity of m and (5.5) we obtain

$$\lim_{s\to\infty} m\Big(E \setminus \bigcup_{i=s}^{\infty} F_{n_i}\Big) = m(E)\,. \tag{5.6}$$

Since m is monotone we have

$$m(E \setminus F_{n_i}) \geq m\Big(E \setminus \bigcup_{i=s}^{\infty} F_{n_i}\Big).$$

Hence by (5.3) and (5.6)

$$m(E) - \epsilon > m(E).$$

Contradiction. □

Theorem (5.3) and Theorem (5.2) imply:

Corollary 5.1 *If m is a finite fuzzy measure, then the autocontinuity from above and the autocontinuity from below are equivalent for m.*

Definition 5.2 A set function $m, m: \mathcal{R} \to [-\infty, +\infty]$ is called uniformly autocontinuous from above (resp. from below), if for every $\epsilon > 0$, there exists $\delta = \delta(\epsilon) > 0$, such that, whenever $A \in \mathcal{R}$, $B \in \mathcal{R}$, $A \cap B = \emptyset$ (resp. $B \subset A$), $|m(B)| \leq \delta$), then

$$m(A) - \epsilon \leq m(A \cup B) \leq m(A) + \epsilon$$
$$(\text{resp. } m(A) - \epsilon \leq m(A \setminus B) \leq m(A) + \epsilon)$$

holds. m is called uniformly autocontinuous, if it is both uniformly autocontinuous from above and from below.

If m is a monotone set function, then '$A \cap B = \emptyset$' and '$B \subset A$' may be omitted in the statement of Definition 5.2.

Theorem 5.3 *If m is a positive monotone set function, then the following statements are equivalent:*

(i) m is uniformly autocontinuous;

(ii) m is uniformly autocontinuous from above;

(iii) m is uniformly autocontinuous from below;

(iv) for every $\epsilon > 0$, there exists $\delta = \delta(\epsilon) > 0$, such that, whenever $A \in \Sigma$, $B \in \Sigma$, $m(B) \leq \delta$, then

$$m(A \Delta B) - \epsilon \leq m(A) \leq m(A \Delta B) + \epsilon$$

holds.

Proof. (i) $\Rightarrow$ (ii) and (iv) $\Rightarrow$ (i) are obvious.

(ii) $\Rightarrow$ (iii): Since $m(B \cap A) \leq m(B) \leq \delta$, the desired conclusion follows from $m(A) = m((A \setminus B) \cup (B \cap A)) \leq m(A \setminus B) + \epsilon$.

(iii) $\Rightarrow$ (iv): Since $m(A \cap B) \leq m(B) \leq \delta$, we have

$$m(A \Delta B) = m((A \cup B) \setminus (B \cap A)) \geq m(A \cup B) - \epsilon \geq m(A) - \epsilon .$$

On the other hand, since $m(B \setminus A) \leq m(B) \leq \delta$, we obtain

$$m(A) \geq m(A \setminus B) = m((A \Delta B) \setminus (B \setminus A)) \geq m(A \Delta B) - \epsilon . \qquad \square$$

5.1.2 Examples

Example 5.1 Let $\mathcal{R}$ be a ring of subsets of a set X and let $v \colon \mathcal{R} \to [0, +\infty)$ be a countable additive measure, or a monotone autocontinuous set function. Further, let $f \colon [0, +\infty) \to [0, +\infty)$ be an increasing continuous function $f(0) = 0$. Then, the set function m on $\mathcal{R}$ defined by the equality $m(A) = f(v(A)), A \in \mathcal{R}$ is a monotone autocontinuous set function (see Subsection 2.2.11). Moreover, if v is a monotone uniformly autocontinuous set function and function f is uniformly continuous on the range of v, then m is a monotone uniformly autocontinuous set function. Specially, suppose that $\mathcal{R}$ is a σ-ring and that $v \colon \mathcal{S} \to [0, +\infty)$ is a countable additive measure. Then, since the range of v is a compact subset of $[0, +\infty)$, (see also Chapter 9) and since the continuous function f is uniformly continuous on compacts, then m defined above is a monotone uniformly autocontinuous set function.

Example 5.2 Suppose now that $\lambda \colon \mathcal{S} \to [-\infty, +\infty]$ is the usual Lebesgue measure on the real line. Let $\mathcal{P} = \{A \in \mathcal{P},\ \lambda(A) < +\infty\}$ and $\mathcal{P} = \{A \in \mathcal{P}, A \subset [0,1]\}$. Then $\mathcal{P}$ is a δ-ring (a ring closed with respect to countable intersection), $\mathcal{P}_1$ is a σ-ring, the monotone autocontinuous set function m_1 on $\mathcal{P}$ defined by the equality $m_1(A) = (\lambda(A))^2$ is not uniformly autocontinuous, while its restriction to $\mathcal{P}_1$ is uniform but not subadditive. Finally, the monotone autocontinuous set function m_2 on $\mathcal{P}$ defined by equality $m_2(A) = \lambda(A)/(1 + \lambda(A))$ is subadditive but not additive.

Example 5.3 Important examples of subadditive monotone autocontinuous and continuous set functions are the continuous semivariation $\hat{m}$ of an operator-valued measure m, and the continuous L_p-norm $\hat{m}_p(f, \cdot)$ of a measurable function f.

In the next example we give a set function m which is autocontinuous, but which is not uniformly autocontinuous.

Example 5.4 Let $X = X^- \cup X^+$, where $X^- = -\mathbf{N}$, $X^+ = \mathbf{N}$ and let $\Sigma = \mathcal{P}(X)$. A set function m is defined on Σ as

$$m(A) = |A^*| + \sum_{i \in A \setminus A^*} \frac{1}{2^{-i}} \qquad (A \in \Sigma),$$

where

$$A^* = \{i : i \in X^-,\ |i| \leq \sup\{j : j \in A \cap X^+\}\} \cup (A \cap X^+)$$

is obviously autocontinuous, but it is not uniformly autocontinuous. Namely, for every $\delta > 0$ and $0 < \epsilon < 1$, there exist $i \in X^-$ and $j = -i \in X^+$, such that $m(\{i\}) = 1/2^{-i} < \delta$ and

$$m(\{j, i\}) - m(\{j\}) = 1 > \epsilon\,.$$

In the following example we construct a finite monotone autocontinuous set function on a σ-algebra which is not uniformly autocontinuous.

Example 5.5 Let $X_0 = \{0, 1\}$, $\Sigma_0 = \mathcal{P}(X_0)$, $v_0(\{0\}) = v_0(\{1\}) = 1/2$, $(X_i, \Sigma_i, v_i) = (X_0, \Sigma_0, v_0)$ for every $i \in \mathbf{N}$ and take the product $(X, \Sigma, v) = (\times_{i=1}^{\infty} X_i, \times_{i=1}^{\infty} \Sigma_i, \times_{i=1}^{\infty} v_i)$ (see Example 5, after §38 in [Halmos 1950]). Further let $Y = [0, 1]$ and let $\mathcal{B}$ be the σ-algebra of Borel measurable subset of Y and let λ be the Lebesgue measure on $\mathcal{B}$.

For $n \in \mathbf{N}$ we take

$$A_n = \{x = (x_1, \ldots, x_n, \ldots) \in X : x_n = 0\}$$

and

$$\mathcal{U}_n = \left\{A \in \Sigma : v(A \Delta A_n) \leq \frac{1}{12}\right\}.$$

We have $A_n \in \Sigma$, $v(A_n) = 1/2$, and $v(A_n \Delta A_k) = \frac{1}{2}$ and $\mathcal{U}_n \cap \mathcal{U}_k = \emptyset$ for $n \neq k$, $(n, k \in \mathbf{N})$.

We introduce $(T, \mathcal{S}, m)$ as follows $T = X \cup Y$,

$$\mathcal{S} = \Sigma \oplus \mathcal{B} = \{C \subset T : C = A \cup B, \text{ where } A \in \Sigma \text{ and } B \in \mathcal{B}\},$$

and for such $C \in \mathcal{S}$ we define

$$m(C) = v(A) + \lambda(B) + \sum_{n=1}^{\infty} \Big(0 \vee \Big(\frac{1}{12} - v(A \Delta A_n)\Big)\Big)(n\lambda(B) \wedge 1).$$

We introduce for $E, F \in \Sigma$ the notation ERF if and only if $E \subset F$ or $E \supset F$. We will prove that $n \neq k$, $E \in \mathcal{U}_n$, $F \in \mathcal{U}_k$ implies $E(\text{non } R)F$.

Let $E \supset F$. Then $v((A_k \backslash A_n) \cap E) \geq v((A_k \backslash A_n) \cap F)$ by the monotonicity of v. We have further

$$\begin{aligned} v((A_k \setminus A_n) \cap F) &= v(A_k \setminus A_n) - v((A_k \setminus A_n) \setminus F) \\ &\geq \frac{1}{4} - v(A_k \setminus F) \geq \frac{1}{4} - v(A_k \Delta F) \geq \frac{1}{4} - \frac{1}{12} = \frac{2}{12}, \end{aligned}$$

and

$$\begin{aligned} v((A_k \setminus A_n) \cap E) &= v((A_k \setminus A_n) \cap (A_n \Delta (A_n \Delta E))) \\ &= v((A_k \setminus A_n) \cap ((A_n \setminus (A_n \Delta E)) \cup ((A_n \Delta E) \setminus A_n))) \\ &= v((A_k \setminus A_n) \cap ((A_n \Delta E) \setminus A_n)) \\ &\leq v((A_k \setminus A_n) \cap (A_n \Delta E)) \leq v(A_n \Delta E) \leq \frac{1}{12}, \end{aligned}$$

a contradiction. By symmetry, the inclusion $E \subset F$ is also false.

We will now prove that m is monotone on $\mathcal{S}$.

Let $E, F \in \mathcal{S}$ and let $E \supset F$. Then we have $E = E_1 \cup E_2$, $F = F_1 \cup F_2$, where $E_1, F_1 \in \Sigma$, $E_2, F_2 \in \mathcal{B}$, $F_1 \subset E_1$ and $F_2 \subset E_2$.

If $E_1, F_1 \in \Sigma \setminus \cup_{n=1}^{\infty} \mathcal{U}_n$, the inequality $m(F) \leq m(E)$ follows by the definition of m.

If $E_1 \in \mathcal{U}_n$ for some n, then by the previously proved property of the relation R, $F_1 \in \Sigma \setminus \cup_{k=1, k\neq n}^{\infty} \mathcal{U}_k$. Suppose first that $F_1 \in \Sigma \setminus \cup_{k=1,k\neq n}^{\infty} \mathcal{U}_k$. Then we have

$$\begin{aligned} m(E) - m(F) &= v(E_1) - v(F_1) + \lambda(E_2) - \lambda(F_2) \\ &\quad + \Big(\frac{1}{12} - v(E_1 \Delta A_n)\Big)(n\lambda(E_2) \wedge 1) \geq 0. \end{aligned}$$

Let $F_1 \in \mathcal{U}_n$. Then

$$\begin{aligned} m(E) - m(F) &= v(E_1) - v(F_1) + \lambda(E_2) - \lambda(F_2) \\ &\quad + \Big(\frac{1}{12} - v(E_1 \Delta A_n)\Big)(n\lambda(E_2) \wedge 1) - \Big(\frac{1}{12} - v(F_1 \Delta A_n)\Big)(n\lambda(F_2) \wedge 1) \\ &\geq v(E_1) - v(F_1) + (n\lambda(E_2) \wedge 1)\Big(\frac{1}{12} - v(E_1 \Delta A_n) - \frac{1}{12} + v(F_1 \Delta A_n)\Big) \\ &\geq v(E_1) - v(F_1) - v(E_1 \Delta F_1) = 0, \end{aligned}$$

where we used in the last inequality the fact that

$$v(E_1 \Delta A_n) \leq v(E_1 \Delta F_1) + v(F_1 \Delta A_n)).$$

Finally if $F_1 \in \mathcal{U}_n$ for some n, then by the proved fact at the beginning $E_1 \in \Sigma \setminus \cup_{k=1,k\neq n}^{\infty} \mathcal{U}_k$.

The case $E_1 \in \mathcal{U}_n$ has already been considered above. Thus let $E_1 \in \Sigma \setminus \cup_{k=1}^{\infty} \mathcal{U}_k$. Therefore

$$\begin{aligned} m(E) - m(F) &= v(E_1) - v(F_1) + \lambda(E_2) - \lambda(F_2) \\ &\quad - \Big(\frac{1}{12} - v(F_1 \Delta A_n)\Big)(n\lambda(F_2) \wedge 1) \\ &\geq v(E_1) - v(F_1) - \frac{1}{12} + v(F_1 \Delta A_n) \\ &> v(E_1) - v(F_1) - (v(E_1 \Delta A_n) - v(F_1 \Delta A_n)) \\ &\geq v(E_1 - v(F_1) - v(E_1 \Delta F_1) = 0. \end{aligned}$$

Thus the monotonicity of m on $\mathcal{S}$ is proved.

It is easy to verify that m is continuous and autocontinuous on $\mathcal{S}$. However, m is not uniformly autocontinuous on $\mathcal{S}$, since for every $\delta > 0$ and every $B \in \mathcal{B}$, with $\lambda(B) = \delta$, we have

$$m(A_n \cup B) - m(A_n) > \frac{1}{12} \quad \text{for} \quad n \geq \frac{1}{\delta}.$$

5.1.3 Countable Case

Although Example 5.5 shows that autocontinuity does not generally imply the uniformly autocontinuity of a finite monotone continuous from above set function, we shall prove in this section that for the countable basic set X this implication is true.

In this section m will be a positive monotone continuous set function.

Lemma 5.1 *Let X be a countable set, and let m be null-additive, continuous from below with $m(X) < \infty$. Then for every $A \in \Sigma$, $m(A) = 0$ if and only if $A \cap A_0 = \emptyset$, where $A_0 = \{x \in X : m(A) > 0\}$, specially, if $A_0 = \emptyset$, then $m(A) = 0$, for every $A \in \Sigma$.*

Proof. The necessity is obvious. Now we show the sufficiency. Let $A = \{x\}$ and I is a finite or countable set. Suppose $m(\{x_i\}) = 0$ $(i \in I)$, then, if I is a finite set, by the null-additivity, we have $m(A) = 0$, if I is a countable set, let $A_s = \cup_{k=1}^{s}\{x_k\}$, then, $A_s \uparrow A$, but $m(A_s) = 0$, so, $m(A) = \lim_{s\to\infty} m(A_s) = 0$. □

Lemma 5.2 *With the same assumption of Lemma 5.1 on X and m, if $B_s \in \Sigma$ and $m(B_s) \to 0$, then $m(\limsup_s B_s) = 0$.*

Proof. By Lemma 5.1, it suffices to show that $\limsup B_s \cap A_0 = \emptyset$. Let $x \in \limsup B_s \cap A_0$, then $m(\{x\}) \leq m(B_s) \to 0$, this is contradiction. □

Theorem 5.4 *Let X be a countable set and let m be a non-negative monotone continuous set function with $m(X) < \infty$, then the null-additivity of m is equivalent to uniform autocontinuity.*

Proof. It suffices to show the necessity. Suppose that the statement is not true, then there exists $\epsilon_0 > 0$ and $A_s, B_s \in \Sigma$ $(s \in \mathbf{N})$, with $m(B_s) \to 0$, such that

$$m(A_s \cup B_s) \geq m(A_s) + \epsilon_0 \qquad (s \in \mathbf{N}) \tag{5.7}$$

By Lemma 5.2, we have $m(\limsup_s B_s = 0$. Hence by null-additivity and (5.7), we obtain

$$\begin{aligned} m(\limsup_s A_s) &= m(\limsup_s A_s \cup \limsup_s B_s) \\ &= m(\limsup_s (A_s \cup B_s)) = \lim_s m\Big(\bigcup_{k=s}^{\infty}(A_k \cup B_k)\Big) \\ &\geq \limsup_s \sup_{k\geq s} m(A_k \cup B_k) = \overline{\lim}_s m(A_s \cup B_s)) \\ &\geq \overline{\lim}_s (m(A_s) + \epsilon_0) = \overline{\lim}_s m(A_s) + \epsilon_0 \end{aligned} \tag{5.8}$$

By Lemma 5.1, we may assume that $A_0 \neq \emptyset$, and $A_0 = \{x_1, x_2, \ldots, x_s \ldots\}$ may be finite or countable.

If $A_0 \cap \limsup_s A_s = \emptyset$ then, by Lemma 5.2, we have $m(\limsup_s A_s) = 0$, and this is a contradiction with (5.8).

If $A_0 \cap \limsup_s A_s \neq \emptyset$, let $s_1 = \inf\{i : x_i \in A_0 \cap \limsup_s A_s\}$. Therefore there exists a subsequence $\{A_s^{(1)}\}$ of $\{A_s\}$ such that $x_{s1} \in A_s^{(1)}$ $(s \in \mathbf{N})$. Let $s_2 = \inf\{i : x_i \in A_0 \cap \limsup_s A_s^{(1)}, x_i \neq x_{s_1}\}$. Then there exists a subsequence $\{A_s^{(2)}\}$ of $\{A_s^{(1)}\}$, such that $x_{s_2} \in A_s^{(2)}$ $(s \in \mathbf{N})$ and inductively either there exists some positive integer k and some subsequence $\{A_s^{(k)}\}$ of $\{A_s\}$ with $\{x_{s1}, x_{s2}, \dots, x_{sk}\} \subset A_s^{(k)}$ $(s \in \mathbf{N})$ such that $\limsup_s A_s^{(k)} \cap A_0 = \{x_{s1}, \dots, x_{sk}\}$, or there exist subsequences $\{A_s^{(k)}\}$ of $\{A_s\}$ with $\{A_s^{(k+1)}\} \subset \{A_s^{(k)}\}$ $(k \in \mathbf{N})$ and $x_{sk} \in A_s^{(k)}$ $(s, k \in \mathbf{N})$, such that $\limsup_s A_s^{(k)} \cap A_0 \supseteq \{x_{s1}, \dots, x_{sk}\}$.

In the first case, due to Lemma 5.2, we have $m(\limsup_s A_s^{(k)} \setminus A_0) = 0$. By the null-additivity we have

$$\begin{aligned} m(\{x_{s1}, \dots, x_{sk}\}) &= m((A_0 \cap \limsup_s A_s^{(k)}) \\ &= m((A_0 \cap \limsup_s A_s^{(k)}) \cup (\limsup_s A_s^{(k)} \setminus A_0)) \\ &= m(\limsup_s A_s^{(k)}) \geq \overline{\lim}_s m(A_s^{(k)}) + \epsilon_0 \\ &\geq m(\{x_{s1}, \dots, x_{sk})\} + \epsilon_0 \,. \end{aligned}$$

Contradiction. In the second case, we have $A_0 \cap \limsup_k A_k^{(k)} = \{x_{s1}, x_{s2}, \dots\}$. We have $A_0 \cap \limsup_s A_k^{(k)} = \{x_{s1}, x_{s2}, \dots\}$. Namely, suppose the contrary, i.e., let $x \in A_0 \cap \limsup_k A_k^{(k)}$, $x \neq x_{sk}$ $(k \in \mathbf{N})$. Then there exists s_0 such that $x = x_{s0}$, and $x_{s0} \in \limsup_k A_k^{(k)}$, and further, there exists a subsequence $\{A_{k_i}^{(k_i)}\}$ of $\{A_k^{(k)}\}$ such that $x_{s0} \in A_{k_i}^{(k_i)}$ $(i \in \mathbf{N})$. It is easy to see that for every positive integer j, $\{A_{k_i}^{(k_i)}\}_{i=j}^{\infty}$ is a subsequence of $\{A_s^{(s_0)}\}_{s=1}^{\infty}$, that is $x_{s0} \in A_0 \cap \limsup_s A_s^{(k_j)}$. Hence $s_0 > s_{kj+1}$ $(j \in \mathbf{N})$, but $\lim_{j\to\infty} s_{kj+1} = \infty$. This is a contradiction, thus $x \in \{x_{s1}, x_{s2}, \dots\}$. Therefore $A_0 \cap \limsup_k A_s^{(k)} = \{x_{s1}, x_{s2}, \dots\}$. Analogously, since $m(\limsup_k A_k^{(k)} \setminus A_0) = 0$, and since $\{x_{s1}, \dots, x_{sn}\} \subset A_n^{(n)}$, $(n \in \mathbf{N})$, we have

$$\begin{aligned} m(\{x_{s1}, x_{s2}, \dots\}) &= m(\limsup_s A_k^{(k)} \cap A_0) = m(\limsup_k A_k^{(k)}) \\ &\geq \overline{\lim}_k m(A_k^{(k)}) + \epsilon_0 \geq \overline{\lim}_k m(\{x_{s1}, \dots, x_{sk}\}) + \epsilon_0 \\ &= m(\{x_{s1}, x_{s2}, \dots\}) + \epsilon_0 . \end{aligned}$$

Contradiction. In a quite similar way we can show that m is uniformly autocontinuous from below. □

Combining Theorem 5.4 and the property that uniform autocontinuity implies autocontinuity and autocontinuity implies null-additivity, we obtain the following two corollaries.

Corollary 5.2 *If X is a countable set and m is a non-negative monotone continuous set function, then autocontinuity is equivalent to uniform autocontinuity.*

Corollary 5.3 *With the same assumption of Corollary 5.2 on X and m, null-additivity is equivalent to autocontinuity.*

5.2 Submeasures and Frechet-Nikodym Topologies

In this section we will consider special non-additive set function — submeasures, to which we will match special kind of topologies. Such a consideration enables us to characterize in the topological sense a more general class of non-additive set functions and then to transfer some results from submeasures on this general class of set functions.

5.2.1 Topological Ring

First we will give some known facts about topological rings (for more details see [Bourbaki 1955]).

Let Γ be a topology on a ring $\mathcal{R}$ such that the operations $(A, B) \mapsto A\Delta B$ and $(A, B) \mapsto A \cap B$, from $\mathcal{R} \times \mathcal{R}$ (with the product topology) to $\mathcal{R}$, are continuous. The pair $(\mathcal{R}, \Gamma)$ is then said to be a topological ring of sets. The topology Γ, consistent with the ring structure of $(\mathcal{R}, \Delta, \cap)$ will be shortly called an r-topology on $\mathcal{R}$. The operations $(A, B) \mapsto A \cup B$ and $(A, B) \mapsto A \setminus B$ are also continuous.

Definition 5.3 A family $\mathcal{W}$ of subsets of X is a filter if

(i) $\emptyset \in \mathcal{W}$

(ii) If $F \subset A \subset X$ for $F \in \mathcal{W}$, then $A \in \mathcal{W}$.

(iii) A finite intersection of elements from $\mathcal{W}$ belongs to $\mathcal{W}$.

Theorem 5.5 *(i) If $(\mathcal{R}, \Gamma)$ is a topological ring of sets and Ω a base of (the filter of all) neighbourhood system of $\emptyset$ in $\mathcal{R}$ then, for each $A \in \mathcal{R}$, $A\Delta\Omega = \{A\Delta\mathcal{U} : \mathcal{U} \in \Omega\}$ is a base of (the filter of all) neighbourhood system of A in $\mathcal{R}$.*

(ii) In order that a filter (or filter base) of neighbourhood system of $\emptyset$ determines an r-topology in $\mathcal{R}$ it is necessary that the following conditions

be satisfied:
(a) For each $\mathcal{U} \in \Omega$ *there exists* $\mathcal{V} \in \Omega$ *such that* $\mathcal{V}\Delta\mathcal{V} \subset \mathcal{U}$,
(b) For each $\mathcal{U} \in \Omega$ *there exists* $\mathcal{V} \in \Omega$ *such that* $\mathcal{V} \cap \mathcal{V} \subset \mathcal{U}$,
(c) For each $A \in \mathcal{R}$ *and* $\mathcal{U} \in \Omega$ *there exists* $\mathcal{V} \in \Omega$ *such that* $A \cap \mathcal{V} \subset \mathcal{U}$.

(iii) If $(\mathcal{R}, \Gamma)$ *is a topological ring of sets and* $\mathcal{P}$ *is a subring (subalgebra) of the ring (algebra)* $\mathcal{R}$, *then* $\overline{\mathcal{P}}^{\Gamma}$ *(closure in* Γ*) is a subring (subalgebra) of* $\mathcal{R}$.

An ideal I of a ring $\mathcal{R}$ is a subset of $\mathcal{R}$ such that, if A_1, $A_2 \in I$ then $A_1 \cup A_2 \in I$ and if $B \in \mathcal{R}$ is such that $B \subset A$ for some $A \in I$ then $B \in I$. If additionally for every sequence $\{A_n\}$ in I we have $\cup_{n=1}^{\infty} A_n \in I$ then we call I a σ-ideal.

Definition 5.4 Let $(\mathcal{R}, \Gamma)$ be a topological ring of sets. A set $E \in \mathcal{R}$ is called Γ-null if $E \in \mathcal{N}_\Gamma(\mathcal{R}) = \overline{\{\emptyset\}}^{\Gamma}$ (closure in Γ). $\mathcal{N}_\Gamma(\mathcal{R})$ is a closed ideal of the ring $\mathcal{R}$, named in the sequel as the ideal of Γ-null sets of the ring $\mathcal{R}$.

Definition 5.5 An r-topology Γ in ring $\mathcal{R}$ is said to be monotone or a Frechet-Nikodym (shortly FN-topology) if, for each neighbourhood $\mathcal{U}$ of $\emptyset$, there is a neighbourhood $\mathcal{V}$ of $\emptyset$ such that $\mathcal{V} \cap \mathcal{R} \subset \mathcal{U}$, i.e., such that $B \in \mathcal{U}$ whenever $B \in \mathcal{R}$ and $B \subset A \in \mathcal{V}$. A ring equipped with an FN-topology is called a FN-ring.

Let $(X, \mathcal{R}, \Gamma)$ be a topological ring of sets, Ω a base of neighbourhood system of $\emptyset$ in $\mathcal{R}$, and let $\mathcal{S}$ be a ring of subsets of X containing $\mathcal{R}$. For every $\mathcal{U} \in \Omega$ take $\mathcal{U}^* = \{B \in \mathcal{S} : B \subset A \text{ for some } A \in \mathcal{U}\}$, and let $\Omega^* = \{\mathcal{U}^* \in \Omega\}$. By (i) in Theorem 5.5 we have that Ω^* is a base on neighbourhood system of $\emptyset$ for a FN-topology in $\mathcal{S}$, denoted $\Gamma^*(\mathcal{S})$ and called the natural extension of Γ on $\mathcal{S}$, which does not depend on the choice of Ω. Clearly, $\Gamma = \Gamma^*(\mathcal{R})$ if and only if Γ is monotone. A base $\mathcal{N}$ at $\emptyset$ in $(\mathcal{R}, \Gamma)$ is called a normal base of neighbourhood system of $\emptyset$ if every $\mathcal{U} \in \mathcal{N}$ is a normal subfamily of $\mathcal{R}$, i.e., $B \in \mathcal{U}$ provided $B \in \mathcal{R}$ and $B \subset A$ for some $A \in \mathcal{U}$.

Theorem 5.6 *(i) If* $(\mathcal{R}, \Gamma)$ *is an* FN*-ring then, for each base* Ω *of neighbourhood system of* $\emptyset$, Ω^* *(defined as above for* $\mathcal{S} = \mathcal{R}$*) is a normal base of neighbourhood system of* $\emptyset$ *in* $(\mathcal{R}, \Gamma)$.

(ii) Let Ω *le a filter base of neighbourhood system of* $\emptyset$ *in* $\mathcal{R}$ *and suppose that:*
(a) Each class $\mathcal{U} \in \Omega$ *is normal in* $\mathcal{R}$.
(b) For each $\mathcal{U} \in \Omega$ *there is* $\mathcal{V} \in \Omega$ *such that* $\mathcal{V} \cup \mathcal{V} \subset \mathcal{U}$. *Then* Ω *is a normal base of neighbourhood system of* $\emptyset$ *for some* FN*-topology in* $\mathcal{R}$.

(iii) Let $(X, \mathcal{R}, \Gamma)$ be an FN-ring and let a ring $\mathcal{S} \subset \mathcal{P}(X)$ contains $\mathcal{R}$. Then $\Gamma^*(\mathcal{S})$ is the strongest FN-topology on $\mathcal{S}$ for which $\Gamma^*(\mathcal{S})|_{\mathcal{R}} = \Gamma$. Hence, if $\mathcal{N}$ is an FN-topology on $\mathcal{S}$ such that $\mathcal{N}|_{\mathcal{R}} = \Gamma$, then $\overline{\mathcal{R}}^{\Gamma^*(\mathcal{S})} \subset \overline{\mathcal{R}}^{\mathcal{N}}$.

(iv) For any family $\{\Gamma_i\}_{i \in I}$ of r-topologies on a ring $\mathcal{R}$ the topology $\Gamma = \sup(\Gamma_i : i \in I)$ (the coarsest topology stronger than each Γ_i) is an r-topology on $\mathcal{R}$. Moreover, if all Γ_i are monotone then Γ is monotone, too.

Proof.
(iii) If $\mathcal{H}$ is a normal base of neighbourhood system of $\emptyset$ for an FN-topology $\mathcal{N}$ on $\mathcal{S}$ such that $\mathcal{N}|_{\mathcal{R}} = \Gamma$, then $\mathcal{H}$ is a normal base of neighbourhood system of $\emptyset$ in $(\mathcal{R}, \Gamma)$. By $(\mathcal{V} \cap \mathcal{R})^* \subset \mathcal{V}$ $(\mathcal{V} \in \mathcal{H})$ and the fact that $(\mathcal{H}|_{\mathcal{R}})^*$ is a base at $\emptyset$ in $(\mathcal{S}, \Gamma(\mathcal{S}))$ we obtain $\mathcal{N} \subset \Gamma^*(\mathcal{S})$.

(iv) For each $i \in I$ let Ω_i be a base at $\emptyset$ in $(\mathcal{R}, \Gamma_i)$. Then $\Omega = \bigcup\{\Omega_{i_1} \bigcap \cdots \bigcap \Omega_{i_n} : \{i_1, \ldots, i_n\} \subset I\}$ is a base at $\emptyset$ for which Γ, by (iii) in Theorem 5.5, is an r-topology on $\mathcal{R}$. If all Ω_t are normal then so is Ω, and Γ is an FN-topology. □

5.2.2 Topology Induced by Submeasures

Now we match the Frechet- Nikodym topologies to submeasures.

If η is a submeasure on $\mathcal{R}$ then the classes $\mathcal{U}_\epsilon = \{A \in \mathcal{R} : \eta(A) \leq \epsilon\}$, $\epsilon > 0$, constitute a normal base of neighbourhood system at $\emptyset$ for a FN-topology, denoted by $\Gamma(\eta)$ and called the FN-topology generated by η, on $\mathcal{R}$. The topology $\Gamma(\eta)$ is pseudometrizable; the functions $(A, B) \mapsto \min(1, \eta(A)\Delta B))$, $(A, B) \mapsto \eta(A\Delta B)/(1+\eta(A\Delta B))$ $(= 1$ if we have $\infty/\infty)$, $(A, B) \mapsto \eta(A\Delta B)$ (if $\eta(A) < \infty$ for each $A \in \mathcal{R}$) are examples of (equivalent) pseudometrics on $\mathcal{R}$ generating $\Gamma(\eta)$. A set $E \in \mathcal{R}$ is $\Gamma(\eta)$-null if and only if $\eta(E) = 0$.

If m is a subadditive set function on $\mathcal{R}$ then the semivariation of m is given by (see Section 2.1)

$$E \mapsto \hat{m}(E) = \sup\{m(F) : F \subset E, F \in \mathcal{R}\}$$

is a submeasure on $\mathcal{R}$.

The terms η-null, η-continuous and symbols $\mathcal{N}_\eta(\mathcal{R})$, $\mathcal{N}_\eta$, $m \ll \eta$ are substitutes for $\Gamma(\eta)$-null, $\Gamma(\eta)$-continuous, ... When considering the topology

$\Gamma(\eta)$ it may be always assumed that η is bounded (replacing, if necessary, η by the submeasure $A \mapsto \min(1, \eta(A))$.

If η is a submeasure on $\mathcal{R}$ then the outer submeasure induced by η, denoted η^*, is a submeasure defined on $\mathcal{P}(X)$ as follows: $\eta^*(B) = \inf\{\eta(A) : B \subset A \in \mathcal{R}\}$, if there is $A \in \mathcal{R}$ covering B, and equal ∞ otherwise (see Section 2.1).

We have $\eta^*|_{\mathcal{R}} = \eta$ and $\Gamma(\eta^*)$ is a FN-topology on $\mathcal{P}(X)$ equivalent with the natural extension $(\Gamma(\eta))^*$ on $\mathcal{P}(X)$. By (iii) Theorem 5.5 $\mathcal{R}^*$ equal to the closure of $\mathcal{R}$ in $(\mathcal{P}(X), \Gamma(\eta^*))$ is a subring of $\mathcal{P}(X)$. Clearly, every η^*-null set is in $\mathcal{R}^*$ and a subset of η^*-null set is η^*-null itself. It is easy to verify that a set $F \subset R$ is in $\mathcal{R}^*$ if and only if there are sequences $\{H_n\}$, $\{K_n\}$ in $\mathcal{R}$ such that $K_n \subset F \subset H_n$, $K_n \nearrow$, $H_n \searrow$ and $\lim_{n\to\infty} \eta(H_n \setminus K_n) = 0$. Hence, assuming $\mathcal{R}$ to be a σ-ring, $F \in \mathcal{R}^*$ if and only if there are sets H, $K \in \mathcal{R}$ such that $K \subset F \subset H$ and $\eta(H \setminus K) = 0$, i.e., $\mathcal{R}^*$ consists of all sets of the form $F = E \cup Z$, where $E \in \mathcal{R}$ and $\eta^*(Z) = 0$. Therefore for $\mathcal{R}^* = \mathcal{R}$ subsets of η-null sets are η-null; if it is the case we shall say that $\mathcal{R}$ (or η) is null-complete. If it is not so $\mathcal{R}^*$ is called the null-completion on $\mathcal{R}$. Note that if $\mathcal{R}$ is a σ-ring and $\mathcal{N}_\eta(\mathcal{R})$ is a σ-ideal then $\mathcal{R}^*$ is a σ-ring and $\mathcal{N}_{\eta^*}(\mathcal{R}^*)$ is a σ-ideal.

Theorem 5.7 *If an FN-topology Γ in a ring $\mathcal{R}$ has a countable base of neighbourhood system of $\emptyset$, then there exists a submeasure η on such that $\Gamma = \Gamma(\eta)$.*

Proof. Let d be a translation invariant pseudometric on $\mathcal{R}$, i.e., $d(A\Delta E, B\Delta E) = d(A, B)$, that generates Γ. Let η be the semivariation for the subadditive set function $A \mapsto a(A) = d(A, \emptyset)$, and let, for each $\epsilon > 0$, $\mathcal{U}_\epsilon = \{A \in \mathcal{R} : a(A) \le \epsilon\}$, $\mathcal{V}_\epsilon = \{A \in \mathcal{R} : \eta(A) \le \epsilon\}$. The collections of these classes are bases at $\emptyset$ in $(\mathcal{R}, \Gamma)$ and $(\mathcal{R}, \Gamma(\eta))$, respectively. Let us observe that $\mathcal{V}_\epsilon = \{A \in \mathcal{R} : a(B) \le \epsilon$ for each $B \in \mathcal{R}, B \subset A\}$. By (i) in Theorem 5.6 the classes $(\mathcal{U}_\epsilon)^* = \{A \in \mathcal{R} : A \subset B$ for some $B \in \mathcal{R}$ such that $a(B) \le \epsilon\}$ consist a normal base at $\emptyset$ in $(\mathcal{R}, \Gamma)$. Evidently, $\mathcal{V}_\epsilon \subset (\mathcal{U}_\epsilon)^*$ for each $\epsilon > 0$. On the other hand, since Γ is monotone, for each $\epsilon > 0$ there is $\delta > 0$ such that $a(A) \le \epsilon$ whenever $A \subset B$ and $a(B) \le \delta$. Therefore $(\mathcal{U}_\delta)^* \subset \mathcal{V}_\epsilon$. Hence $\Gamma = \Gamma(\eta)$. □

Propositon 5.1 *Let $(\mathcal{R}, \Gamma)$ be a topological ring of sets and let η be a finite submeasure on $\mathcal{R}$. Then the following statements are equivalent:*

(a) η is uniformly continuous.

(b) η is continuous.

(c) η is continuous at the point $\emptyset$ of $\mathcal{R}$.

(d) $\Gamma(\eta) \subset \Gamma$.

Proof. It is enough to prove (c) $\Rightarrow$ (a). If $A, B \in \mathcal{R}$ then $\eta(A) \leq \eta(A\Delta B) + \eta(B)$, $\eta(B) \leq \eta(B\Delta A) + \eta(A)$ and hence $|\eta(A) - \eta(B)| \leq \eta(A\Delta B)$. This gives (c) $\Rightarrow$ (a). □

The preceding Proposition remains valid also for a $[0,\infty]$-valued submeasure η if uniform continuity (and similarly, continuity) is understood in a slightly wider sense: for each $\epsilon > 0$ there is a neighbourhood $\mathcal{U}$ of $\emptyset$ in $\mathcal{R}$ such that

(1) if $A, B \in \mathcal{R}$, $A\Delta B \in \mathcal{U}$ and $\eta(A) < \infty$, then $\eta(B) < \infty$ and $|\eta(A) - \eta(B)| < \epsilon$, and

(2) if $\eta(A) = \infty$ then also $\eta(B) = \infty$.

In the following 'continuity of a submeasure' (finite or not) will always be used in the wider sense.

Let Γ be monotone. A subadditive set function m on $\mathcal{R}$ is Γ-continuous at the point $\emptyset$ in $\mathcal{R}$ if and only if $\hat{m} \ll \Gamma$. If η is a submeasure on $\mathcal{R}$ then $m \ll \eta$ if and only if $\hat{m} \ll \eta$. Similarly, equi-Γ-continuity of a family of subadditive set functions is equivalent to equi Γ-continuity of the family of corresponding semivariations.

Theorem 5.8 *(i) For any family $H = \{\eta_i\}_{i\in I}$ of submeasures on $\mathcal{R}$ there exists the coarsest FN-topology on $\mathcal{R}$, denoted $\Gamma(H) \equiv \Gamma(\eta_i : i \in I)$ and called the FN-topology generated (or determined) by the family H, such that $\eta_i \ll \Gamma(H)$ for all $i \in I$. A set $E \in \mathcal{R}$ is $\Gamma(H)$-null if and only if $\eta_i(E) = 0$ for each $i \in I$.*

If $\eta_1, \ldots, \eta_n$ are submeasures on $\mathcal{R}$ then

$$\Gamma(\eta_i : 1 \leq i \leq n) = \Gamma\Big(\sup_{1\leq i\leq n} \eta_i\Big) = \Gamma(\eta_1 + \cdots + \eta_n).$$

(ii) For each FN-topology Γ on $\mathcal{R}$ there is a family $(\eta_i : i \in I)$ of submeasures on $\mathcal{R}$ such that $\Gamma = \Gamma(\eta_i : i \in I)$.

Proof. (i) Follows by (iv) in Theorem 5.6.

(ii) Let $\{d_i : i \in I\}$ be a family of continuous invariant pseudometrics on $\mathcal{R}$ generating Γ. Setting $a_i(A) = d_i(A, \emptyset)$ and then $\eta_i = \hat{a}_i$ we obtain the needed family. □

5.2.3 Non-Negative Monotone Set Functions

Suppose that m, $m: \mathcal{R} \to [0, +\infty)$, is monotone and $m(\emptyset) = 0$. Then we may associate with it a uniformly $\mathcal{U}(m)$ on $\mathcal{R}$, a base of $\emptyset$ which is formed by the classes

$$\mathcal{U}_\delta = \{(A, B) \in \mathcal{R} \times \mathcal{R} : m(A \Delta B) < \delta\}, \qquad \delta > 0.$$

$\mathcal{U}(m)$ is pseudometrizable, and since each $\mathcal{U}_\delta$ is invariant, i.e.,

$$(A, B) \in \mathcal{U}_\delta \Rightarrow (A \Delta C, B \Delta C) \in \mathcal{U}_\delta,$$

there exists a pseudometric d on $\mathcal{R}$ generating $\mathcal{U}(m)$, which is also invariant:

$$d(A, B) = d(A \Delta C, B \Delta C) \qquad (A, B, C \in \mathcal{R}).$$

Let $\Gamma(m)$ denote the topology on $\mathcal{R}$ associated with $\mathcal{U}(m)$ (or with d). $\Gamma(m)$ is the weakest invariant topology on $\mathcal{R}$ with respect to which m is continuous as $\emptyset$, that is

$$(E_n \to \emptyset \text{ in } \Gamma(m)) \Longleftrightarrow (m(E_n) \to 0).$$

Note also that $E_n \to E$ in $\Gamma(m)$ if and only if $m(E_n \Delta E) \to 0$.

Theorem 5.9 *Let $m: \mathcal{R} \to R_+$ be monotone. Then*

(i) $\Gamma(m)$ is an FN-topology, i.e., there exists a submeasure η on $\mathcal{R}$ such that

$$m(E_n) \to 0 \Leftrightarrow \eta(E_n) \to 0,$$

if and only if m satisfies the following condition:

(ac) $$m(A_n) + m(B_n) \to 0, \text{ then } m(A_n \cup B_n) \to 0.$$

(ii) m is continuous on $(\mathcal{R}, \Gamma(m))$ if and only if it is autocontinuous and continuous from above.

(iii) m is uniformly continuous on $(\mathcal{R}, \Gamma(m))$ if and only if m is uniformly autocontinuous and continuous from above.

Proof. The classes

$$\mathcal{U}_\delta(\emptyset) = \{A \in \mathcal{R} : m(A) < \delta\}$$

form a base of $\Gamma(m)$-neighbourhoods of $\emptyset$, and each of them is normal:

$$B \subset A \in \mathcal{U}_\delta(\emptyset) \Rightarrow B \in \mathcal{U}_\delta(\emptyset).$$

Hence, by (iii) in Theorem 5.6, a necessary and sufficient condition for $\Gamma(m)$ to be an FN-topology is that for every $\epsilon > 0$ there exists a $\delta > 0$ such that

$$A, B \in \mathcal{U}_\delta(\emptyset) \Rightarrow A \cup B \in \mathcal{U}_\delta(\emptyset).$$

This is evidently equivalent to (ac). Since $\Gamma(m)$ is pseudometrizable, $\Gamma(m)$ is an FN-topology if and only if there is a submeasure η for which $\Gamma(m) = \Gamma(\eta)$, by Theorem 5.7, and such a submeasure can be defined by

$$\eta(E) = \sup\{d(F, \emptyset) : F \subset E,\ F \in \mathcal{R}\},$$

where d is any invariant semimetric determining $\Gamma(m)$ (see the proof of Theorem 5.7. □

If m is a set function on $\mathcal{R}$, we write $m \ll \eta$ to indicate that m is continuous on $(\mathcal{R}, \eta)$ and then say that m is η-continuous. Suppose that m_1 and m_2 are set functions on $\mathcal{R}$ determined FN-topologies $\Gamma(m_1)$, and Γ_2, respectively; then we say that m_1 and m_2 are equivalent and write $m_1 \sim m_2$ if $\Gamma(m_1) = \Gamma(m_2)$.

5.3 Extensions

First we will prove an extension theorem for submeasures and then by Theorem 5.9 which gives the topological correspondence with some non-additive set functions, we prove an extension theorem for these general set functions. We remark that for an arbitrary fuzzy measure does not exist always an extension and when it exists it is not always unique.

First we have

Theorem 5.10 *Let $\mathcal{R}$ be a ring generating the σ-ring $\mathcal{S}$ and let η be an order continuous submeasure on $\mathcal{S}$. Then $\mathcal{R}$ is a dense subset of $(\mathcal{S}, \Gamma(\eta))$.*

Proof. By (iii) in Theorem 5.5 $\overline{\mathcal{R}} = \overline{\mathcal{R}}^{\Gamma(\eta)}$ is a subring of $\mathcal{S}$. Let $A_n \in \overline{\mathcal{R}}$ ($n \in \mathbf{N}$) be disjoint sets and let $A = \cup_{n=1}^{\infty} A_n$. Since $\cup_{k=1}^{n} A_k \in \overline{\mathcal{R}}$ and $\eta(A \Delta \cup_{k=1}^{n} A_k) \searrow 0$, we have $A \in \overline{\mathcal{R}}$. Therefore $\overline{\mathcal{R}}$ is a σ-ring and hence $\mathcal{S} = \overline{\mathcal{R}}$. □

Theorem 5.11 *Let $\mathcal{R}$ be a ring of subsets of a set X and let η be an exhaustive and order continuous submeasure on $\mathcal{R}$. Then there is a σ-ring $\mathcal{R}_0$ of subsets of X and an order continuous submeasure η' on $\mathcal{R}_0$ such that:*

(a) $\mathcal{R}_0 \supset \mathcal{R}$ and $\mathcal{R}_0 = \overline{\mathcal{R}}^{\Gamma(\eta')}$;

(b) $\eta'|_{\mathcal{R}} = \eta$;

(c) The σ-ring $\mathcal{R}_0$ is null complete with respect to η', i.e., if $F \subset E \in \mathcal{R}_0$ and $\eta'(E) = 0$ then $F \in \mathcal{R}_0$ and $\eta'(F) = 0$;

(d) η' is the unique order continuous submeasure on $\mathcal{R}_0$ which satisfies the condition (b);

(e) The σ-ring $\mathcal{R}_0$ is the null-completion of the σ-ring $\mathcal{S}$ generated by $\mathcal{R}$.

Proof. Let $\mathcal{R}_\sigma$ be the class of all sets of the form $A_1 \cup A_2 \cup \ldots$, where $A_n \in \mathcal{R}$, ($n \in \mathbf{N}$). Let us note that $\mathcal{R}_\sigma$ is closed with respect to countable unions and finite intersections and that $E \setminus F \in \mathcal{R}_\sigma$ if $E \in \mathcal{R}_\sigma$, $F \in \mathcal{R}$. For each $E \in \mathcal{R}_\sigma$ set $\eta''(E) = \sup\{\eta(F) : F \subset E, F \in \mathcal{R}\}$. It is easy to see that if $\{F_n\} \subset \mathcal{R}$ and $F_n \nearrow E$ then $\eta''(E) = \lim \eta(F_n)$. Some properties of the set function η'' on $\mathcal{R}_\sigma$ are of importance in the following.

(1) $\eta''|_{\mathcal{R}} = \eta$, η'' is nondecreasing.

(2) η'' is σ-subadditive. Namely, let $E_n \in \mathcal{R}_\sigma$, $E_k^n \in \mathcal{R}$ ($k, n \in \mathbf{N}$) and $E_k^n \nearrow E_n$ ($k \to \infty$). Then, denoting $F = \cup_{n=1}^{\infty} E_n$ and

$$F_r = E_r^1 \cup E_r^2 \cup \cdots \cup E_r^r,$$

we have $F_r \in \mathcal{R}$, $F_r \nearrow F$ and $\eta''(F) = \lim_{r\to\infty} \eta(F_r)$. Since $\eta(F_r) \le \sum_{k=1}^{r} \eta''(E_k)$ for $r \in \mathbf{N}$, taking r to infinity we obtain

$$\eta''(F) \le \sum_{n=1}^{\infty} \eta''(E_n).$$

(3) If $\mathcal{R} \ni E_n \nearrow E$ then $\eta''(E \setminus E_n) \searrow 0$. Indeed, for given $\epsilon > 0$ there is, by Proposition 2.8, $k \in \mathbf{N}$ such that $\eta(E_m \setminus E_n) \le \epsilon$ if $m \ge n \ge k$. Since $E_m \setminus E_n \nearrow E \setminus E_n$ as $m \to \infty$, we have

$$\eta''(E \setminus E_n) = \lim_{m\to\infty} \eta(E_m \setminus E_n) \le \epsilon$$

for $n \ge k$.

(4) η'' is exhaustive on $\mathcal{R}_\sigma$.

(5) η'' is order continuous on $\mathcal{R}_\sigma$. Suppose that $F_n \in \mathcal{R}_\sigma$, $F_n \searrow \emptyset$ and $\eta''(F_n) > 2a > 0$, $(n \in \mathbf{N})$. By (3), for each $n \in \mathbf{N}$ there is a set $E_n \in \mathcal{R}$, such that $E_n \subset F_n$ and $\eta''(F_n \setminus E_n) < a/2^n$. Setting $D_n = E_1 \cap \cdots \cap E_n$, we have $D_n \in \mathcal{R}$, $D_n \subset F_n$ and, as easily seen, $\eta''(F_n \setminus D_n) < a$, $(n \in \mathbf{N})$. On the other hand, from the inequality $\eta''(F_n) \leq \eta''(D_n) + \eta''(F_n \setminus D_n)$ we obtain $\eta(D_n) > a$ for $n \in \mathbf{N}$, and this is impossible since $D_n \searrow \emptyset$ and η is order continuous on $\mathcal{R}$.

Let
$$\widehat{\mathcal{R}} = \{E : E \subset X,\ E \subset F \ \text{ for some } \ F \in \mathcal{R}_\sigma\},$$
and let $\eta'(E) = \inf\{\eta''(F) : E \subset F \in \mathcal{R}_\sigma\}$ for each $E \in \hat{\mathcal{R}}$. It is obvious that $\hat{\mathcal{R}}$ is a σ-ring and that $\eta'|_{\mathcal{R}_\sigma} = \eta''$. Let $\mathcal{R}_0 = \overline{\mathcal{R}_\sigma}^{\Gamma(\eta')}$. Thus $E \in \mathcal{R}_0$ if and only if there is a sequence $\{E_n\} \subset \mathcal{R}_0$ such that $\eta'(E \Delta E_n) \to 0$. Now we are going to prove that $\mathcal{R}_0$ and η' restricted to $\mathcal{R}_0$ satisfy all conditions of the theorem. First of all, let us note that (3) implies that $\mathcal{R}_0 = \overline{\mathcal{R}}^{\Gamma(\eta')}$. Therefore, by (iii) in Theorem 5.5 $\mathcal{R}_0$ is a ring.

We will need the following fact.

(6) If $E \in \mathcal{R}_0$, then there is a sequence $\{H_n\} \subset \mathcal{R}_0$ such that $E \subset H_n (n \in \mathbf{N})$ and $\eta'(H_n \setminus E) \to 0$; it may be also required that $H_n \searrow$. Namely, let $E_n \in \mathcal{R}$ be such that $\eta'(E \Delta E_n) \to 0$. Then, for each $n \in \mathbf{N}$, there is a set $G_n \in \mathcal{R}_\sigma$ such that $E \Delta E_n \subset G_n$ and $\eta'(G_n) \leq \eta'(E \Delta E_n) + n^{-1}$. It suffices to set $H_n = (E_1 \cup G_1) \cap \cdots \cap (E_n \cup G_n)$.

(7) The submeasure η' is σ-subadditive on $\widehat{\mathcal{R}}$. Indeed, let $E_n \in \widehat{\mathcal{R}}$ $(n \in \mathbf{N})$. We may suppose that $\sum_{n=1}^{\infty} \eta'(E_n) < \infty$. Given $\epsilon > 0$ there is, for each $n \in \mathbf{N}$, a set $F_n \in \mathcal{R}_\sigma$ such that $E_n \subset F_n$ and $\eta'(F_n) \leq \eta'(E_n) + \epsilon/2^n$. Therefore
$$\eta'\Big(\bigcup_{n=1}^{\infty} E_n\Big) \leq \eta'\Big(\bigcup_{n=1}^{\infty} F_n\Big) \leq \sum_{n=1}^{\infty} \eta'(F_n) \leq \epsilon + \sum_{n=1}^{\infty} \eta'(E_n).$$
We will prove that $\mathcal{R}_0$ is a σ-ring. Let $E = \cup_{n=1}^{\infty} E_n$, where $E_n \in \mathcal{R}_0$ $(n \in \mathbf{N})$. From (6) it follows that, given $\epsilon > 0$, there is, for each $n \in \mathbf{N}$, a set $H_n \in \mathcal{R}_\sigma$, such that $E_n \subset H_n$ and $\eta'(H_n \setminus E_n) < 2^n$. Let $H = \cup_{n=1}^{\infty} H_n$. We have $H \in \mathcal{R}_\sigma$, $E \subset H$ and $\eta'(H \Delta E) < \epsilon$. Therefore $E \in \mathcal{R}_0$.

(8) The submeasure η' is order continuous on $\mathcal{R}_0$. Let $E_n \in \mathcal{R}_0$ and $E_n \searrow \emptyset$. By (6) there is a sequence $\{H_n\} \subset \mathcal{R}_\sigma$ such that
$$E_n \subset H_n, \quad \eta'(H_n \setminus E_n) < 2^{-n} \qquad (n \in \mathbf{N}).$$

Let $H = \bigcap_{n=1}^{\infty} H_n$. Since $\bigcap_{n=1}^{\infty} E_n = \emptyset$, we have

$$H \subset \bigcap_{n=1}^{\infty} \bigcup_{k=n}^{\infty} (H_k \setminus E_k).$$

Hence $\eta'(H) = 0$. We claim that $\eta'(H_n) \searrow 0$. Given $\epsilon > 0$, similarly as in the proof of (5) we can find sets $D_n \in \mathcal{R}$ such that

$$D_n \subset H_n, \quad D_n \searrow, \quad \eta'(H_n \setminus D_n) < \epsilon \qquad (n \in \mathbf{N}).$$

Let a set $F \in \mathcal{R}_\sigma$ be such that $H \subset F$, $\eta'(F) < \epsilon$, and let sets $F_n \in \mathcal{R}$ be so chosen that $F_n \nearrow F$. Let $G_n = D_n \setminus F_n$ $(n \in \mathbf{N})$. Since $G_n \searrow \emptyset$, $H_n \subset (H_n \setminus D_n) \cup G_n \cup F_n$, we have $\overline{\lim}_{n\to\infty} \eta'(H_n) \leq 2\epsilon$. Therefore $\lim_n \eta'(H_n) = 0$, and hence also $\lim_n \eta'(E_n) = 0$.

Thus we have already proved that $\mathcal{R}_0$ is a σ-ring containing $\mathcal{R}$, η' is an order continuous submeasure on $\mathcal{R}_0$, and the conditions (a), (b) are satisfied. The condition (c) is trivial. In order to prove (d), let an order continuous submeasure γ on $\mathcal{R}_0$ be such that $\gamma|_{\mathcal{R}} = \eta$. Then also $\gamma|_{\mathcal{R}_\sigma} = \eta''$. Let $E \in \mathcal{R}_0$, and let a nonincreasing sequence $\{H_n\}$ be as in (6). Then

$$\eta'(E) = \lim \eta''(H_n) = \lim \gamma(H_n).$$

Let $G_n \in \mathcal{R}_0$ $(n \in \mathbf{N})$ be such that $H_n \setminus E \subset G_n$ and $\eta''(G_n) \to 0$. Then $\gamma(H_n \setminus E) \leq \gamma(G_n)$. Hence $\gamma(E) = \lim \gamma(H_n)$. Thus $\gamma(E) = \eta'(E)$ for each set $E \in \mathcal{R}_0$. Now, let $H = \bigcap_{n=1}^{\infty} H_n$. We have $E \subset H \in \mathcal{S}$ and $\eta'(H \setminus E) = 0$. Further, let sets $K_n \in \mathcal{R}_\sigma$ be such that $K_n \searrow, H \setminus E \subset K_n$ and $\eta'(K_n) \searrow 0$. Then $H \setminus E \subset K = \bigcap_{n=1}^{\infty} K_n \in \mathcal{S}$ and $\eta'(K) = 0$. Therefore $\mathcal{R}_0$ satisfies (e). □

Now we have the general extension theorem.

Theorem 5.12 *Let m_0, $m_0 \colon \mathcal{R} \to [0, +\infty)$, be a monotone autocontinuous and continuous set function and let $\mathcal{S}$ be the σ-ring generated by $\mathcal{R}$. Then m_0 can be extended to a monotone autocontinuous and continuous set function $m \colon \mathcal{S} \to [0, +\infty)$ if and only if the conditions are fulfilled:*

(i) m_0 is exhaustive.

(ii) m_0 has property (ac) *from Theorem* 5.9;

(iii) If $\{A_n\} \subset \mathcal{R}_\sigma$ and $A_n \searrow$, then for every $\epsilon > 0$ there is a $\delta > 0$ such that $m^+(A_n \setminus B) \geq m^+(A_n) - \epsilon$ for all $n \in \mathbf{N}$ and all $B \in \mathcal{R}$ with $m_0(B) < \delta$.

The extension m, if it exists, is unique.

Here $\mathcal{R}_\sigma$ denotes the class of sets which are limits of increasing sequences in $\mathcal{R}$, and $m_\sigma : \mathcal{R}_\sigma \to [0, +\infty)$ is defined by

$$m_\sigma(A) = \lim_{n\to\infty} m_0(A_n),$$

where $\{A_n\}$ is any sequence in $\mathcal{R}$ such that $A_n \nearrow A$. Then m_σ is well defined, finite valued and extends m_0.

Proof. Necessity. Let m be a required extension of m_0 to $\mathcal{S}$. Then, by Theorem 5.9, there exists an order continuous submeasure η on $\mathcal{S}$ such that $m \sim \eta$ and $m \ll \eta$. It is obvious that m is exhaustive, and hence so is m_0.

Suppose that condition (ii) is not satisfied. Then there exist an increasing sequence $\{A_n\}$ in $\mathcal{R}$, and $\epsilon > 0$ and a sequence $\{B_n\}$ in $\mathcal{R}$ with $m_0(B_n) \to 0$ such that

$$m_0(A_n \cup B_n) > m_0(A_n) + \epsilon \qquad (n \in \mathbf{N}).$$

Let

$$A = \bigcup_{n=1}^{\infty} A_n ;$$

then

$$\eta((A_n \cup B_n)\Delta A) \leq \eta(A \backslash A_n) + \eta(B_n) \to 0.$$

Since $m \ll \eta$, we obtain

$$m_0(A_n \cup B_n) \to m(A) \text{ and } m_0(A_n) \to m(A),$$

but this contradicts our assumption.

In quite an analogous way, we may verify that m_0 must satisfy (iii).

Sufficiency. Theorem 5.9(i) and (ii) implies that there exists an order continuous submeasure η_0 on $\mathcal{R}$ such that $m_0 \sim \eta_0$ and $m_0 \ll \eta_0$. Let η be the unique order continuous extension of η_0 to $\mathcal{S}$ by Theorem 5.11.

If there exists a required extension m of m_0, and if v is an order continuous submeasure on $\mathcal{S}$ equivalent to m, then

$$v(E_n) \to 0 \iff m(E_n) \to 0,$$

whence $(v|_\mathcal{R}) \sim m_0 \sim \eta_0$ and so $v \sim \eta$. Therefore m, if it exists, is a continuous (hence unique) extension of m_0 from $(\mathcal{R}, \eta_0)$ to $(\mathcal{S}, \eta)$. Therefore, as $\mathcal{R}$ is dense in $(\mathcal{S}, \eta)$ by Theorem 5.10 and $(\mathcal{S}, \eta)$ is complete by

Theorem 2.8, to prove the existence of such a continuous extension m it is necessary and sufficient to show that

(**C**) If $\{A_n\} \subset \mathcal{R}$ is such that $m_0(A_n \Delta A_m) \to 0$ (which is equivalent to $\eta_0(A_n \Delta A_m) \to 0$), then $\lim_{n\to\infty} m_0(A_n)$ exists in $[0, +\infty)$.

If m_0 is uniformly autocontinuous, then this is trivially true.
Suppose that a sequence $\{B_n\} \subset \mathcal{R}$ is such that

$$m_0\Big(\bigcup_{i=n}^{k-1}(B_{i+1}\Delta B_i)\Big) < \frac{1}{n} \tag{5.9}$$

holds for all $n \in \mathbf{N}$ and $k > n$.

Take $\epsilon > 0$ and then, for each $n \in \mathbf{N}$, choose $\delta_n > 0$ such that for each $C \in \mathcal{R}$, $m_0(C) < \delta_n$ we have $m_0(B_n \cup C) \leq m_0(B_n) + \epsilon$, which is possible by autocontinuity. Then, if $C \in \mathcal{R}_\sigma$ and $m_\sigma(C) < \delta_n$, we also have

$$m_\sigma(B_n \cup C) \leq m_0(B_n) + \epsilon\,.$$

For each $n \in \mathbf{N}$ and $k \geq n$ let

$$C_{nk} = \bigcup_{i=n}^{k} B_i \qquad \text{and} \qquad C_n = \bigcup_{i=n}^{\infty} B_i\,.$$

From the exhaustivity of m_0 it follows that

$$m_\sigma(C_n \backslash C_{nk}) \to 0 \ \text{ as } \ k \to \infty\,.$$

Therefore for each $n \in \mathbf{N}$ there exists $k_n \geq n$ such that

$$m_\sigma(C_n \backslash C_{nk}) < \delta_n \ \text{ for all } \ k \geq k_n\,.$$

We have

$$m_\sigma(B_n \cup (C_n \backslash C_{nk})) \geq m_0(B_n) + \epsilon \ \text{ for all } \ n \in \mathbf{N} \ \text{ and } \ k \geq k_n. \tag{5.10}$$

By condition (iii) there exists $\delta > 0$ such that

$$m_\sigma(C_n \backslash B) \geq m^+(C_n) - \epsilon$$

for all $n \in \mathbf{N}$ and all $B \in \mathcal{R}$ with $m_0(B) < \delta$. Now, since

$$C_{nk} \backslash B_n \subset \bigcup_{i=n}^{k-1} (B_{i+1} \Delta B_i),$$

we have $m_0(C_{nk} \backslash B_n) < n^{-1}$ for all n and $k \geq n$. Choose $r \in \mathbf{N}$ with $r^{-1} < \delta$; then, if $n \geq r$ and $k \geq n$, we have

$$m_\sigma(C_n \backslash (C_{nk} \backslash B_n)) \geq m_\sigma(C_n) - \epsilon. \tag{5.11}$$

Since

$$C_n \backslash (C_{nk} \backslash B_n) = B_n \cup (C_n \backslash C_{nk}),$$

combining (5.10) and (5.11) (for $k = k_n$) we obtain

$$0 \leq m_\sigma(C_n) - m_0(B_n) \leq 2\epsilon \quad \text{for all} \quad n \geq r.$$

Since $C_n \searrow$, $\lim m_\sigma(C_n)$ exists, and so does $\lim m_0(B_n)$.

Now let $\{A_n\}$ be as in (C), and let $\{A'_n\}$ and $\{A''_n\}$ be two arbitrary subsequences of $\{A_n\}$. Then a simple argument using the σ-subadditivity of η shows that we may select a subsequence $\{B_n\}$ of $\{A_n\}$ which satisfies (5.9) and contains infinitely many terms of each of $\{A'_n\}$ and $\{A_n''\}$. By the preceding part of the proof, $q = \lim m_0(B_n)$ exists, and our argument implies easily that every subsequence of $\{m_0(A_n)\}$ has a subsequence converging to q. Hence $\lim m_0(A_n)$ exists and equals q.

The required extension m of m_0 is thus defined as follows: given $A \in \mathcal{S}$, choose $\{A_n\}$ in $\mathcal{R}$ such that $\eta(A \Delta A_n) \to 0$ and then take

$$m(A) = \lim_n m_0(A_n).$$

Since m extends m_0, and m is η-continuous m is order continuous. It remains only to verify that m is monotone. But this is an easy consequence of monotonicity of m_σ and of the fact that for every $A \in \mathcal{S}$ there exists $\{A_n\} \subset \mathcal{R}_\sigma$ with $A \subset A_n$ for each n, $A_n \searrow$ and $\eta(A_n \setminus A) \to 0$. □

5.4 Linearly Indexed Families of Submeasures

We shall introduce a family of set functions as a generalization of a submeasure, which induces a modular. For modular function space a Lebesgue dominated convergence type theorem will be proved. All these results will be used for nonlinear approximation of functions in modular spaces, and thus in generalized Orlicz spaces by some nonlinear singular integral operators.

5.4.1 The Modular Space

Let X be a non-empty set, Σ a σ-algebra of subsets of X and V a real vector space.

Definition 5.6 A map η, $\eta\colon V \times \Sigma \to [0, +\infty]$, denoted by $\{\eta_f(A)\}_{f\in V} = \eta(f, A)$, is called a linearly indexed family of submeasures, if the following conditions hold for all $f \in V$, $A \in \Sigma$:

(S1) $\eta_f(X) = 0$, if and only if $f = 0$;

(S2) $\eta_{-f}(A) = \eta_f(A)$;

(S3) $\eta_{\alpha f+\beta g}(A) \le \eta_f(A) + \eta_g(A)$ for $f, g \in V$, $\alpha, \beta \ge 0$, $\alpha + \beta = 1$;

(S4) $\eta_f(A) \le \eta_f(B)$ if $A, B \in \Sigma$, $A \subset B$;

(S5) $\eta_f(\bigcup_{i=1}^{n} A_i) \le \sum_{i=1}^{n} \eta_{nf}(A_i)$ if $A_i \in \Sigma$, A_i are pairwise disjoint and $n \in \mathbf{N}$.

Example 5.6 If $\eta_f, f \in V$ is independent of the variable f, then $\eta_f = \eta$ $(f \in V)$ is a submeasure on Σ, i.e., $\eta(A \cup B) \le \eta(A) + \eta(B)$ for any $A, B \in \Sigma$, $\eta(\emptyset) = 0$ and $\eta(A) \le \eta(B)$ for $A \subset B$.

Example 5.7 Let $\eta_f(A) = \int_A f\, d\mu$ $(A \in \Sigma)$ for $f \ge 0$, $f \in L^1(X)$ then $\{\eta_f\}_{f\in V}$ is a linearly indexed family of submeasures, where

$$V = \{f : f \ge 0,\ f \in L^1(X)\}.$$

Definition 5.7 Let μ be a σ-finite measure on Σ. The family $\{\eta_f\}_{f\in V}$ is called absolutely continuous with respect to μ, if there exists a number $\alpha > 0$ such that for every $f \in V$ satisfying $\eta_f(X) < \infty$, the following two conditions hold:

(i) For each $\epsilon > 0$ there exists $A \in \Sigma$ such that $\mu(A) < \infty$ and

$$\eta_{\alpha f}(X \setminus A) < \epsilon.$$

(ii) For each $\epsilon > 0$ there exists $\delta > 0$ such that for each $\mu(B) < \delta$ implies $\eta_{\alpha f}(B) < \epsilon$.

Let $L^0(X) = L^0(X, \mu)$ be the space of all extended real-valued, Σ-measurable functions f on X, finite μ-almost everywhere with equality μ almost everywhere.

Let $\mathbf{F}$ be a vector subspace of $L^0(X)$ having the following properties:

(1) simple functions on X belong to $\mathbf{F}$,

(2) if $f \in \mathbf{F}$ then $|f| \in \mathbf{F}$.

Let χ_A be the characteristic function of the set $A \subset X$. We make an additional assumption:

(S6) If $f \in \mathbf{F}$, $A \in \Sigma$, then $f\chi_A \in \mathbf{F}$ and $\eta_f(A) = \eta_{f\chi_A}(X)$.

Propositon 5.2 *The map ρ: $\mathbf{F} \to [0, +\infty]$ defined by $\rho(f) = \eta_f(X)$ is a modular on* $\mathbf{F}$, *i.e.,*

(i) $\rho(f) = 0$ if and only if $f = 0$,

(ii) $\rho(-f) = \rho(f)$,

(iii) $\rho(\alpha f + \beta g) \leq \rho(f) + \rho(g)$ for $f, g \in V$, $\alpha, \beta \geq 0$, $\alpha + \beta = 1$.

Remark 5.1 Property (iii) implies $\rho(f + g) \leq \rho(2f) + \rho(2g)$.

Definition 5.8 The vector subspace of $\mathbf{F}$

$$\mathbf{F}_\rho = \{f \in \mathbf{F} : \rho(\lambda f) \to 0 \text{ as } \lambda \to 0+\}$$

is called the modular space generated by ρ.

Definition 5.9 The modular ρ is called:

(i) monotone, if $f, g \in V$ and $|f| \leq |g|$ imply $\rho(f) \leq \rho(g)$.

(ii) finite, if for every $A \in \Sigma$, $\mu(A) < \infty$ implies $\chi_A \in \mathbf{F}_\rho$,

(iii) absolutely finite, if ρ is finite and there exists $\lambda_0 > 0$ such that for each $\epsilon > 0$ there exists $\delta > 0$ that for any $B \in \Sigma$, $\mu(B) < \delta$ implies $\rho(\lambda_0 \chi_B) < \epsilon$,

(iv) absolutely continuous, if for every $f \in \mathbf{F}$ such that $\rho(f) < \infty$ the conditions hold

(a) for every $\epsilon > 0$ there exists a set $A \in \Sigma$ such that $\mu(A) < \infty$ and $\rho(f\chi_{X \setminus A}) < \epsilon$,

(b) for every $\epsilon > 0$ there exists $\delta > 0$ such that for every $B \in \Sigma$ with $\mu(B) < \delta$ there holds $\rho(f\chi_B) < \epsilon$.

(v) J-convex, if for any $p(t) \geq 0$, $\int_X p(t)\, d\mu(t) = 1$ and every measurable function $F\colon X \times X \to \mathbf{R}$ there holds:

$$\rho\Big(\int_X p(t)|F(t,\cdot)|\, d\mu(t)\Big) \leq \int_X p(t)\rho(F(t,\cdot))\, d\mu(t)\,.$$

5.4.2 Lebesgue Dominated Convergence Type Theorem

Theorem 5.13 *Let $\{\eta_f\}_{f\in F}$ be a linearly indexed family of submeasures on $F \subset L^0(\Omega)$, absolutely continuous with respect to μ and satisfying (S6). Let $\rho(f) = \eta_f(X)$ be a monotone and finite modular. Let $f_n \in \mathbf{F}_\rho$ for $n \in \mathbf{N}$, $f_n(t) \to 0$ μ-a.e. in X as $n \to \infty$ and let there exist a function $g \in \mathbf{F}_\rho$ such that $\rho(3g) < \infty$ and $|f_n(t)| \leq g(t)$ μ-a.e. in X for $n \in \mathbf{N}$. Then $\rho(f_n) \to 0$ as $n \to \infty$.*

Proof. For an arbitrary but fixed $\epsilon > 0$ we choose by absolutely continuity of ρ, the set $A \in \Sigma$ and the number $\delta > 0$ (replacing f by $3g$). Classical Egoroff's theorem (see also Subsection 7.1.3) implies the existence of a set $A_\epsilon \in \Sigma$, $A_\epsilon \subset A$, such that $\mu(A_\epsilon) < \delta$ and $\lim_{n\to\infty} f_n(t) = 0$ uniformly in $A \setminus A_\epsilon$. Using the representation

$$|f_n| = \frac{1}{3}(3|f_n|\chi_{X\setminus A} + 3|f_n|_{\chi_{A\setminus A_\epsilon}} + 3|f_n|_{\chi_{A_\epsilon}})$$

and the monotonicity of modular ρ we have

$$\begin{aligned}\rho(f_n) &\leq \rho(3f_n\chi_{X\setminus A}) + \rho(3f_n\chi_{A\setminus A_\epsilon}) + \rho(3f_n\chi_{A_\epsilon})\\ &\leq \rho(3g\chi_{X\setminus A}) + \rho(3f_n\chi_{A\setminus A_\epsilon}) + \rho(3g\chi_{A_\epsilon}) \qquad (5.12)\\ &\leq \frac{2}{3}\epsilon + \rho(3f_n\chi_{A\setminus A_\epsilon}) \qquad (n \in \mathbf{N})\,.\end{aligned}$$

By $\chi_A \in \mathbf{F}$ there exists a number $\lambda_\epsilon > 0$ such that $\rho(\lambda\chi_A) < \frac{1}{3}\epsilon$ for every $\lambda \in (0, \lambda_\epsilon)$. Using the uniform convergence $\lim_{n\to\infty} f_n(t) = 0$ in $A \setminus A_\epsilon$, we obtain $n_0 \in \mathbf{N}$ such that $3|f_n(t)| < \lambda_\epsilon$ for every $t \in A \setminus A_\epsilon$ and $n \geq n_0$. This fact implies

$$\rho(3f_n\chi_{A\setminus A_\epsilon}) \leq \rho(\lambda_\epsilon\chi_{A\setminus A_\epsilon})) \leq \rho(\lambda_\epsilon\chi_A) < \frac{1}{3}\epsilon$$

for $n > n_0$. Hence by (5.12)

$$\rho(f_n) \leq \frac{2}{3}\epsilon + \rho(3f_n\chi_{A\setminus A_\epsilon}) < \epsilon$$

for $n \geq n_0$. □

5.4.3 Monotone Type Sequence of Functions

Let $\phi = \{\phi_n\}$, where $\phi_n\colon [0,\infty) \to [0,+\infty]$, ϕ_n is continuous and nondecreasing

$$\phi_n(0) = 0,\ \phi_n(u) > 0 \ \text{ for } \ u > 0,\ \lim_{n\to\infty} \phi_n(u) = +\infty\,.$$

Definition 5.10 ϕ is called of monotone type if there are an index n_0 and sets $R_1, R_2 \subset [0,+\infty)$, $R_1 \cap R_2 = \emptyset$, $R_1 \cup R_2 = [0,+\infty)$ such that the sequence $\phi_{n_0}(u), \phi_{n_0+1}(u), \ldots$ is nonincreasing for $u \in R_1$, nondecreasing for $u \in R_2$, and $\tilde{\phi}(u) = \lim_{n\to\infty} \phi_n(u) < +\infty$ for all $u \geq 0$.

Remark 5.2 If $\phi_n(u) = u^{p_n}$, $p_n > 0$, then ϕ is of monotone type if and only if $\{p_n\}$ is monotone.

We define

$$I_{\phi_n}(f) = \int_X \phi_n(|f(t)|)\,d\mu \quad \text{and} \quad I_\phi(f) = \sup_n I_{\phi_n}(f)$$

for $f \in L^0(X)$, $n \in \mathbf{N}$.

Then $I_{\phi_1}, I_{\phi_2}, \ldots$ and I_ϕ are modulars on $\mathbf{F} = L^0(X)$. Modulars I_{ϕ_n} generate Orlicz spaces $L^{\phi_n}(X)$ (as the respective modular spaces) and modular I_ϕ generates the modular space

$$L_{I_\phi}(X) = \{f \in L^0(X) : I_\phi(\lambda f) \to 0 \ \text{ as } \ \lambda \to 0+\}\,.$$

Let

$$\eta_f(A) = I_\phi(f\chi_A) \qquad (f \in L^0(X),\ A \in \Sigma)\,.$$

Supposing ϕ to be of monotone type, we have $\{\eta_f\}_{f\in L^0(X)}$ is a linearly indexed family of submeasures satisfying (S6) and absolutely continuous with respect to μ. If, moreover, $\phi_n(u)$ are equicontinuous at $u = 0$, then I_ϕ is absolutely finite. We shall prove these facts in the next section, which is more general.

5.4.4 Matrix Transformation

Let $\phi = \{\phi_j\}$ and $I_{\phi_j}(f)$ be as in the previous subsection. Further, let $\mathbf{A} = (a_{ij})$ be an infinite, nonnegative matrix, no column of which consists of zeros only, such that:

$$\lim_{i\to\infty} a_{ij} = 0 \quad \text{for} \quad j \in \mathbf{N},$$

$$\sum_{j=1}^{\infty} a_{ij} \le 1 \quad \text{for} \quad i \in \mathbf{N} \quad \text{and} \quad \lim_{j\to\infty} \sum_{j=1}^{\infty} a_{ij} = 1\,.$$

We introduce $\mathbf{A}_\phi \colon L^0(X) \to [0, +\infty]$ by the formula:

$$\mathbf{A}_\phi(f) = \sup_i \sum_{j=1}^{\infty} a_{ij} I_{\phi_j}(f) \qquad (f \in L^0(X))\,.$$

$\mathbf{A}_\phi$ is a modular on $L^0(X)$ and $\eta_f(A) = \mathbf{A}_\phi(f\chi_A)$ is a linearly indexed family of submeasures on $\mathbf{F} = L^0(X)$ satisfying (S6).

Theorem 5.14 *Let* $\mathbf{A} = (a_{ij})$ *satisfy the above conditions and let* $\phi = \{\phi_j\}$ *be of monotone type. Then the family* $\eta_f(A) = \mathbf{A}_\phi(f\chi_A)$ *is absolutely continuous with respect to* μ. *If, moreover,* $\phi_j(u)$ *are equicontinuous at* $u = 0$, *then* $\mathbf{A}_\phi(f) = \eta_f(X)$ *is absolutely finite.*

Proof. By $\mathbf{A}_\phi(f) < \infty$ we have $I_{\phi_j}(f) = \int_X \phi_j(|f(t)|)\, d\mu < \infty$ and each of the set functions $I_{\phi_j}(f\chi_A)$ $(j \in \mathbf{N})$ is absolutely continuous with respect to μ.

Hence for every $\epsilon > 0$ there exist a set $A_1 \in \Sigma$, an index n_0 and a number $\delta_1 > 0$ such that

$$\begin{aligned} &\sup_i \sum_{j=1}^{n_0-1} a_{ij} I_{\phi_j}(f\chi_{X\setminus A_1}) < \frac{1}{2}\epsilon \quad (\mu(A_1) < \infty), \\ &\sup_i \sum_{j=1}^{n_0-1} a_{ij} I_{\phi_j}(f\chi_A) < \frac{1}{2}\epsilon \quad (A \in \Sigma,\ \mu(A) < \delta_1)\,. \end{aligned} \tag{5.13}$$

In order to estimate expressions for the sum with indices $j \ge n_0$ we introduce

$$A_0 = \{t \in X : |f(t)| \in R_1\} \quad \text{and} \quad B_0 = \{t \in X : |f(t)| \in R_2\}\,.$$

Then we obtain

$$I_{\phi_j}(f\chi_{A_0}\chi_C) \le I_{\phi_{n_0}}(f\chi_{A_0}\chi_C)$$

for $j \geq n_0$ and an arbitrary $C \in \Sigma$. Consequently there exists a set $A_2 \in \Sigma$ and a number $\delta_2 > 0$ such that

$$\begin{aligned} \sup_i \sum_{j=n_0}^{\infty} a_{ij} I_{\phi_j}(f\chi_{X\setminus A_2}\chi_{A_0}) &< \frac{1}{4}\epsilon, \quad (u(A_2) < \infty), \\ \sup_i \sum_{j=n_0}^{\infty} a_{ij} I_{\phi_j}(f\chi_A\chi_{A_0}) &< \frac{1}{4}\epsilon, \quad (A \in \Sigma,\ u(A) < \delta_2). \end{aligned} \tag{5.14}$$

The definition of the set B_0 implies

$$\phi_j(|f(t)|\chi_{B_0}(t)\chi_C(t)) \nearrow \tilde{\phi}(|f(t)|\chi_{B_0}(t)\chi_C(t)) \text{ as } j \nearrow \infty$$

for every set $C \in \Sigma$. Hence

$$I_{\phi_j}(f\chi_{B_0}\chi_C) \nearrow I_{\tilde{\phi}}(f\chi_{B_0}\chi_C) \text{ as } j \nearrow \infty .$$

Then we have

$$I_{\tilde{\phi}}(f\chi_{B_0}\chi_P) \leq \sup_C \sum_{j=1}^{\infty} a_{ij} I_{\phi_j}(f\chi_{B_0}\chi_C) \leq \mathbf{A}_\phi(f) < \infty .$$

Hence one can find a set $A_3 \in \Sigma$ and a number $\delta_3 > 0$ such that

$$\begin{aligned} \sup_C \sum_{j=n_0}^{\infty} a_{ij} I_{\phi_j}(f\chi_{X\setminus A_3}\chi_{B_0}) &< \frac{1}{4}\epsilon \quad (u(A_3) < \infty), \\ \sup_C \sum_{j=n_0}^{\infty} a_{ij} I_{\phi_j}(f\chi_A\chi_{B_0}) &< \frac{1}{4}\epsilon, \quad (A \in \Sigma,\ u(A_3) < \delta_3). \end{aligned} \tag{5.15}$$

Taking $X_\epsilon = A_1 \cup A_2 \cup A_3$ and $\delta = \min(\delta_1, \delta_2, \delta_3)$ and collecting the inequalities (5.13)–(5.14) we easily obtain the absolute continuity of $\mathcal{A}_\phi(f\chi_A)$ with a constant $\alpha = \frac{1}{2}$.

Absolute finiteness of $\mathbf{A}_\phi$ is obtained in an obvious way. □

Obviously, if $a_{nj} = \delta_j$ $(j \in \mathbf{N})$ is the Kronecker symbol, then $\mathbf{A}_\phi(f) = I_\phi(f)$ reduces in the case of the previous subsection.

5.4.5 Functions of Monotone Type

Let $[a,b) \subset \mathbf{R}$ (b may be also $+\infty$) and $\phi\colon [a,b) \times [0,+\infty) \to [0,+\infty)$, be a function such that

$$\phi(x,0) = 0, \ \phi(x,u) > 0 \ \text{ for } \ u > 0,$$
$$\phi(x,u) \nearrow +\infty \ \text{ as } \ u \nearrow +\infty, \ \text{ for all } \ x \in [a,b),$$

$\phi(x,u)$ is a continuous function of x for all $u \geq 0$.

Definition 5.11 ϕ is called of monotone type, if there are sets $R_1, R_2 \subset [0,+\infty)$, $R_1 \cap R_2 = \emptyset$, $R_1 \cup R_2 = [0,+\infty)$ such that the function $\phi(x,u)$ of variable x is nonincreasing if $u \in R_1$, nondecreasing if $u \in R_2$, and $\tilde{\phi}(u) = \lim_{x \to b-0} \phi(x,u) < \infty$ for all $u \geq 0$.

We define

$$J_\phi(x,f) = \int_X \phi(x,|f(t)|)\, d\mu(t), \quad I_\phi(f) = \sup_{x \in [a,b)} J_\phi(x,f)$$

for $f \in L^0(X)$ and $x \in [a,b)$.

Then I_ϕ is a modular on $L^0(X)$. Taking $\eta_f(A) = I_\phi(f\chi_A)$, we have the following theorem.

Theorem 5.15 *Let* $\phi\colon [a,b)\times[0,+\infty) \to [0,+\infty)$ *be of monotone type. Then the family* $\eta_f(A) = I_\phi(f\chi_A)$ *is absolutely continuous with respect to* μ*. If, moreover, the family of functions* $\phi(x,u)$ *of variable* u *is equicontinuous (with respect to* x*) at* $u = 0$*, then* $I_\phi(f) = \eta_f(X)$ *is absolutely finite.*

Proof. The proof is similar to that of Theorem 5.14 with the exception of the part concerning indices $j < n_0$ which is unnecessary now. We define sets A_0 and B_0 as before, i.e., given $f \in L^0(X)$ such that $I_\phi(f) < \infty$, we take

$$A_0 = \{t \in X : |f(t)| \in R_1\} \quad \text{and} \quad B_0 = \{t \in X : |f(t)| \in R_2\}.$$

We choose for a given $\epsilon > 0$ sets $A_1, A_2 \in \Sigma$ of finite measure μ and numbers $\delta_1, \delta_2 > 0$ in such a manner that

$$I_\phi(f\chi_{X\setminus A_1}\chi_{A_0}) < \frac{1}{2}\epsilon, \quad I_\phi(f\chi_A\chi_{A_0}) < \frac{1}{2}\epsilon \ \text{ if } \ \delta \in \Sigma, \ \mu(A) < \delta_1 \tag{5.16}$$

and

$$I_\phi(f\chi_{X\setminus A_2}\chi_{B_0}) < \frac{1}{2}\epsilon,\ I_\phi(f\chi_A\chi_{B_0}) < \frac{1}{2}\epsilon \text{ if } A \in \Sigma,\ \mu(A) < \delta_2\,. \quad (5.17)$$

Taking $X_\epsilon = A_1 \cup A_2$ and $\delta = \min(\delta_1, \delta_2)$ we see easily combining (5.16) and (5.17), that $I_\phi(f\chi_A)$ is absolutely continuous with respect to μ, where $\alpha = 1$. □

5.4.6 ρ-Modulus of Continuity

Let G be a locally compact Abelian group with Haar measure $\mu_{\mathbf{H}}(A)$ of a Haar-measurable set $A \subset G$, i.e., $\mu_{\mathbf{H}}$ is a Borel measure defined on the σ-ring generated by the class of all compact sets and $\mu_{\mathbf{H}}(C) < \infty$, C compact set, such that $\mu_{\mathbf{H}}(U) > 0$ for every non-empty Borel open set U, and such that $\mu_{\mathbf{H}}(t + A) = \mu_{\mathbf{H}}(A)$ for $t \in G$, and A is a Borel set. We suppose that $\mu_{\mathbf{H}}$ has the property

(P) $\mu_{\mathbf{H}}(A) < \infty$ implies $\mu_{\mathbf{H}}(A\Delta(A-t)) \to \mathbf{0}$ as $t \to 0$, where Δ is the symmetric difference.

For example, the groups $(\mathbf{R}^n, +)$, $(T, \cdot)$, where T is the unit circle in the complex plain, any discrete group G, and any compact group satisfy the property (P).

We denote by $\int_G f\,dt$ the integral of f, $f\colon G \to \mathbf{R}$, with respect to Haar measure and by $\mathcal{U}$ the family of neighbourhoods of $\mathbf{0}$ in G.

Let ρ be a modular on a subspace $\mathbf{F}$ of $L^0(G)$.

Definition 5.12 The function ω_ρ, $\omega_\rho\colon \mathbf{F} \times G \to [0, +\infty)$, defined by

$$\omega_\rho(f, U) = \sup_{t\in U} \rho(f(t+\cdot) - f(\cdot))$$

for $f \in \mathbf{F}$ and $U \in \mathcal{U}$ is called the ρ-modulus of continuity.

Theorem 5.16 *Let η_f be a linearly indexed family of submeasures on translation invariant subspace $\mathbf{F}$ of $L^0(G)$, i.e., if $f \in \mathbf{F}$ then $f(\cdot + t) \in \mathbf{F}$ for every $t \in G$, is absolutely continuous with respect to the Haar measure $\mu_{\mathbf{H}}$ on G and satisfying* (S6).

Let ρ be defined by $\rho(f) = \eta_f(X)$, this is a monotone and absolutely finite modular on $\mathbf{F}$ which is translation invariant, i.e.,

$$\rho(f(t+\cdot)) = \rho(f) \qquad (f \in \mathbf{F},\ t \in G)\,.$$

If $f \in \mathbf{F}_\rho$, then there is a number $\lambda_0 > 0$ such that for every $\epsilon > 0$ there exists a $U_\epsilon \in \mathcal{U}$ for which

$$\omega_\rho(\lambda_0 f, U_\epsilon) < \epsilon .$$

Proof. We take first the special case $f = \chi_A$, when $A \in \Sigma$, $\mu_{\mathbf{H}}(A) < \infty$. Then we have

$$\omega_\rho(\lambda_0 f, U) = \sup_{t \in U} \rho(\lambda_0(\chi_A(t+\cdot) - \chi_A(\cdot))) = \sup_{t \in U} \rho(\chi_{A\Delta(A-t)})$$

for $u \in \mathcal{U}$, $x_0 > 0$. The theorem follows by absolute finiteness of ρ and the property (P) of G.

As a second case we take that f is an integrable simple function, i.e., $f = \sum_{k=1}^n c_k \chi_{A_k}$, where $A_k \in \Sigma$, $\mu_{\mathbf{H}}(A) < \infty$, A_k are pairwise disjoint for $k = 1, 2, \dots, n$. Taking $c = \max\{|c_k| : k = 1, 2, \dots, n\}$, we have $|f| \leq c\chi_{\cup_{k=1}^n A_k}$. Since we have for every $f_1, f_2, \dots, f_n \in \mathbf{F}$ and $U \in \mathcal{U}$

$$\omega_\rho\Big(\sum_{i=1}^n f_i, U\Big) \leq \sum_{i=1}^n \omega_\rho(n f_i, U),$$

we obtain for any $\lambda > 0$ that $\omega_\rho(\frac{\lambda}{c} f, U) \leq \omega_\rho(\lambda \chi_{\cup_{k=1}^n A_k}, U) \to 0$ as $U \to \mathbf{0}$.

As third case let $f \in \mathbf{F}_\rho$, $f \geq 0$. We take a sequence $\{f_n\}$ of simple functions such that $0 \leq f_n(t) \uparrow f(t)$ for every $t \in G$. We obtain $f_n \in \mathbf{F}$ and so $0 \leq f - f_n \in \mathbf{F}$, $(f - f_n)(t) \downarrow 0$ for $t \in G$. Moreover, $|(f - f_n)(t)| \leq f(t)$ and $f \in \mathbf{F}_\rho$, whence $\rho(3\lambda f) < \infty$ for some $\lambda > 0$. We apply the Lebesgue type Theorem 5.13 with $g = \lambda f$, obtaining $\rho(\lambda(f_n - f)) \to 0$ as $n \to \infty$. By absolute continuity of ρ, for every $\gamma > 0$ there is a set $B_{(\gamma)} \in \Sigma$ such that $\mu_{\mathbf{H}}(B_{(\gamma)}) < \infty$ and $\rho(\lambda f \chi_{G \setminus B_{(\gamma)}}) < \gamma$. Taking $\gamma = \frac{1}{n}$, where $n \in \mathbf{N}$ and writing $A_n = B_{(1/n)}$, we have $A_n \in \Sigma$, $\mu_{\mathbf{H}}(A_n) < \infty$ and $\rho(\lambda f \chi_{G \setminus A_n}) < \frac{1}{n}$ for $n \in \mathbf{N}$. Let $g_n = f_n \chi_{A_n}$, then

$$\begin{aligned}
\rho\Big(\frac{1}{2}\lambda(f - g_n)\Big) &= \rho\Big(\frac{1}{2}\lambda(f - f_n) + \frac{1}{2}\lambda(f_n - g_n)\Big) \\
&\leq \rho(\lambda(f - f_n)) + \rho(\lambda f_n \chi_{G \setminus A_n}) \\
&\leq \rho(\lambda(f - f_n)) + \rho(\lambda f \chi_{G \setminus A_n}) \leq \rho(\lambda(f - f_n)) + \frac{1}{n}
\end{aligned}$$

for $n \in \mathbf{N}$, due to monotonicity of ρ. Since $\rho(\lambda(f - f_n)) \to 0$ as $n \to \infty$, so we get $\rho(\frac{1}{2}\lambda(f - g_n)) \to 0$ as $n \to \infty$. Now, take an arbitrary $c_0 > 0$. Then we have

$$\begin{aligned}\rho(c_0(f(t+\cdot) - f(\cdot))) &\le \rho(2c_0((f - g_n)(t+\cdot) - (f - g_n)(\cdot))) \\ &\quad + \rho(2c_0(g_n(t+\cdot) - g_n(\cdot))) \\ &\le \rho(4c_0(f - g_n)(t+\cdot)) + \rho(4c_0(f - g_n)) \\ &\quad + \rho(2c_0(g_n(t+\cdot) - g_n(\cdot))).\end{aligned}$$

Since $\rho(\frac{1}{2}\lambda(f - g_n)) < \infty$ for sufficiently large n, say for $n \ge n_0$, so taking $c_0 = \frac{\lambda}{8}$, we obtain

$$\begin{aligned}\rho\Big(\frac{\lambda}{8}(f(t+\cdot) - f(\cdot))\Big) &\le \rho\Big(\frac{1}{2}\lambda(f - g_n)\Big) + \rho\Big(\frac{\lambda}{2}(f - g_n)\Big) \\ &\quad + \rho\Big(\frac{\lambda}{4}(g_n(t+\cdot) - g_n(\cdot))\Big) \\ &\le 2\rho\Big(\frac{1}{2}\lambda(f - g_n)\Big) + \rho\Big(\frac{\lambda}{4}(g_n(t+\cdot) - g_n(\cdot))\Big)\end{aligned}$$

for $n \ge n_0$, where we have used that ρ is translation invariant. Hence

$$\omega_\rho\Big(\frac{\lambda}{8}f, U\Big) \le 2\rho\Big(\frac{1}{2}\lambda(f - g_n)\Big) + \omega_\rho\Big(\frac{\lambda}{4}g_n, U\Big) \tag{5.18}$$

for $n \ge n_0$ and $U \in \mathcal{U}$. Now, let us take an arbitrary $\epsilon > 0$. Since $\rho(\frac{1}{2}\lambda(f - g_n)) \to 0$, there exists an $n_1 \ge n_0$ such that

$$\rho\Big(\frac{1}{2}\lambda(f - g_n)\Big) < \frac{1}{4}\epsilon \quad \text{for} \quad n \ge n_1\,. \tag{5.19}$$

Moreover, since g_n is an integrable simple function, we have $\omega_\rho(\lambda_n g_n, U) \to 0$ as $U \to \mathbf{0}$ for some $\lambda_n > 0$. Taking $n = n_1$ we thus have

$$\omega_\rho\Big(\frac{\lambda_1}{4}g, U\Big) \to 0, \ U \to \mathbf{0}$$

for λ_1 which satisfies the condition

$$0 < \lambda_1 \le \min(\lambda, 4\lambda_{n_1})\,.$$

This implies the existence of a neighbourhood U_1 of $\mathbf{0}$ such that

$$\omega_\rho\Big(\frac{\lambda_1}{4}g_{n_1}, U_1\Big) < \frac{1}{2}\epsilon\,. \qquad (5.20)$$

The inequality (5.18) by the inequality (5.19) and (5.20) obtains the following form

$$\omega_\rho\Big(\frac{\lambda_1}{8}f, U\Big) \le \omega_\rho\Big(\frac{\lambda_1}{8}f, U_1\Big) < \epsilon$$

for any $U \subset U_1$, $U \in \mathcal{U}$. So we have the conclusion of the theorem for positive functions for $\lambda_0 = \frac{\lambda_1}{8}$.

Now we will prove that the conclusion of theorem is true also for the general case, i.e., for on arbitrary $f \in \mathbf{F}_\rho$. Then we can represent f in the form $f = f_+ - f_-$, where f_+, f_- are the positive and the negative part of f, respectively. The monotonicity of ρ implies $f_+, f_- \in \mathbf{F}_\rho$. Since we have proved the theorem for positive case we can apply it on f_+ and f_-. So we have

$$\omega_\rho(\lambda_0 f, U) \le \omega_\rho(2\lambda_0 f_+, U) + \omega_\rho(2\lambda_0 f_-, U) \to 0$$

as $U \to \mathbf{0}$ for sufficiently small $\lambda_0 > 0$. □

5.4.7 Application to Approximation Theory

Let $\mathbf{K} = \{K_n\}$ be a kernel ,i.e., $K_n\colon G \times \mathbf{R} \to \mathbf{R}$, $K_n(t, 0) = 0$, and $K_n(\cdot, u) \in L^1(G)$ for all $u \ge 0$. Let

$$L_n(t) = \sup_{u \ne v} \frac{|K_n(t,u) - K_n(t,v)|}{|u - v|} \qquad (t \in G,\ n \in \mathbf{N}).$$

We suppose the Lipschitz condition, i.e., $\overline{L}_n = \int_G L_n(t)\,dt < +\infty$.

Definition 5.13 The kernel $\mathbf{K}$ is called singular if:
(1) for every $U \in \mathcal{U}$, $\frac{1}{\overline{L}_n} \int_{G\setminus U} L_n(t) \to 0$ as $n \to \infty$;
(2) the sequence $\{\overline{L}_n\}$ is bounded;
(3) $r_n = \sup_{u \ne 0} | \int_G K_n(t, u)\,dt - 1| \to 0$ as $n \to \infty$.

Now, we consider a sequence $\{T_n\}$ of nonlinear integral operators:

$$(T_n f)(s) = \int_G K_n(t - s, f(t))\,dt = \int_G K_n(t, f(t + s))\,dt \qquad (n \in \mathbf{N}).$$

Then the following estimation holds.

Theorem 5.17 *Let G be a locally compact abelian group and let K be a singular kernel. Let ρ be a monotone translation invariant and J-convex modular in a translation invariant subspace* **F** *defined by $\rho(t) = \eta_f(X)$ ($f \in \mathbf{F}$). Then for all $\lambda > 0$, $U \in \mathcal{U}$, $n \in \mathbf{N}$ and $f \in \mathbf{F}_\rho$ there holds the inequality*

$$\rho(\lambda(T_n f - f)) \leq \omega_\rho(2\lambda \overline{L}_n f, U) + 2\rho(4\lambda \overline{L}_n f) \times \frac{1}{\overline{L}_n} \int_{G \setminus U} L_n(t)dt + \rho(2\lambda r_n f). \tag{5.21}$$

Proof. Using Remark 5.1 on ρ we have

$$\rho(\lambda(T_n f - f)) \leq C_1 + C_2 \,, \tag{5.22}$$

where

$$C_1 = \rho\Big(2\lambda \int_G |K_n(t, f(t+\cdot)) - K_n(t, f(\cdot))|\, dt\Big),$$

$$C_2 = \rho\Big(2\lambda \int_G K_n(t, f(\cdot))\, dt - f(\cdot)\Big).$$

J-convexity of ρ implies

$$\begin{aligned} C_1 &\leq \rho\Big(\int_G L_n(t) 2\lambda \overline{L}_n |f(t+\cdot) - f(\cdot)|\, dt\Big) \\ &\leq \int_U \frac{L_n(t)}{\overline{L}_n} \rho(2\lambda \overline{L}_n |f(t+\cdot) - f(\cdot)|)\, dt \\ &\quad + \int_{G\setminus U} \frac{L_n(t)}{\overline{L}_n} \rho(2\lambda \overline{L}_n |f(t+\cdot) - f(\cdot)|)\, dt \,. \end{aligned}$$

By the definition of ω_ρ, the first of the integrals at the right-hand side of the last inequality is estimated by

$$\int_U \frac{L_n(t)}{\overline{L}_n} \omega_\rho(2\lambda \overline{L}_n f, U)\, dt \leq \omega_\rho(2\lambda L_n f, U) \,.$$

In order to estimate the second integral we observe that $|f(t+s)-f(s)| \leq |f(t+s)|+|f(s)|$ $(t, s \in G)$ and so, by translation invariantness and monotonicity of ρ, we obtain

$$\begin{aligned}\int_{G\setminus U} & L_n(t)\rho(2\lambda\overline{L}_n|f(t+\cdot)-f(\cdot)|)\,dt \\ &\leq \int_{G\setminus U} \frac{L_n(t)}{\overline{L}_n}\rho(4\lambda\overline{L}_n f)\,dt + \int_{G\setminus U} L_n(t)\rho(4\lambda\overline{L}_n f)\,dt \\ &\leq 2\rho(4\lambda L_n f)\int_{G\setminus U} \frac{L_n(t)}{\overline{L}_n}\,dt\,.\end{aligned}$$

Hence

$$C_1 \leq \omega_\rho(2\lambda\overline{L}_n, U) + 2\rho(4\lambda\overline{L}_n f)\int_{G\setminus U}\frac{L_n(t)}{\overline{L}_n}\,dt. \tag{5.23}$$

Since, we have

$$2\lambda\left|\int_G K_n(t, f(s))\,dt - f(s)\right| \leq 2\lambda r_n|f(s)|$$

for $s \in G$, we obtain

$$C_2 \leq \rho(2\lambda r_n f)\,. \tag{5.24}$$

Taking (5.23) and (5.24) in (5.22) we obtain the inequality (5.21). □

Theorem 5.18 *Let G be a locally compact abelian group possessing property* (P) *and let* **K** *be a singular kernel. Let ρ be a monotone, J-convex absolutely finite, absolutely continuous, and translation invariant modular in a translation invariant subspace* **F**. *Then for every $f \in \mathbf{F}_\rho$ there exists a $\lambda > 0$ such that*

$$\rho(\lambda(T_n f - f)) \to 0, \quad as \quad n \to \infty\,.$$

Proof. In order to estimate the first sum let the right-hand side of the inequality in (5.21) in Theorem 5.17 we have that for every $\epsilon > 0$ there exist $\lambda_1 > 0$ and $U \in \mathcal{U}$ such that

$$\omega_\rho(2\lambda L' f, U) < \frac{1}{3}\epsilon \ \text{ for } \ 0 < \lambda \leq \lambda_1\,. \tag{5.25}$$

where $L' = \sup_{1\leq n<\infty} \overline{L}_n$.

Let $M(\lambda) = \rho(4\lambda cf)$. Since $f \in \mathbf{F}_\rho$, there is a $\lambda_2 > 0$ such that $M(\lambda) \leq M(\lambda_2) < \infty$ for $0 < \lambda \leq \lambda_2$. By singularity of $\mathbf{K}$, there exists $n_0 \in \mathbf{N}$ such that

$$2M(\lambda_2)\frac{1}{L_n}\int_{G\setminus U} L_n(t)\,dt < \frac{1}{3}\epsilon \quad \text{for } n \geq n_1\,. \tag{5.26}$$

Let $\lambda_2 = \min(\lambda_1, \lambda_2)$. Since $f \in \mathbf{F}_\rho$, there exists an $\gamma > 0$ such that $\rho(2\lambda_0\gamma) < \frac{\epsilon}{3}$. By singularity of γ there exists $n_2 \in \mathbf{N}$ such that $r_n < \gamma$ for $n \geq n_2$. Hence

$$\rho(2\lambda r_n f) < \frac{1}{3}\epsilon \quad \text{for } n \geq n_2\,. \tag{5.27}$$

Taking (5.25)–(5.27) in the inequality (5.21) we obtain

$$\rho(\lambda(T_n f - f)) < \epsilon \quad \text{for } 0 < \lambda \leq \lambda_0 \text{ and } n \geq \max(n_1, n_2)\,. \qquad \square$$

Remark 5.3 Theorem 5.17 and 5.18 hold also with respect to an abstract nonempty set of indices instead of the set of natural numbers and convergence with respect to a filter of subsets of the set.

Example 5.8 Let G be a locally compact abelian group. Let $\varphi\colon G\times[0,+\infty) \to [0,+\infty)$ be a measurable function of t for every $u \neq 0$, and $\varphi(t,u)$ is a nondecreasing continuous function of u for a.e. $t \in G$, $\varphi(t,0) = 0$, $\varphi(t,u) > 0$ for $u > 0$. Let $\mathbf{F}$ be a family of all measurable, finite almost everywhere functions. Then $\{\eta_f\}_{f\in\mathbf{F}}$

$$\eta_f(A) = \int_A \varphi(t, |f(t)|)\,dt$$

is a linearly indexed family of submeasure with the property (S6), and so

$$\rho(f) = \eta_f(G)$$

is a monotone, absolutely continuous modular in the space $\mathbf{F}$. The space $\mathbf{F}_\rho$ generated by the modular ρ is the generalized Orlicz space $L^\varphi(G)$.

Example 5.9 If additionally in the previous example

$$\int_A \varphi(t,u) < \infty$$

for sufficiently small $u > 0$ for $A \in \Sigma$, $\mu_{\mathbf{H}}(A) < \infty$, then ρ is finite.

5.5 Notes and References

Early traces of the notion of autocontinuity were in the papers [Aleksjuk 1968], [Aleksjuk 1970], [Dobrakov 1974] and [Drewnowski 1978]. The name and further properties related to autocontinuity were given in [Z. Wang 1984]. Examples in 5.1.2 are from [Dobrakov 1974], [Dobrakov 1972], [Z. Wang 1984]. The result for countable case in 5.1.3 was obtained in [Congxin and Minghu 1993] and [Z. Wang and Klir 1992].

Submeasures appear very often in classical measure theory as a tool for investigation of additive and σ-additive measures: [Dunford and Swartz 1958], [Halmos 1950], [M. M. Rao 1987]. The Nikodym-Frechet topology is based on [Drewnowski 1972a], [Orlicz 1968]. The Frechet-Nikodym topology approach to measure theoretical problems was further developed in many papers, for example see [H. Weber 1984], [H. Weber 1986].

The topological relation of submeasures and autocontinuous monotone set functions was obtained in [Drewnowski 1978]. A version of the extension of Theorem 5.11 was obtained for first time in [Aleksjuk 1970] and then in the present form in [Drewnowski 1972a]. The general theorem on the extension was obtained in [Dobrakov 1974] and the proof with submeasures was given in [Drewnowski 1978].

Linearly indexed families of submeasures were introduced by J. Musielak, and this notion is implicitly contained in [Musielak 1987], [Musielak 1991] and [Musielak 1993], where also the application in approximation theory can be found.

Chapter 6

Decompositions of Set Functions

In this chapter we adapt some classical decomposition theorems to null-additive set functions such as: Hahn decomposition on positive and negative sets, Jordan decomposition, Lebesgue decomposition on continuous and singular parts, and Saks decomposition related to atoms. At the end we will prove a version of the Hewitt-Yosida decomposition theorem for decomposable measures defined on lattice (generalized measures).

6.1 Hahn Decomposition Theorem

6.1.1 Signed Fuzzy Measures

Σ always denotes a σ-algebra of subsets of the given set X.

Definition 6.1 A signed fuzzy measure (revised monotone continuous from above and below set function) $m, m\colon \Sigma \to [-\infty, \infty]$, is an extended real-valued set function m defined on a σ-algebra Σ and with the properties:

(SFM_1) $m(\emptyset) = 0$;

(SFM_2) if $E, F \in \Sigma$, $E \cap F = \emptyset$, then

(a) $m(E) \geq 0$, $m(F) \geq 0$, $\max(m(E), m(F)) > 0$ implies

$$m(E \cup F) \geq \max(m(E), m(F));$$

(b) $m(E) \leq 0$, $m(F) \leq 0$, $\min(m(E), m(F)) < 0$ implies

$$m(E \cup F) \leq \min(m(E), m(F));$$

(c) $m(E) > 0$, $m(F) < 0$ implies $m(E) \geq m(E \cup F) \geq m(F)$.

(SFM_3) $(E_1 \subset E_2 \subset \ldots,\ E_n \in \Sigma) \Rightarrow m(\bigcup_{n=1}^{\infty} E_n) = \lim_{n\to\infty} m(E_n)$;

(SFM_4) $(E_1 \subset E_2 \subset \ldots,\ E_n \in \Sigma$, there exists $n_0 \in \mathbf{N}$, $|m(E_{n_0})| < \infty) \Rightarrow m(\bigcap_{n=1}^{\infty} E_n) = \lim_{n\to\infty} m(E_n)$.

The property (SFM_2) of m is called the revised monotonicity.

Example 6.1 Any non-negative signed fuzzy measure is a usually fuzzy measure. Namely, the condition (SFM_2)(a) implies the monotonicity of m. But also each fuzzy measure is a signed fuzzy measure.

Example 6.2 The classical signed measure is a signed fuzzy measure.

Definition 6.2 A set $A \subset X$ is called a positive set (resp. negative set) for signed fuzzy measure m on (Σ, X) if for every subset E of A which belongs to Σ we have $m(E) \geq 0$ (resp. $m(E) \leq 0$).

Theorem 6.1 *If m is a null-additive revised monotone set function defined on Σ, then the union of two negative sets with respect to m is again a negative set and the union of two positive sets with respect to m is again a positive set.*

Proof. Let $A, B \in \Sigma$ be two negative sets. Then for any subset D of $A \cup B$ which belongs to Σ, since

$$D = (A \cap D) \cup ((B \cap D) \setminus A), \quad A \cap D \subset A \quad \text{and} \quad (B \cap D) \setminus A \subset B,$$

we obtain by $m(A \cap D) \leq 0$ and $m((B \cap D) \setminus A) \leq 0$ for the case

$$\min\{m(A \cap D), m((B \cap D) \setminus A)\} < 0$$

by the revised monotonicity

$$m(D) \leq \min\{m(A \cap D), m((B \cap D) \setminus A)\} < 0,$$

and if at least one of $m(A \cap D)$ or $m((B \cap D) \setminus A)$ is zero by the null-additivity of m that $m(D) \leq 0$.

The proof is analogous for positive sets. □

The previous theorem is not true for every set function with the property of revised monotonicity.

Example 6.3 Let $X = \{a, b, c, d\}$ and $\Sigma = \mathcal{P}(X)$. We define the set function $m: \Sigma \to [-1, 1]$ in the following way: $m(\emptyset) = m(\{b\}) = m(\{d\}) = 0$, $m(\{b, d\}) = 1$ and $m(A) = -1$ for other cases. Then the sets $\{a, b\}$ and $\{c, d\}$ are negative sets, but their union, the set X, is not a negative set. This set function is not null-additive, although it is revised monotone.

6.1.2 Hahn Decomposition

We have the following Hahn decomposition type theorem.

Theorem 6.2 *Let m be a null-additive revised monotone set function on (Σ, X), which is continuous from above and from below (i.e., a null-additive signed fuzzy measure). If m takes at most one of the values $-\infty$ or $+\infty$, and if*

$$E \in \Sigma,\ |m(E)| < +\infty \Rightarrow |m(F)| < +\infty \qquad (F \subset E,\ F \in \Sigma),$$

then there exist two disjoint sets A and B from Σ such that $A \cup B = X$ and A is a positive set and B is a negative set.

To prove Theorem 6.2 we need the following lemma.

Lemma 6.1 *If m is a signed fuzzy measure with the conditions given in Theorem 6.2 and $E \in \Sigma$, $m(E) < 0$, then there exists a negative set $E_0 \in \Sigma$ such that $E_0 \subset E$ and $m(E_0) \leq m(E)$.*

Proof. If E is a negative set, then the statement follows taking $E_0 = E$.

If E is not a negative set, then there exists $A \in \Sigma$, $A \subset E$ such that $m(A) > 0$. Let

$$n_0 = \min\{n : \text{ there exists } A \in \Sigma,\ A \subset E,\ m(A) > 1/(n+1)\},$$

and take $A_0 \in \Sigma$, $A_0 \subset E$ with $m(A) > 1/(n_0 + 1)$.

If $E \setminus A_0$ is a negative set, take $E_0 = E \setminus A_0$. From the revised monotonicity of a signed fuzzy measure and the facts that $m(A_0) > 0$, $m(E) < 0$, we have $m(E \setminus A_0) < 0$. Then the revised monotonicity of the signed fuzzy measure and the facts that $m(E \setminus A_0) < 0$ and $m(A_0) > 0$ imply $m(E_0) = m(E \setminus A_0) \leq m(E)$, and the conclusion follow.

If $E \setminus A_0$ is not a negative set, then there exists $A \in \Sigma$, $A \subset E \setminus A_0$ such that $m(A) > 0$. Let

$$n_1 = \min\{n : \text{ there exists } A \in \Sigma,\ A \subset E \setminus A_0,\ m(A) > 1/(n+1)\},$$

and take $A_1 \in \Sigma$, $A_1 \subset E \setminus A_0$ with $m(A_1) > 1/(n_1 + 1)$.

If $E \setminus (A_0 \cup A_1)$ is a negative set, take $E_0 = E \setminus (A_0 \cup A_1)$. By the revised monotonicity of a signed fuzzy measure and the fact that $m(A_0 \cup A_1) \geq m(A_0) \vee m(A_1) > 0$ and $m(E) > 0$, we have $m(E \setminus (A_0 \cup A_1)) < 0$. Then the revised monotonicity of the set function and the facts that $m(E\setminus(A_0\cup A_1))<0$, $m(A_0\cup A_1)>0$ imply $m(E_0)=m(E\setminus(A_0\cup A_1))\leq m(E)$, and the conclusion follow.

If $E\setminus(A_0\cup A_1)$ is a not negative set, then there exists $A\in\Sigma$, $A\subset E\setminus(A_1\cup A_1)$ such that $m(A)>0$. Let

$$n_2=\min\{n: \text{ there exists } A\in\Sigma,\ A\subset E\setminus(A_0\cup A_1),\ m(A)>1/(n+1)\}$$

and take $A_2\in\Sigma$, $A_2\subset E\setminus(A_0\cup A_1)$ with $m(A_2)>1/(n_2+1)$.

Repeating the above processes, we can obtain two cases:

(i) there exists $s\in\{0,1,2,\ldots\}=\mathbf{N}'$ such that $E\setminus\bigcup_{n=0}^{s}A_n$ is a negative set, and the conclusion follows by taking $E_0=E\setminus\bigcup_{n=0}^{s}A_n$; or

(ii) $E\setminus\bigcup_{n=0}^{s}A_n$; $(s\in\mathbf{N}')$ is not a negative set.

If (ii) occurs, we take $E_n\in\Sigma$, $E_n\subset E$ as $E_0=E\setminus\bigcup_{n=1}^{\infty}A_n$, then E_0 is a negative set and $m(E_0)\leq m(E)$.

Namely, we will prove that $\lim_{s\to\infty}1/n_s=0$.

Since $\bigcup_{n=s}^{\infty}A_n\subset E$, $-\infty<m(E)<0$, then $|m(\bigcup_{n=s}^{\infty}A_n)|<+\infty$ $(s\in\mathbf{N}')$. Furthermore, the fact that $A_0,A_1,A_2,\ldots$ are pairwise disjoint implies $\bigcup_{n=s}^{\infty}A_n\searrow\emptyset$, then

$$\lim_{s\to\infty}m\Big(\bigcup_{n=s}^{\infty}A_n\Big)=0\,.$$

We have

$$m\Big(\bigcup_{n=s}^{\infty}A_n\Big)\geq\bigvee_{n=s}^{\infty}m(A_n)\geq m(A_s)>0\qquad(s\in\mathbf{N}'),$$

we obtain

$$\lim_{s\to\infty}m(A_s)=0\,.$$

Therefore by $m(A_s)>1/(n_s+1)>0$, we obtain $\lim_{s\to\infty}1/n_s=0$.

Now, suppose $A\in\Sigma$, $A\subset E\setminus\bigcup_{n=0}^{\infty}A_n$, then $A\subset E\setminus\bigcup_{n=0}^{s-1}A_n$ $(s\in\mathbf{N})$, and hence $m(A)\leq 1/n_s$. Therefore $m(A)\leq\lim_{s\to\infty}1/n_s=0$, and then E_0 is a negative set.

Since $m(E)<0$,

$$m\Big(\bigcup_{n=0}^{\infty}A_n\Big)\geq\bigvee_{n=0}^{\infty}m(A_n)>0\,,$$

then, from the revised monotonicity of signed fuzzy measures, $m(E\setminus\bigcup_{n=0}^{\infty} A_n) < 0$, and the revised monotonicity of signed fuzzy measure m and the facts that $m(E \setminus \bigcup_{n=0}^{\infty} A_n) < 0$ and $m(\bigcup_{n=0}^{\infty} A_n) > 0$ we obtain

$$m(E_0) = m\Big(E \setminus \bigcup_{n=0}^{\infty} A_n\Big) \leq m(E). \qquad \square$$

Proof of Theorem 6.2. For $m(A) \geq 0$ $(A \in \Sigma)$, then m we can take $A = X$ and $B = \emptyset$, and the conclusion follows.

If there exists $H \in \Sigma$ such that $m(H) < 0$, put

$$n_0 = \min\{n : \text{ there exists } H \in \Sigma,\ m(H) < -1/(n+1)\},$$

and choose $H_0 \in \Sigma$ such that $m(H_0) < -1/(n_0 + 1)$. By Lemma 6.1 there exists a negative set $E_0 \in \Sigma$, $E_0 \subset H_0$ such that $m(E_0) \leq m(H_0)$.

If $X \setminus E_0$ is a positive set, then we can take $A = X \setminus E_0$ and $B = E_0$ and the conclusion follows.

If $X \setminus E_0$ is not a positive set, then we can find $H \in \Sigma$, $H \subset X \setminus E_0$ such that $m(H) < 0$. Let

$$n_1 = \min\{n : \text{ there exists } H \in \Sigma,\ H \subset X \setminus E_0,\ m(H) < -1/(n+1)\},$$

and choose $H_1 \in \Sigma$, $H_1 \subset X \setminus E_0$ such that $m(H_1) < -1(n_1 + 1)$. By Lemma 6.1, there exists a negative set $E_1 \in \Sigma$, $E_1 \subset H_1$ such that $m(E_1) \leq m(H_1)$.

If $X \setminus (E_0 \cup E_1)$ is a positive set, then we can take $A = X \setminus (E_0 \cup E_1)$ and $B = E_0 \cup E_1$, and the conclusion follows.

If $X \setminus (E_0 \cup E_1)$ is not a positive set, then there exists $H \in \Sigma$, $H \subset X \setminus (E_0 \cup E_1)$ such that $m(H) < 0$. Let

$$n_2 = m\{n : \text{ there exists } H \in \Sigma,\ H \subset X\setminus(E_0 \cup E_1),\ m(H) < -1/(n+1)\},$$

and choose $H_2 \in \Sigma$, $H_2 \subset X \setminus (E_0 \cup E_1)$ such that $m(H_2) < -1(n_2 + 1)$. By Lemma 6.1 there exists a negative set $E_2 \in \Sigma$, $E_2 \subset H_2$ such that $m(E_2) \leq m(H_2)$.

Repeating the above processes, one of the following cases will occur

(a) there exists $s \in \mathbf{N}$ such that $X \setminus \bigcup_{n=0}^{s} E_n$ is a positive set;

(b) For each $s \in \mathbf{N}$ we have $X \setminus \bigcup_{n=0}^{s} E_n$ is not a positive set.

If (a) occurs, we can take $A = X \setminus \bigcup_{n=0}^{s} E_n$ and $B = \bigcup_{n=0}^{s} E_n$, and the conclusion follows.

If (b) occurs, we can take $A = X \setminus \bigcup_{n=0}^{\infty} E_n$ and $B = \bigcup_{n=0}^{\infty} E_n$. Then A is a positive set and B a negative set.

Namely, since m is null-additive B is a negative set by Theorem 6.1 as a union of negative sets. Now we will prove that A is a positive set.

For all $E \in \Sigma$, $E \subset A = X \setminus \bigcup_{n=0}^{\infty} E_n$, we have

$$E \subset X \setminus \bigcup_{n=0}^{s-1} E_n \qquad (s \in \mathbf{N}),$$

and hence

$$m(E) \geq -1/n_s \qquad (s \in \mathbf{N}). \tag{6.1}$$

In addition, since $E_0, E_1, E_2, \ldots$ are pairwise disjoint $\bigcup_{n=s}^{\infty} E_n \searrow \emptyset$, and since

$$-\infty < m\Big(\bigcup_{n=s}^{\infty} E_n\Big) \leq \bigwedge_{n=s}^{\infty} m(E_n) < 0 \qquad (s \in \mathbf{N}')$$

we have

$$\lim_{s\to\infty} m\Big(\bigcup_{n=s}^{\infty} E_n\Big) = 0,$$

which, using the fact that $m(\bigcup_{n=s}^{\infty} E_n) \leq m(E_n) < 0$ implies

$$\lim_{s\to\infty} m(E_s) = 0 \qquad (s \in \mathbf{N}').$$

Therefore by $m(E_s) \leq m(H_s) < \frac{-1}{n_s+1} < 0$ $(s \in \mathbf{N})$ follows $\lim_{s\to\infty} 1/n_s = 0$. Hence by (6.1)

$$m(E) \geq \lim_{s\to\infty} (-1/n_s) = 0,$$

which implies that $A = X \setminus \bigcup_{n=0}^{\infty} E_n$ is a positive set. □

Remark 6.1 The classical signed measure satisfies the conditions of the previous theorem, so in this case the previous theorem reduces on the classical Hahn decomposition theorem.

The previous Hahn decomposition of m in Theorem 6.2 (which is not unique!) we shall denote by (A, B).

6.2 Jordan Decomposition Theorem

In this section we shall suppose, without loss of generality, that the signed fuzzy measure m satisfies the condition

$$-\infty < m(E) \leq +\infty \qquad (E \in \Sigma).$$

Now we can prove the following version of the Jordan decomposition theorem

Theorem 6.3 *Let m be a null-additive signed fuzzy measure. Then there exist uniquely determined null-additive fuzzy measures m^+ and m^- such that*

$$m^+ \geq m \geq -m^- .$$

Moreover, if m has a representation in the form

$$m = v_1 - v_2 ,$$

where v_1 and v_2 are null-additive fuzzy measures, then $v_1 \geq m^+$ and $v_2 \geq m^-$.

Proof. By Theorem 6.2 there exist a positive set A and a negative set B for m, both from Σ. Then we define

$$m^+(E) = m(E \cap A) \qquad \text{and} \qquad m^-(E) = -m(E \cap B).$$

The previous definition is correct. Namely, if we take two different Hahn decompositions (A_1, B_1) and (A_2, B_2) of m, i.e., $X = A_i \cup B_i$ with $A_i \cap B_i = \emptyset$ $(i = 1, 2)$ and A_i are positive sets and B_i are negative sets with respect to m, then we have

$$m(E \cap A_1) = m(E \cap A_2) \qquad \text{and} \qquad m(E \cap B_1) = m(E \cap B_2).$$

We shall prove only the first equality since the second can be proved in a quite similar way.

Since $E \cap (A_2 \setminus A_1)$ is a subset of $E \cap A_2$ we have

$$m(E \cap (A_2 \setminus A_1)) \geq 0 .$$

Since $E \cap (A_2 \setminus A_1)$ is a subset also of the set $E \cap B_1$ we obtain

$$m(E \cap (A_2 \setminus A_1)) \leq 0 .$$

The last two inequalities imply

$$m(E \cap (A_2 \setminus A_1)) = 0 .$$

Exchanging A_1 and A_2 in the last equality we obtain

$$m(E \cap (A_1 \setminus A_2)) = 0 .$$

Finally, by the null-additivity of m we have

$$\begin{aligned} m(E \cap A_1) &= m(E \cap (A_1 \cap A_2) \cup (E \cap (A_1 \setminus A_2)) \\ &= m(E \cap (A_1 \cap A_2) \cup (E \cap (A_1 \setminus A_2) \cup (E \cap (A_2 \setminus A_1))) \\ &= m(E \cap (A_1 \cap A_2) \cup (E \cap (A_2 \setminus A_1))) = m(E \cap A_2) . \end{aligned}$$

It is obvious that m^+ and m^- are fuzzy measures. We shall prove that m^+ is null-additive. Namely, if $C \in \Sigma$ is such that $m^+(C) = 0$ we have $m(C \cap A) = 0$ and so we obtain

$$m^+(E \cup C) = m((E \cup C) \cap A) = m((E \cap A) \cup (C \cap A)) = m(E \cap A) = m^+(E) .$$

In quite an analogous way we can prove that m^- is also null-additive.

By the revised monotonicity of m (condition (c) in (SFM_2)) and definitions of m^+ and m^- we obtain

$$m^+ \geq m \geq -m^- .$$

Now suppose that m is represented in the form $v_1 - v_2$, where v_1 and v_2 are null-additive fuzzy measures. Then by the inequality $m \leq v_1$ we have

$$m^+(E) \leq m(E \cap A) \leq v_1(E \cap A) \leq v_1(E) \qquad (E \in \Sigma) .$$

In an analogous way, we obtain by the inequality $m \leq v_2$ that

$$m^-(E) = -m(E \cap B) \leq v_2(E \cap A) \leq v_2(E) \qquad (E \in \Sigma). \qquad \square$$

Since for the null-additive signed fuzzy measure m the set functions m^+ and m^- are uniquely determined the following definition makes sense.

Definition 6.3 The total variation $|m|$ of a null-additive signed fuzzy measure m is given by

$$|m| = m^+ + m^- .$$

If $m \in BV$, then the total variation of m coincides with its chain variation.

We have the following analogous property as for the classical signed measure.

Theorem 6.4 *The total variation $|m|$ of a null-additive signed fuzzy measure m is a null-additive fuzzy measure which satisfies the inequality*

$$|m(E)| \le |m|(E) \qquad (E \in \Sigma) .$$

Proof. It is easy to check that $|m|$ is a fuzzy measure. We shall prove the inequality. For that purpose we shall use the inequality

$$m^+ \ge m \ge -m^- .$$

Suppose first that $m(E) > 0$. Then by the last inequality $m^+(E) \ge m(E)$. Hence

$$|m|(E) = m^+(E) + m^-(E) \ge m(E) = |m(E)| .$$

Suppose now that $m(E) < 0$. Then we have $-m(E) \le m^-(E)$. Hence

$$|m(E)| \le m^-(E) \le |m|(E) .$$

If $m(E) = 0$, then since $|m|$ is always non-negative we have also in this case the desired inequality. □

Definition 6.4 Let m and v be two null-additive signed fuzzy measures on Σ. m is absolutely continuous with respect to v, $m \ll v$, if $m(E) = 0$ for every $E \in \Sigma$ such that $|v|(E) = 0$.

We have the following characterization of the notion 'absolute continuity' which is analogous to that in the classical measure theory.

Theorem 6.5 *If m and v are null-additive signed fuzzy measures, then the following conditions are equivalent to each other*
(i) $m \ll v$,
(ii) $m^+ \ll v$ *and* $m^- \ll v$,
(iii) $|m| \ll |v|$.

6.3 Lebesgue Decomposition Theorem

6.3.1 Null-Additive Set Functions

Let m and v be two non-negative monotone set functions defined on a ring $\mathcal{R}$.

Definition 6.5 Let m and v be two finite monotone set functions. If $E \in \mathcal{R}$, $v(E) = 0$ implies $m(E) = 0$, then we say that m is absolutely continuous with respect to v.

Definition 6.6 Let m and v be two finite monotone set functions. If for every $\epsilon > 0$ there is a $\delta > 0$ such that $E \in \mathcal{R}$, $v(E) < \delta$ implies $m(E) < \epsilon$, then we say that m is absolutely ϵ–continuous with respect to v.

Theorem 6.6 *Let m and v be two finite monotone set functions defined on a σ-ring $\mathcal{R}$ such that they are continuous from above and continuous from below. If v is autocontinuous from above, then m is absolutely continuous with respect to v if and only if m is absolutely ϵ-continuous with respect to v.*

Proof. It is obvious that if m is absolutely ϵ-continuous with respect to v, then m is absolutely continuous with respect to v.

Suppose now that $E \in \mathcal{R}$, $v(E) = 0$ implies $m(E) = 0$. If the theorem would not be true, then there would exist $\epsilon > 0$ and a sequence $\{E_n\}$ from $\mathcal{R}$ such that

$$v(E_n) < \frac{1}{n} \qquad \text{and} \qquad m(E_n) > \epsilon \qquad (n \in \mathbf{N}). \tag{6.2}$$

Since v is autocontinuous from above there exists a subsequence $\{E_{n_k}\}$ of the sequence $\{E_n\}$ such that

$$v\Big(\bigcup_{i=s}^{k} E_{n_i}\Big) < \frac{1}{s} \quad \text{for} \quad s = 1, 2, \ldots, k. \tag{6.3}$$

By the continuity from above of v we have

$$\lim_{s \to \infty} v\Big(\bigcup_{i=s}^{\infty} E_{n_i}\Big) = v\Big(\bigcap_{s=1}^{\infty} \bigcup_{i=s}^{\infty} E_{n_i}\Big). \tag{6.4}$$

Since v is continuous from below we obtain by (6.3)

$$v\Big(\bigcup_{i=s}^{\infty} E_{n_i}\Big) = \lim_{k\to\infty} v\Big(\bigcup_{i=s}^{k} E_{n_i}\Big) \leq \frac{1}{s}\,.$$

Hence by (6.4)

$$v\Big(\bigcap_{s=1}^{\infty}\bigcup_{i=s}^{\infty} E_{n_i}\Big) = 0\,,$$

which implies

$$m\Big(\bigcap_{s=1}^{\infty}\bigcup_{i=s}^{\infty} E_{n_i}\Big) = 0\,.$$

On the other hand, we obtain by the continuity from above and continuity from below of the monotone set function m and (6.2)

$$m\Big(\bigcap_{s=1}^{\infty}\bigcup_{i=s}^{\infty} E_{n_i}\Big) = \lim_{s\to\infty} m\Big(\bigcup_{i=s}^{\infty} E_{n_i}\Big) = \lim_{s\to\infty}\lim_{k\to\infty} m\Big(\bigcup_{i=s}^{k} E_{n_i}\Big) \geq m(E_{n_s}) > \epsilon\,.$$

Contradiction. □

Theorem 6.7 *Let m be a finite null-additive monotone set function defined on a σ-ring $\mathcal{R}$ which is continuous from above and continuous from below. Then there exists a set A from a σ-ring $\mathcal{R}$ such that*

$$m(A) = \sup\{m(E) : E \in \mathcal{R}\}, \tag{6.5}$$

$m(E\setminus A) = 0$ and $m(E) = m(E\cap A)$ $(E \in \mathcal{R})$.

Proof. We shall choose a sequence $\{A_n\}$ from $\mathcal{R}$, which will generate the desired set A. Let $A_0 = \emptyset$. We take A_1 from $\mathcal{R}$ such that

$$m(A_1) = \sup\{m(E) : E \in \mathcal{R}\}.$$

This is possible by the continuity from below of m. We choose A_2 from $\mathcal{R}$ such that

$$m(A_2) = \sup\{m(E) : E \subset X \setminus A_1\}.$$

Repeating this procedure, we choose a sequence $\{A_n\}$ such that

$$m(A_n) = \sup\Big\{m(E) : E \subset X \setminus \bigcup_{i=0}^{n-1} A_i,\ E \in \mathcal{R}\Big\} \tag{6.6}$$

holds. We take $A = \bigcup_{i=0}^{\infty} A_i$. Then the construction (6.5) holds. The continuity from above of m implies

$$\lim_{n\to\infty} m\Big(E \setminus \bigcup_{i=0}^{n} A_i\Big) = m(E \setminus A). \tag{6.7}$$

By (6.6) we obtain

$$\overline{\lim_{n\to\infty}}\, m(A_n) \geq \lim_{n\to\infty} m\Big(E \setminus \bigcup_{i=0}^{n} A_i\Big).$$

Hence by the exhaustivity of m and (6.7) $m(E \setminus A) = 0$. Hence by the null-additivity of m

$$m(E) = m((E \cap A) \cup (E \setminus A)) = m(E \cap A). \qquad \square$$

Definition 6.7 Let m and v be two finite monotone set functions defined on $\mathcal{R}$. The monotone set function m is called singular with respect to v, denoted by $m \perp v$, if there exists a set A from $\mathcal{R}$ such that

$$m(E \setminus A) = v(E) = 0 \qquad (E \in \mathcal{R}).$$

Remark 6.2 By Theorem 6.7, if for null-additive monotone set functions m and v, which are continuous from above and continuous from below, $m \perp v$ holds, then we have $v \perp m$ too.

Now we have the following two theorems of the Lebesgue decomposition type for null-additive monotone set functions.

Theorem 6.8 *Let m and v be two finite null-additive monotone set functions on a σ-ring $\mathcal{R}$. Then there exist two null-additive monotone set functions m_c and m_s such that $m_c(E) = m(E \setminus A)$ and $m_s(E) = m(E \cap A)$ for a set $A \in \mathcal{R}$ and m_c is absolutely continuous with respect to v and m_s is singular with respect to v.*

Proof. The family

$$\mathcal{R}_1 = \{E \in \mathcal{R} : v(E) = 0\}$$

is a σ-subring of the σ-ring $\mathcal{R}$. By Theorem 6.7 the restriction of m on $\mathcal{R}_1$ has a set $A \in \mathcal{R}_1$ such that $m(E \setminus A) = 0$ and $m(E) = m(E \cap A)$ for $E \in \mathcal{R}_1$. We take

$$m_c(E) = m(E \setminus A)$$

and

$$m_s(E) = m(E \cap A)$$

for each $E \in \mathcal{R}$. It is easy to check that m_c and m_s are null-additive monotone set functions and that m_c is absolutely continuous with respect to v and m_s is singular with respect to v. □

Theorem 6.9 *Let m and v be two finite autocontinuous from above monotone set functions on $\mathcal{R}$, which are continuous from above and from below. Then there exist two autocontinuous from above fuzzy measures m_c and m_s such that $m_c(E) = m(E \setminus A)$ and $m_s(E) = m(E \cap A)$ for a set $A \in \mathcal{R}$ and m_c is absolutely ϵ-continuous with respect to v and m_s is singular with respect to v.*

Proof. We take the same m_c and m_s as in the proof of Theorem 6.8. Then by Theorem 6.6 m_c is absolutely ϵ-continuous. □

6.3.2 σ-Null-Additive Set Functions

Let $\mathcal{R}$ be a ring, and m a function from $\mathcal{R}$ into $[0, \infty]$. We do not suppose the monotonicity of m. We write

$$\mathcal{N}(m) = \{A \in \mathcal{R} : \ m(A \cap Y) = 0, \ \forall Y \in \mathcal{R}\}.$$

Theorem 6.10 *Let $\mathcal{R}$ be a ring. If $m \colon \mathcal{R} \to [0, \infty]$ is a null-additive set function then the set $\mathcal{N}(m)$ is an ideal in $\mathcal{R}$.*

Proof. By definition we have for $B \in \mathcal{R}$ and $A \in \mathcal{N}(m)$ such that $B \subset A$ that $m(B) = 0$, i.e., $B \in \mathcal{N}(m)$.

For $A_1, A_2 \in \mathcal{N}(m)$ and for an arbitrary but fixed subset B of $A_1 \cup A_2$ which belongs to $\mathcal{R}$ we have $B \setminus A_1$ and $B \cap A_1$, and so

$$m(B \setminus A_1) = m(B \cap A_1) = 0 .$$

Hence, since $B \setminus A_1$ and $B \cap A_1$ are disjoint sets and m is null-additive,

$$m(B) = m((B \setminus A_1) \cup (B \cap A_1)) = 0 ,$$

i.e., $A_1 \cup A_2 \in \mathcal{N}(m)$. □

Corollary 6.1 *Let $\mathcal{R}$ be a σ-ring. If $m\colon \mathcal{R} \to [0, \infty]$ is σ-null-additive then the set $\mathcal{N}(m)$ is a σ-ideal of $\mathcal{R}$.*

Proof. Consequence of Theorem 6.10 and Proposition 2.3. □

Definition 6.8 Let $\mathcal{R}$ be a ring and $\mathcal{M}$ a subset of $\mathcal{R}$. $\mathcal{M}$ is said to satisfy the countable chain condition (CCC) if every disjoint subset (consisting of disjoint sets from $\mathcal{M}$) of $\mathcal{M} \setminus \{\emptyset\}$ is at most countable. Let m be a function from $\mathcal{R}$ into $[0, \infty]$, m is said to satisfies the (CCC) if $\mathcal{R} \setminus \mathcal{N}(m)$ satisfy the (CCC).

Lemma 6.2 *Let $\mathcal{R}$ be a ring and let m be a function from $\mathcal{R}$ into $[0, \infty]$. If m is exhaustive then it satisfies the (CCC).*

Proof. It suffices to observe that if $\{A_i\}_{i \in I}$ is a disjoint family of $\mathcal{R} \setminus \mathcal{N}(m)$, by the exhaustivity of m it follows that for every $p \in \mathbf{N}$ the set

$$I_p = \left\{ i \in I : \exists Y \in \mathcal{R} \text{ such that } m(A_i \cap Y) > \frac{1}{p} \right\}$$

is finite and that is $I = \cup_{p \in \mathbf{N}} I_p$.

Definition 6.9 Let $\mathcal{R}$ be a ring and let m and v be two functions from $\mathcal{R}$ into $[0, \infty]$. We say that m is v-continuous if $\mathcal{N}(v) \subset \mathcal{N}(m)$. We say that m is v-singular if there exists an element $A \in \mathcal{N}(\mu)$ such that $T \setminus A \in \mathcal{N}(m)$ for every T in $\mathcal{R}$.

Lemma 6.3 *Let $\mathcal{I}$ be a σ-complete ideal of $\mathcal{R}$ and $\mathcal{N}$ a subset of $\mathcal{I}$ such that $\mathcal{I} \setminus \mathcal{N}$ satisfies the (CCC). Then there exists $A \in \mathcal{I}$ such that*

$$\mathcal{I} = \{ Y \in \mathcal{R} : Y \setminus A \in \mathcal{N} \}.$$

Proof. Let $\mathcal{H}$ be a disjoint, maximal subset of $\mathcal{I}\setminus\mathcal{N}$. By the suppositions it follows that $A = \bigcup_{Y\in\mathcal{H}} Y$ belongs to $\mathcal{I}$ and it is easy to verify that

$$\mathcal{I} = \{Y \in \mathcal{R} : Y \setminus A \in \mathcal{I}\}$$

and

$$\{Y \in \mathcal{R} : Y \setminus A \in \mathcal{N}\} \subset \{Y \in \mathcal{R} : Y \setminus A \in \mathcal{I}\}.$$

But if T is an element of $\mathcal{R}$ such that $T\setminus A \in \mathcal{I}\setminus\mathcal{N}$, we have that $\mathcal{H}\cup\{T\setminus A\}$ is a disjoint subset of $\mathcal{I} \setminus \mathcal{N}$ against the maximality of $\mathcal{H}$. □

Theorem 6.11 *Let $\mathcal{R}$ be a σ-ring, m a null-additive set function, and v a finite σ-null-additive set function from $\mathcal{R}$ into $[0,+\infty]$. If $\mathcal{N}(m) \setminus \mathcal{N}(v)$ satisfies (CCC) then there exists $A \in \mathcal{N}(v)$ such that null-additive set functions*

$$m_1 : Y \mapsto m(Y \setminus A), \quad m_2 : Y \mapsto m(Y \cap A)$$

are, respectively, v-continuous and v-singular.

Proof. By Corollary 6.1 it follows that $\mathcal{N}(v)$ is a σ-ideal, then, by Lemma 6.3, there exists $A \in \mathcal{N}(v)$ such that

$$\mathcal{N}(v) = \{Y \in \mathcal{R} : Y \setminus A \in \mathcal{N}(v) \cap \mathcal{N}(m)\} \subset \{Y \in \mathcal{R} : Y \setminus A \in \mathcal{N}(m)\}.$$

Since $\{Y \in \mathcal{R} : Y \setminus A \in \mathcal{N}(m)\} = \mathcal{N}(m_1)$, m_1 is v-continuous. Moreover $T \setminus A$ belongs to $\mathcal{N}(m_2)$ for every $T \in \mathcal{R}$, then m_2 is v-singular. □

Remark 6.3 If in Theorem 6.7 we suppose that v is exhaustive, from Lemma 6.2 it follows that $\mathcal{N}(m) \setminus \mathcal{N}(v)$ satisfies the (CCC).

Corollary 6.2 *Let $\mathcal{R}$ be a σ-ring, m and v two k-triangular set functions on $\mathcal{R}$. If v is σ-k-subadditive monotone and exhaustive, then there exist $A \in \mathcal{N}(v)$ such that*

$$m_1 : Y \mapsto m(Y \setminus A), \quad m_2 : Y \mapsto m(Y \cap A)$$

are two k-triangular set functions on $\mathcal{R}$, the first v-continuous and the second v-singular.

Proof. This follows from σ-null-additivity of v, Lemma 2.3, Remark 6.3 and Theorem 6.11. □

Theorem 6.12 *Let $\mathcal{R}$ be a ring, m and v two set functions from $\mathcal{R}$ into $[0,+\infty]$, the first of them null-additive. If A, B are two elements of $\mathcal{N}(v)$ such that the functions*

$$m_1 : Y \mapsto m(Y \setminus A), \quad m_1' : Y \mapsto m(Y \setminus B)$$

are v-continuous, then $m_1 = m_1'$.

Moreover also the two null-additive, v-singular functions

$$m_2 : Y \mapsto m(Y \cap A), \quad m_2' : Y \mapsto m(Y \cap B)$$

are equal.

Proof. Since $B \in \mathcal{N}(v) \subset \mathcal{N}(m_1)$, for every $Y \in \mathcal{R}$ we have

$$m_1(Y) = m_1((Y \setminus B) \cup (Y \cap B)) = m_1(Y \setminus B) = m(Y \setminus (A \cup B)).$$

Since $A \in \mathcal{N}(v) \subset \mathcal{N}(m_1')$, we have also

$$m_2(Y \cap (A \setminus B)) = m_1'(Y \cap A) = 0 \qquad (Y \in \mathcal{R}),$$

then $A \setminus B \in \mathcal{N}(m_2)$. Therefore, for every $Y \in \mathcal{R}$ we have

$$\begin{aligned} m_2(Y) = m_2(Y \cap A) &= m_2((Y \cap A \cap B) \cup (Y \cap (A \setminus B)) \\ &= m_2(Y \cap A \cap B) = m(Y \cap A \cap B). \end{aligned}$$

□

Theorem 6.13 *Let $\mathcal{R}$ be a ring and m and v two functions from $\mathcal{R}$ into $[0,+\infty]$, the first of them null-additive. Let $\oplus$ be a pseudo-addition. There exist at most two null-additive functions m_1 and m_2 from $\mathcal{R}$ into $[0,+\infty]$ such that*

(i) m_1 is v-continuous;
(ii) m_2 is v-singular;
(iii) $m = m_1 \oplus m_2$.

Proof. By Theorem 6.12 it suffices to prove that if m_1 and m_2 are two null-additive set functions on $\mathcal{R}$ which satisfy (i), (ii) and (iii) then there exists $A \in \mathcal{N}(v)$ such that

$$m_1(Y) = m(Y \setminus A), \quad m_2(Y) = m(Y \cap A) \qquad (Y \in \mathcal{R}).$$

Since m_2 is v-singular, there exists $A \in \mathcal{N}(v)$ such that $T \setminus A \in \mathcal{N}(m_2)$ for every $T \in \mathcal{R}$. Moreover from the v-continuity of m_1 it follows also that $A \in \mathcal{N}(m_1)$. Then we have, for every $Y \in \mathcal{R}$

$$\begin{aligned} m_1(Y) &= m_1((Y \cap A) \cup (Y \setminus A)) = m_1(Y \setminus A) \\ &= m_1(Y \setminus A) \oplus m_2(Y \setminus A) = m(Y \setminus A) \end{aligned}$$

and

$$\begin{aligned} m_2(Y) &= m_2((Y \cap A) \cup (Y \setminus A)) = m_2(Y \setminus A) \\ &= m_1(Y \cap A) \oplus m_2(Y \cap A) = m(Y \cap A). \end{aligned}$$ □

Corollary 6.3 *Let $\mathcal{R}$ be a σ-ring , $\oplus$ a pseudo-addition. Let m be a $\oplus$-decomposable measure on $\mathcal{R}$ and v a σ-null-additive and exhaustive function on $\mathcal{R}$. Then m can be uniquely represented as the $\oplus$-sum of two $\oplus$-decomposable functions m_1 and m_2 such that m_1 is v-continuous and m_2 is v-singular. Moreover m_2 is m_1-singular.*

Proof. This follows from Theorem 6.7, Remark 6.3 and Theorem 6.13.□

Remark 6.4 The main point in Corollary 6.3 is the possibility of supposition of σ-$\oplus$-decomposability of v (which implies σ-null-additivity) instead of order continuity. In this way it can be considered also the possibility measure , i.e., set function $\pi: \mathcal{R} \to [0,1]$ which satisfies

$$\pi(A \cup B) = \pi(A) \vee \pi(B)$$

for $A, B \in \mathcal{R}$ such that $A \cap B = \emptyset$, which in general does not be order continuous although it is σ-$\vee$-decomposable.

6.4 Atoms of Null-Additive Set Functions

Let m be a positive set function defined on a σ-algebra Σ.

Definition 6.10 Let m be a positive monotone set function on Σ. A set $A \in \Sigma$ with $m(A) > 0$ is called an atom provided that if $B \subset A$ then
(i) $m(B) = 0$, or
(ii) $m(A) = m(B)$ and $m(A \setminus B) = 0$.

Remark 6.5 For σ-finite fuzzy measure m, each atom has always finite measure. For null-additive monotone set functions we may suppose in (ii) only $m(A \setminus B) = 0$.

Lemma 6.4 *Let m be a finite null-additive exhaustive set function. Then m is atomless on a given set $E \in \Sigma$ with $m(E) > 0$ if and only if for a given $\epsilon > 0$ there exists $F \subset E$ such that*

$$0 < m(F) \leq \epsilon .$$

Proof. The 'if' part: Suppose the contrary, i.e., that $E \in \Sigma$ is an atom such that $m(E) > 0$ and for each $\epsilon > 0$ there exist $F \subset E$ such that $0 < m(F) \leq \epsilon$. We choose ϵ such that $0 < \epsilon < m(E)$. By the atomness of E we obtain $m(E \setminus F) = 0$. Null-additivity of m implies $m(E) = m(F)$. Hence $m(E) \leq \epsilon$. Contradiction.

'Only if': Suppose that the statement in this part of theorem is not true. Therefore, since E is atomless, there exists a sequence $\{F_n\}$ of pairwise disjoint subsets F_n of $E, F_n \in \Sigma$ $(n \in \mathbf{N})$, such that

$$m(F_n) > \epsilon \qquad (n \in \mathbf{N}).$$

Contradiction with the exhaustivity of m. □

Theorem 6.14 *Let m be a null-additive monotone set function which is exhaustive. If each set $E \in \Sigma$, $m(E) > 0$, contains an atom of m, then for each set $E \in \Sigma$ there exist at most a countable number of atoms E_i $(i \in I)$ of m such that*

$$m\Big(E \setminus \bigcup_{i \in I} E_i\Big) = 0 .$$

Each set $A \in \Sigma$ contains at most a countable number of different atoms of m.

Proof. Let $E \in \Sigma$, $m(E) > 0$. If E is an atom of m, the theorem is proved, taking $E_1 = E$. If E is not an atom of m we take $E_1 \subset E$ from Σ to be an atom of m such that

$$m(E_1) > \frac{1}{2} \sup\{m(A) : A \subset E,\ A \text{ is an atom of } m\}.$$

Then either $E \setminus E_1$ is an atom of m, and then the theorem is proved, or we have a set $E_2 \subset E \setminus E_1$ from Σ, which is an atom of m and such that

$$m(E_2) > \frac{1}{2} \sup\{m(A) : A \subset E \setminus E_1,\ A \text{ is an atom of } m\}.$$

Repeating this procedure we either obtain that for some natural number n we have that $E \setminus \cup_{i=1}^{n-1} E_i$ is an atom of m, or we construct a sequence $\{E_i\}$ of pairwise disjoint atoms of m from E such that

$$m(E_i) > \frac{1}{2} \sup\Big\{m(A) : A \subset E \setminus \bigcup_{k=1}^{i-1} E_k,\ A \text{ is an atom of } m\Big\} \qquad (i \in \mathbf{N}).$$

The second case implies

$$\lim_{n\to\infty} m(E_n) = 0. \tag{6.8}$$

Hence

$$m\Big(E \setminus \bigcup_{i=1}^{\infty} E_i\Big) = 0.$$

Namely, if we suppose the contrary, i.e.,

$$m\Big(E \setminus \bigcup_{i=1}^{\infty} E_i\Big) > 0,$$

then there exists an atom $A \subset E \setminus \cup_{i=1}^{\infty} E_i$ of m such that

$$m(A) < 2m(E_n) \qquad (n \in \mathbf{N}).$$

Hence by (6.8) $m(A) = 0$, which contradicts the atomness of A with respect to m. □

Let m and v be two null-additive set functions.

Propositon 6.1 *If m is v-continuous and null-additive, then each atom A of v is also an atom of m proved $m(A) > 0$.*

Proof. Let $B \subset A$. $m(A) > 0$ implies $v(B) > 0$, and then $v(A\setminus B) = 0$, that is, $m(A\setminus B) = 0$. Thus $m(A) = m(B)$. □

If for m there is no atom, m is called non-atomic.

A 'vacant' atom which is at most a countable union of null sets can exist.

Example 6.4 Let $X = \mathbf{N}$, $\Sigma = \mathcal{P}(X)$, $m(E) = \sup E - \inf E$ for $E \neq \emptyset$, and $m(\emptyset) = 0$. Then every two-point set $A = \{n_1, n_2\}$ is an atom, and each proper subset of A has m-null-measure. Note that the fuzzy measure space (X, Σ, m) is σ-finite, and that X is represented as a countable union of null sets.

For a monotone continuous measure space (X, Σ, m) we define a subclass $\mathcal{N}(m)$ of Σ as: $E \in \mathcal{N}(m)$ if and only if E is at most a countable union $\cup_n E_n$ where $m(E_n) = 0$ $(n \in \mathbf{N})$.

For the measure m in Example 6.4 we have $\mathcal{N}(m) = \Sigma$.

Theorem 6.15 *If m is an μ-continuous monotone set function, where μ is a classical measure, then there exists an element $N = N(m, \mu) \in \mathcal{N}(m)$ which is maximal in the sense that $\mu(D \backslash N) = 0$ for every $D \in \mathcal{N}(m)$.*

Proof. Let $\alpha = \sup\{\mu(D) : D \in \mathcal{N}(m)\}$, and select $D_n \in \mathcal{N}(m)$ $(n \in \mathbf{N})$, to satisfy $D_1 \subset D_2 \subset \dots$ and $\lim_{n\to\infty} \mu(D_n) = \alpha$. Take

$$N = N(m, \mu) = \bigcup_{n=1}^{\infty} D_n .$$

Then $N \in \mathcal{N}(m)$ and $\mu(N) = \alpha$. The inequality $\mu(D \backslash N) > 0$ for some $D \in \mathcal{N}(m)$, would be imply a contradiction

$$\alpha = \mu(N) < \mu(N) + \mu(D \backslash N) = \mu(D \cup N) \leq \alpha .$$

Therefore $\mu(D \backslash N) = 0$. □

Take $P = P(m, \mu) = X \backslash N(m, \mu)$.

Propositon 6.2 *(i) If $m(E \cap P) = m(F \cap P) = 0$ then $m((E \cup F) \cap P) = 0$;*

(ii) if m is null-additive then $m(N) = 0$ and $m(E \cap P) = m(E)$ for any $E \in \Sigma$.

By (i), such a vacant atom A as in Example 6.4 appears nowhere in P. The opposite implication of the statement (ii) is not true.

Example 6.5 Let $X_1 = [0, 1]$, $X_2 = [-1, 0]$, $X = X_1 \cup X_2$, $\Sigma = \mathcal{B} \cap X$ and μ the Lebesgue measure on Σ. Define m by

$$m(E) = \mu(E \cap X_1) \cdot (1 + \mu(E \cap X_2)) \quad \text{for} \quad E \in \Sigma .$$

Then m is monotone and μ-continuous. We have $\mu(N \Delta X_2) = 0$, and therefore $\mu(N \cap X_2) = 0$, or equivalently $m(N) = 0$. But m is not null-additive because $m(X_1 \cup X_2) = 2 > m(X_1) = 1$ although $m(X_2) = 0$.

We suppose to the end of this section that m is μ-continuous for some measure.

We introduce

$$m_P(E) = m(E \cap P) \qquad (E \in \Sigma).$$

Suppose that A is an atom of m with $m_P(A) > 0$. If $B \subset A$ and $m_P(B) > 0$, then $m(B \cap P) = m(A)$ and $m(A \backslash B \cap P) = 0$. Thus $m_P(A) \geq m_P(B) = m(A)$ and $m_P(A \backslash B) \leq m(A \backslash B \cap P)$ imply respectively that $m_P(B) = m_P(A)$ and $m_P(A \backslash B) = 0$. Therefore A is also an atom of m_P. If m is null-additive in addition, then $m_P(E) = m(E)$ for every E. Therefore $X = P$ may be assumed as we concerned with the atoms of null-additive set functions.

Propositon 6.3 *Suppose that A_1 and A_2 are atoms of m_P. Then only one of the relations*

$$m_P(A_1 \cap A_2) = 0, \quad m_P(A_1 \Delta A_2) = 0$$

is possible.

The property of m_P in Proposition 6.3 is common to additive measures, but not to m_N, the restriction of m to N.

Example 6.6 Let $X = \mathbf{N}$ and $\Sigma = \mathcal{P}(\mathbf{N})$. For $k \in \mathbf{N}$ we define $m(E) = 1 \wedge [(|E| - k) \vee 0]$. Hence $N = X$, i.e. $m_N = m$, and any set A such that $k < |A| \leq 2k + 1$ is an atom.

Take $A_1 = \{1, 2, \ldots, 2k+1\}$ and $A_2 = \{k+1, k+2, \ldots, 3k+1\}$. Hence

$$\begin{aligned} A_1 \cap A_2 &= \{k+1, \ldots, 2k+1\} \quad \text{and} \\ A_1 \Delta A_2 &= \{1, \ldots, k\} \cup \{2k+2, \ldots, 3k+1\}, \end{aligned}$$

which implies $m(A_1 \cap A_2) = 1 = m(A_1) = m(A_2)$ while $m(A_1 \Delta A_2) = 1$.

We denote by $\mathcal{A}(m)$ the family of all the finite or countable infinite unions of atoms.

Theorem 6.16 *If m is an μ-continuous positive monotone continuous set function, then there exists an element $\mathbf{A} \in \mathcal{A}(m)$ such that*

(i) $A \in \mathcal{A}(m)$ implies $\mu(A \backslash \mathbf{A}) = 0$ (maximality),

(ii) m is non-atomic on $X \backslash \mathbf{A}$, and

(iii) $P \cap \mathbf{A}$ is represented as a disjoint union of atoms $\{A_n : n \in \mathbf{N}\}$ in the sense of $m((P \cap \mathbf{A}) \backslash (\cup_n A_n)) = 0$.

Proof. We can prove that $\mathbf{A}$ exists with the property (i) in a quite similar way as was proved for N in Theorem 6.15. (ii) is obvious. To prove (iii) it is sufficient to show that for any countable union $\cup_n A_n$ of atoms, there exists a sequence of disjoint atoms $\{A'_k\}$ such that $m_P(\cup_n A_n \setminus \cup_k A'_k) = 0$.

Let $E_1 = A_1$ and $E_n = A_n \setminus \cup_{j=1}^{n-1} A_j$ for $n = 2, 3, \ldots$. Then $\cup_{n=1}^{\infty} A_n = \cup_{n=1}^{\infty} E_n$, and the $\{E_n\}$ consists of mutually disjoint atoms. Take $N_1 = \{n : m(E_n) > 0\}$, $N_2 = N \setminus N_1$, and $D = \cup_{n \in N_2} E_n$, then $m(D \cap P) = 0$ by Theorem 6.15. Now renumber E_n $(n \in N_1)$ to obtain $A'_1, A'_2, \ldots$. Then $m_P(\cup A_n \setminus \cup A'_k) = 0$ follows from

$$\bigcup_n A_n \setminus \bigcup_k A'_k = \Big(\bigcup_{n \in N_1} E_n\Big) \cup \Big(\bigcup_{n \in N_2} E_n\Big) \setminus \bigcup_k A'_k \subset D. \qquad \square$$

Now we will show that for null-additive set monotone function m the elements of $\mathcal{A}(m)$ can be represented by a monotone set function defined on $\mathcal{P}(\mathbf{N})$.

Lemma 6.5 *If m is null-additive, and $\{A_n : n \in \mathbf{N}\}$ a sequence of atoms satisfying Theorem* 6.16(iii), *then for every $E \in \Sigma \cap \mathbf{A}$, there is a subset $K = K(E)$ of $\mathbf{N}$ such that $m(E) = m(\cup_{n \in K} A_n)$. Moreover K satisfies*

$$K(X) = \mathbf{N}, \quad K(X \setminus E) = \mathbf{N} \setminus K(E) \quad \textit{and} \quad K\Big(\bigcup_j E_j\Big) = \bigcup_j K(E_j).$$

Proof. We take $K(E) = \{n \in \mathbf{N} : m(E \cap A_n) > 0\}$ and $K(E) = \infty$ if $m(E) = 0$. If we prove

$$m(E) = m\Big(\bigcup_{n \in K} A_n \cap E\Big) \quad \text{and} \quad m\Big(\bigcup_{n \in K} A_n \setminus E\Big) = 0,$$

then $m(\bigcup_{n \in K} A_n) = m(E)$ follows, by null-additivity of m.

We have that $m(E \setminus \bigcup_{n \in \mathbf{N}} A_n) \leq m(\mathbf{A} \setminus_{n \in \mathbf{N}} A_n) = m(P \cap \mathbf{A} \setminus \cup_n A_n) = 0$ implies $m(E) = m(\bigcup_{n \in \mathbf{N}} A_n \cap E)$, from which the first equality is obtained by iterated use of null-additivity. The second follows from $m(\cup_{n \in K} A_n \setminus E)$ $= m(\cup_{n \in K}(A_n \setminus E))$ and $m(A_n \setminus E) = 0$ for $n \in K$.

The last part of the lemma is obvious by the definition of K. $\square$

Null-additivity is important in Lemma 6.5

Example 6.7 Let $X = [0, +\infty)$, $\Sigma = \mathcal{B} \cap [0, +\infty)$ and

(i) $m(E) = 0$ if $E \cap \mathbf{N} = \emptyset$,

(ii) $m(E) = \Sigma_{j=1}^{k} j^{-1}$ if $E \cap \mathbf{N} = \{x_1, \ldots, x_k\}$, $0 \notin E$,

(iii) $m(E) = 1 + \Sigma_{j=1}^{k} j^{-1}$ if $E \cap \mathbf{N} = \{x_1, \ldots, x_k\}$, $0 \in E$. If $E \cap \mathbf{N}$ is infinite, let $m(E) = \infty$.

m is not null-additive. Namely, for $E = [0, 1]$ and $F = [1, 2)$ we have $m(E) = 0$, $m(F) = 1$ and nevertheless $m(E \cup F) = 2$. If we take a sequence of disjoint atoms $A_n = (n - 1, n]$ ($n \in \mathbf{N}$), and an interval $I = [0, 1]$, then $m(X \backslash \cup A_n) = 0$ and $m(I) = 2$, although every union of those atoms does not have measure 2.

Theorem 6.17 *Let m be an μ-continuous monotone continuous set function defined on Σ. If m is null-additive, then there exists a map $r \colon \Sigma \cap \mathbf{A} \to \mathcal{P}(\mathbf{N})$ and a positive monotone continuous set functions v on $(\mathbf{N}, \mathcal{P}(\mathbf{N}))$ such that*

(i) $r(X) = \mathbf{N}$, $r(X \backslash E) = \mathbf{N} \backslash r(E)$ and $r(\cup_j E_j) = \cup_j r(E_j)$,

(ii) $m(E) = v(r(E))$ for each $E \in \Sigma \cap \mathbf{A}$.

In other words the monotone measure space $(\mathbf{A}, \Sigma \cap \mathbf{A}, m)$ is represented by a simple monotone measure space $(\mathbf{N}, \mathcal{P}(\mathbf{N}), v)$.

Proof. Take $\{A_n \colon n \in \mathbf{N}\}$ as a sequence of atoms satisfying Theorem 6.15(iii). WWe have that $m(P \cap \mathbf{A}) = m(\mathbf{A})$. Define r and v by

$$r(E) = \{n \in \mathbf{N} : m(E \cap A_n) > 0\} \quad \text{for} \quad E \in \Sigma \cap \mathbf{A}\,,$$
$$v(K) = m\Big(\bigcup_{n \subset K} A_n\Big) \quad \text{for} \quad K \in \mathcal{P}(\mathbf{N})\,.$$

Therefore the statement (i) and (ii) follow from Lemma 6.5. □

6.5 Saks Decomposition Theorem

Theorem 6.18 (Saks Decomposition) *Let m be a non-negative null-additive set function, which is finite, exhaustive and order continuous, or let it be a finite fuzzy measure. Then for every $A \in \Sigma$ and every $\epsilon > 0$ there exists a finite number $A_0, A_1, \ldots, A_r$ of pairwise disjoint elements of Σ such that* *(i)* $A = \cup_{i=0}^{r} A_i$;

(ii) each A_i $(i = 0, 1, \ldots, r)$ *is either an atom of* m *and* $m(A_i) > \epsilon$, *or* $m(A_i) \leq \epsilon$.

Proof. Let us first suppose that m is a finite null-additive set function which is continuous from above at the empty set and $m(\emptyset) = 0$. Since m is exhaustive, for arbitrary but fixed $\epsilon > 0$ there is at most a finite number $A_1, \ldots, A_k$ of pairwise disjoint atoms of m in A such that $m(A_i) > \epsilon$ $(i = 1, \ldots, k)$. We shall choose a sequence $\{A_n\}$ of pairwise disjoint sets from Σ such that $m(A_i) \leq \epsilon$, $i = k+1, k+2, \ldots$ and either
(a) for some $s \in \mathbf{N} \cup \{0\}$, $m(A \setminus \cup_{i=1}^{k+s} A_i) = 0$, or
(b) $m(A \setminus \cup_{i=1}^{k+s} A_i) > 0$ for each $s \in \mathbf{N} \cup \{0\}$, and

$$m(A_{k+i}) \geq \frac{1}{2} \sup\Big\{m(E) : E \in \Sigma,\ E \subset A \setminus \bigcup_{s=1}^{k+i-1} A_s \text{ and } 0 < m(E) \leq \epsilon\Big\} \qquad (i \in \mathbf{N}).$$

Namely, if we have

$$m\Big(A \setminus \bigcup_{i=1}^{k} A_i\Big) = 0,$$

then (a) holds for $A_i = \emptyset$ $(i = k+1, k+2, \ldots)$. Otherwise, by Lemma 6.4 we can choose

$$A_{k+1} \subset A \setminus \bigcup_{i=1}^{k} A_i, \quad A_{k+1} \in \Sigma,$$

such that $m(A_{k+1}) \leq \epsilon$ and

$$m(A_{k+1}) > \frac{1}{2} \sup\Big\{m(E) : E \in \Sigma,\ E \subset A \setminus \bigcup_{s=1}^{k} A_s \text{ and } 0 < m(E) \leq \epsilon\Big\}.$$

Then either

$$m\Big(A \setminus \bigcup_{i=1}^{k+1} A_i\Big) = 0, \qquad \text{— case (a)}$$

or

$$m\Big(A\setminus\bigcup_{i=1}^{k+1}A_i\Big)>0\,.$$

In the second case we repeat this procedures now on the set $A\setminus\cup_{i=1}^{k+1}A_i$. Hence again we have either the case (a) or the case (b) for $i=2$. Repeating this procedure s times ($s\in\mathbf{N}$) we obtain either (a) or (b).

The case (a) gives us the assertion of the theorem directly. The case (b) and the exhaustivity of m imply

$$m\Big(A\setminus\bigcup_{s=1}^{\infty}A_s\Big)=0\,. \tag{6.9}$$

By order continuity of m there exists $r\in\mathbf{N}$ such that

$$m\Big(\bigcup_{i=r+1}^{\infty}A_i\Big)\le\epsilon\,. \tag{6.10}$$

Taking $A_0=A\setminus\cup_{i=1}^{r}A_r$ we have by (6.9) and null-additivity of m

$$m(A_0)=m\Big(\Big(A_0\setminus\bigcup_{i=r+1}^{\infty}A_i\Big)\cup\Big(\bigcup_{i=r+1}^{\infty}A_i\Big)\Big)=m\Big(\bigcup_{i=r+1}^{\infty}A_i\Big)\,.$$

Hence by (6.10) $m(A_0)\le\epsilon$.

Since each finite fuzzy measure is exhaustive, we have proved the theorem for the fuzzy measure case too. □

Remark 6.6 It is easy to check that the preceding proof works also for finite submodular exhaustive and order continuous set function, although it may not be null-additive.

6.6 Hewitt-Yosida Decomposition Theorem

In this section we consider functions defined on more general structures than families of sets, with domain a lattice, and with values in some ordered semigroup. Many important special situations are covered by such an approach, which we want only to illustrate here by a decomposition theorem.

6.6.1 Lattice with Relative Complement

A lattice **L** is a partially ordered set in which every pair of elements x and y join $x \vee y$ and meet $x \wedge y$ in **L**. **L** is σ-complete if countable join of elements from **L** belongs also to **L**.

Definition 6.11 A lattice **L** is called lattice with relative complement if for each element x from any interval $[a, b]$ there exits an element y such that

$$x \vee y = b \qquad \text{and} \qquad x \wedge y = a\,.$$

The element y is called the relative complement of the element x on the interval $[a, b]$.

Example 6.8 The complement, in general, is not unique. Let $\mathbf{L} = \{\mathbf{0}, a, b, c, \mathbf{1}\}$ and the order $\leq$ is defined so that a, b and c are incomparable. Then elements b and c are complements of a on the interval $[\mathbf{0}, \mathbf{1}]$.

Example 6.9 An orthocomplementation on a lattice **L** is a unary operation $'$ on **L** such that $x \wedge x' = \mathbf{0}$ and $x \vee x' = \mathbf{1}$ for all $x \in \mathbf{L}$, idempotent and decreasing. An orthomodular lattice is a lattice with **0**, 1 and an orthocomplementation $'$ such that $x \leq y$ implies $y = x \vee (y \wedge x')$.

Example 6.10 For a distributive lattice with relative complement the complement (with respect to the interval $[\mathbf{0}, \mathbf{1}]$) is unique for each element so for Boolean algebras the complement always exists and it is unique.

Remark 6.7 Each lattice **L** can be embedded in a lattice **L**$'$ with **0** and **1**, and in which each element has a complement (on interval $[\mathbf{0}, \mathbf{1}]$), adding no more than three elements to **L**.

Definition 6.12 A lattice **L** is called a sectionally complemented lattice if for each element x from any interval $[\mathbf{0}, b]$, there exits an element y such that

$$x \vee y = b \qquad \text{and} \qquad x \wedge y = \mathbf{0}\,.$$

Two typical examples are:

Example 6.11 σ-algebra Σ of subsets of given set X.

Example 6.12 Take for $\mathbf{L}$ the subset of $[0,1]^X$ which consists of Σ-measurable functions, i.e., a generated fuzzy σ-algebra of all measurable fuzzy subsets of X.

We shall suppose in the whole section that $\mathbf{L}$ is a σ-complete sectionally complemented lattice.

Theorem 6.19 *If $\{x_n\}$ is a sequence from $\mathbf{L}$ such that $x_n \downarrow x$, then*

$$\inf_n y_n = \mathbf{0},$$

where y_n is a relative complement of x on $[\mathbf{0}, x_n]$, and there exists a sequence $\{y_n\}$ such that $y_n \not< y_{n+1}$ $(n \in \mathbf{N})$.

If $\mathbf{L}$ is distributive lattice, then $y_n \downarrow$.

Proof. Let y_n be a relative complement of x on $[\mathbf{0}, x_n]$ $(n \in \mathbf{N})$, i.e., $x \vee y_n = x_n$ and $x \wedge y_n = \mathbf{0}$. Since $x_n \downarrow x$, we have $\inf_n x_n = x$ and so that $x = \inf_n(x \vee y_n)$. $\mathbf{L}$ is a σ-complete lattice and $y_n \geq \mathbf{0}$, so $\inf_n y_n = y$ always exists. We have

$$x = \inf_n(x \vee y_n) \geq x \vee \inf_n y_n = x \vee y,$$

i.e., $x \geq y$. Since $x \wedge y_n = \mathbf{0}$ $(n \in \mathbf{N})$ holds, we obtain

$$\inf_n(x \wedge y_n) = \mathbf{0}.$$

Hence

$$x \wedge y = x \wedge \inf_n y_n = \inf_n(x \wedge y_n) = \mathbf{0}.$$

Since $x \geq y$, we obtain $y = \mathbf{0}$.

Now, let $\mathbf{L}$ be a distributive lattice then we have

$$x \vee (y_n \wedge y_{n+1}) = (x \vee y_n) \wedge (x \vee y_{n+1}) = x_n \wedge x_{n+1} = x_{n+1}.$$

Hence by

$$x \wedge (y_n \wedge y_{n+1}) = \mathbf{0},$$

we obtain that $y_n \wedge y_{n+1}$ is also a relative complement of x on $[\mathbf{0}, x_{n+1}]$. Since $\mathbf{L}$ is a distributive lattice the relative complement is unique and so we have

$$y_{n+1} = y_{n+1} \wedge y_n \leq y_n.$$

□

6.6.2 $\square$-Decomposable Measures

Let $(\mathbf{L}, \wedge, \vee, \mathbf{0}, \mathbf{1})$ be a σ-complete sequentially complemented lattice with smallest and largest elements $\mathbf{0}$ and $\mathbf{1}$, respectively, and let $(\mathbf{S}, \square, \leq, 0, 1)$ be a σ-complete, lattice ordered commutative semigroup with identity 0 and with smallest and largest elements 0 and 1, respectively. $(\mathbf{S}, \square, \leq)$ is an ordered commutative semigroup if $(\mathbf{S}, \square)$ is a commutative semigroup, $\leq$ is a partial odering such that if $x \leq y$, then $x \,\square\, z \leq y \,\square\, z$ for every $z \in \mathbf{S}$.

The following three examples of the semigroup $\mathbf{S}$ are typical.

Example 6.13 An interval $[0, M]$, M is a positive real number or $M = \infty$, the operation $\square$ is a continuous t-conorm S on $[0, M]$.

Example 6.14 The collection $\mathcal{D}_M$ of probability distribution functions on $[0, M]$, with the operation $\square$ defined by

$$(F \,\square\, G)(x) = \sup\{\min(F(u), G(v)) : u, v \in [0, M],\ u \, S \, v < x\},$$

where S is the same as in Example 6.13, and the partial order $\leq$ is the stochastic order $\preceq$ defined by

$$F \preceq G \Leftrightarrow F(x) \geq G(x) \ \text{ for all } \ x \in [0, \infty] .$$

Example 6.15 The collection $\mathcal{D}_M^q$ of all quasi-inverses $F^q \colon [0,1] \to [0, M]$ of probability distribution functions $F \in \mathcal{D}_M$ given by

$$F^q(\alpha) = \sup\{x \in [0, M] : F(x) < \alpha\}.$$

The operation S on $[0, M]$ is from Example 6.13 and the natural order $\leq$ in $[0, M]$ are pointwise extended on $\mathcal{D}_M^q$.

Definition 6.13 A mapping $m \colon \mathbf{L} \to \mathbf{S}$ satisfying

$$m(\mathbf{0}) = 0 ,$$
$$m(x \wedge y) \,\square\, m(x \vee y) = m(x) \,\square\, m(y) ,$$

is called an $\mathbf{S}$-valued modular function on $\mathbf{L}$.

Modular functions with the property

$$\{x_n\}_{n \in \mathbf{N}} \uparrow \Rightarrow \sup_{n \in \mathbf{N}} m(x_n) = m\Big(\bigvee_{n \in \mathbf{N}} x_n\Big),$$

are called generalized measures.

Combining the previous examples we obtain many important generalized measures.

An **S**-valued generalized measure has the following properties:

(Is) $x \le y \Rightarrow m(x) \le m(y)$,

(□-d) $m(x \vee y) = m(x) \,\square\, m(y)$ for $x \wedge y = \mathbf{0}$,

(σ□-d) $m(\bigvee_{n\in\mathbf{N}} x_n) = \sup_{k\in\mathbf{N}} m(\square_{n=1}^{k} x_k)$ for any sequence $\{x_n\}$ from **L** such that $x_n \wedge y_m$ for $n \neq m$.

Definition 6.14 A mapping $m\colon \mathbf{L} \to \mathbf{S}$ satisfying

-
$$m(\mathbf{0}) = 0\,,$$
$$m(x \vee y) = m(x) \,\square\, m(y)$$

for $x, y \in \mathbf{L}$ such that $x \wedge y = \mathbf{0}$ is called (finitely) □-decomposable measure on **L**.

It is known that a real valued function on a relatively complemented lattice is +-decomposable if and only if it is a modular function (valuation).

Definition 6.15 A lattice semigroup **S** is lower complete if every majorised increasing net $\{x_\alpha\}$ in **S** has the least upper bound, i.e. $\bigvee_\alpha x_\alpha \in \mathbf{S}$.

We suppose in the whole section that **S** is a lower complete lattice and that

$$\inf(A \,\square\, x) = \inf A \,\square\, x \qquad (A \subset \mathbf{S},\ x \in \mathbf{S}) \tag{6.11}$$

holds.

Theorem 6.20 *Let m, $m\colon \mathbf{L} \to \mathbf{S}$, be a non-trivial □-decomposable measure in the sense that $y \square v = y$ for each $y \in m(\mathbf{L})$, $y \neq \mathbf{0}$, implies $v = 0$. Then m is continuous from above, i.e., $x_n \downarrow x$ implies* $\inf_n m(x_n) = m(x)$ *if and only if* $\inf_n m(y_n) = 0$, *where y_n is a relative complement of x on $[\mathbf{0}, x_n]$.*

Proof. Let $\{x_n\}$ be a sequence from **S** such that $x_n \downarrow x$. Let y_n be a relative complement of x on $[\mathbf{0}, x_n]$, i.e.,

$$x \vee y_n = x_n \quad \text{and} \quad x \wedge y_n = \mathbf{0}\,.$$

We suppose $\inf_n m(y_n) = 0$. Then we obtain

$$\inf_n m(x_n) = \inf_n m(y_n \vee x) = \inf_n m(y_n) \,\Box\, m(x) = m(x)\,.$$

Suppose now that $x_n \downarrow x$ implies $\inf_n m(x_n) = m(x)$. If y_n $(n \in \mathbf{N})$ is a relative complement of x on $[\mathbf{0}, x_n]$, then we have

$$m(x) = \inf_n m(x_n) = \inf_n m(x \vee y_n) = \inf_n m(y_n) \,\Box\, m(x)\,.$$

Hence $\inf_n m(y_n) = 0$. □

The set of all non-trivial (in the sense of Theorem 6.20) □-decomposable mappings on $\mathbf{L}$ onto $\mathbf{S}$ will be denoted by $D(\mathbf{L}, \mathbf{S})$ and the subset of all continuous from above mappings will be denoted by $D_{\downarrow}(\mathbf{L}, \mathbf{S})$. We endow the set $D(\mathbf{L}, \mathbf{S})$ with the usual pointwise order, i.e., for $m_1, m_2 \in D(\mathbf{L}, \mathbf{S})$

$$m_1 \leq m_2 \quad \text{iff} \quad m_1(x) \leq m_2(x) \qquad (x \in \mathbf{L}),$$

and with the operation □ defined by

$$(m_1 \,\Box\, m_2)(x) = m_1(x) \,\Box\, m_2(x) \qquad (x \in \mathbf{L})\,.$$

Example 6.16 Let S be an Archimedean t-conorm. Then taking $\mathbf{S} = [0, 1]$, $\Box = S$ and $\mathbf{L} = \Sigma$ a σ-algebra of subsets of a set X, we obtain that $D(\Sigma, [0, 1])$ is the set of S-decomposable measures and that $D_{\downarrow}(\Sigma, [0, 1])$ is a subset of the set $D_{\sigma}(\Sigma, [0, 1])$ of σ-S-decomposable measures.

Example 6.17 Let S be the t-conorm max on $[0, 1]$. Then the elements of $D_{\downarrow}(\Sigma, [0, 1])$ are continuous from below and so

$$D_{\downarrow}(\Sigma, [0, 1]) \subset D_{\sigma}(\Sigma, [0, 1]),$$

with strict inclusion.

6.6.3 Hewitt-Yosida Decomposition

We suppose in this section that $\mathbf{S}$ satisfies (6.11) and also these conditions:

(a) $\mathbf{S}$ is of the countable type, i.e., every subset A of $\mathbf{S}$ that has a supremum in $\mathbf{S}$, and contains a countable subset A_1 such that $\sup A = \sup A_1$.

(b) $x \square \sup A = \sup(x \square A)$, whenever a supremum exists.

(c) If $a_{in} \in \mathbf{S}$ such that $\inf_n a_{in} = 0$ $(i \in \mathbf{N})$ and $a_{in} \not< a_{i(n+1)}$ $(i, n \in \mathbf{N})$, then

$$\inf_n \sup_i a_{in} = 0 ;$$

or instead of (c)

(c_1) $\mathbf{L}$ is distributive lattice and if $a_{in} \in \mathbf{S}$ such that $a_{in} \downarrow 0$ $(i \in \mathbf{N})$, then

$$\inf_n \sup_i a_{in} = 0 .$$

Definition 6.16 A non-empty subset S_1 of $\mathbf{S}$ is a band of $\mathbf{S}$ if it satisfies the following conditions:

(i) $x, y \in S_1$ implies $x \square y \in S_1$;

(ii) $x \leq y$ and $y \in S_1$ imply $x \in S_1$;

(iii) For any increasing net in S_1 its least upper bound (if it exists) belongs to S_1.

Theorem 6.21 *Let $D(\mathbf{L}, \mathbf{S})$ be a lattice. Then $D_{\downarrow}(\mathbf{L}, \mathbf{S})$ is a band of the lower complete lattice semigroup $D(\mathbf{L}, \mathbf{S})$.*

Proof. We shall prove that $D(\mathbf{L}, \mathbf{S})$ is a lower complete lattice semigroup. Let $\{m_i\}_{i \in I}$ be an increasing net in $D(\mathbf{L}, \mathbf{S})$ such that for every $x \in \mathbf{L}$ there is $b \in \mathbf{S}$ such that $m_i(x) \leq b$ and

$$m = \sup_{i \in I} m_i, \quad \text{i.e.,} \quad m(x) = \sup_i m_i(x) \qquad (x \in \mathbf{L}) .$$

Then for all $x, y \subset \mathbf{L}$ such that $x \wedge y = \mathbf{0}$ we have

$$m(x \vee y) = \sup_i m_i(x \vee y) = \sup_i (m_i(x) \square m_i(y)) \leq \sup_i m_i(x) \square \sup_i m_i(y) .$$

Let i, j be any pair of indices from I. There exists $k \in I$ such that $i < k$ and $j < k$. Then we have

$$m_i(x) \square m_j(y) \leq m_k(x) \square m_k(y) = m_k(x \vee y) \leq m(x \vee y) .$$

Hence

$$m(x) \square m(y) = m(x \vee y) .$$

We shall prove that $D_\downarrow(\mathbf{L},\mathbf{S})$ is a band of $D(\mathbf{L},\mathbf{S})$. It is obvious that $m_1, m_2 \in D_\downarrow(\mathbf{L},\mathbf{S})$ implies

$$m_1 \,\square\, m_2 \in D_\downarrow(\mathbf{L},\mathbf{S}).$$

Suppose $m \in D(\mathbf{L},\mathbf{S})$ and $v \in D_\downarrow(\mathbf{L},\mathbf{S})$ such that $m \leq v$. Let $\{x_n\}$ be a sequence in $\mathbf{L}$ such that $x_n \downarrow x$, and y_n a relative complement of x on $[\mathbf{0}, x_n]$, then by Theorem 6.19 $\inf_n y_n = \mathbf{0}$ and since $v \in D_\downarrow(\mathbf{L},\mathbf{S})$ by Theorem 6.20 $\inf_n v(y_n) = 0$. Hence $\inf_n m(y_n) = 0$. Then by Theorem 6.20 $m(x_n) \downarrow m(x)$, i.e., $m \in D_\downarrow(\mathbf{L},\mathbf{S})$. Let $\{m_i\}_{i\in I}$ be an increasing net in $D_\downarrow(\mathbf{L},\mathbf{S})$ and

$$m = \sup_{i\in I} m_i \in D(\mathbf{L},\mathbf{S}).$$

Let $x_n \downarrow x$. Since $m_i \in D_\downarrow(\mathbf{L},\mathbf{S})$ we have by Theorem 6.20

$$\inf_n m_i(y_n) = 0,$$

where y_n is a relative complement of x on $[\mathbf{0}, x_n]$. Now we have by properties (a) and (c) or (c_1) (if $\mathbf{L}$ is distributive lattice), using Theorem 6.19

$$\inf_n m(y_n) = \inf_n \sup_i m_i(y_n) = \inf_n \sup_{i\in C} m_i(y_n) = 0,$$

where C is a countable subset of I. Using properties (6.11) and (b) we obtain

$$\begin{aligned}\inf_n \sup_i m_i(x_n) &= \inf_n \sup_{i\in C} m_i(x \vee y_n) = \inf_n \sup_{i\in C} (m_i(x) \,\square\, m_i(y_n)) \\ &= \inf_n \sup_{i\in C} m_i(x) \,\square\, \inf_n \sup_{i\in C} m_i(y_n) = \inf_n \sup_i m_i(x) = \inf_n m(x) = m(x).\end{aligned}$$

Then by the monotonicity of m we have $m \in D_\downarrow(\mathbf{L},\mathbf{S})$. □

We suppose further that the operation □ and the partial ordering $\leq$ in $\mathbf{S}$ satisfy the conditions:

(i) $x, y \in \mathbf{S}$, $x \,\square\, y = x \,\square\, z$ imply $y = z$;

(ii) $x, y \in \mathbf{S}$, $x \,\square\, y = 0$ imply $x = y = 0$;

(iii) $u \leq v$ if there exists an element $w \in \mathbf{S}$ such that $v = u \,\square\, w$;

(iv) if $x_\alpha \uparrow$, then

$$\sup_\alpha (x_\alpha \wedge y) = (\sup_\alpha x_\alpha) \wedge y\,.$$

Remark 6.8 By the Nakada theorem conditions (i) and (iii) imply that $\mathbf{S}$ is a positive cone of a partially ordered group.

Definition 6.17 For $S_1 \subset \mathbf{S}$, $S_1 \neq \emptyset$, we define

$$S_1^\perp = \{x : x \in \mathbf{S},\ x \wedge y = \mathbf{0} \ \text{ for every } \ y \in S_1\}.$$

Lemma 6.6 *$S_1^\perp$ is a band of $\mathbf{S}$.*

We can easily check the following theorem.

Theorem 6.22 *If S_1 is a band of a lower complete lattice semigroup $\mathbf{S}$, then for any $u \in \mathbf{S}$, there exists a unique $u' \in S_1$ and $u'' \in S_1^\perp$ such that $u = u' \,\square\, u''$, and*

$$u' = \sup\{v : v \in S_1, 0 \le v \le u\},$$
$$u'' = \sup\{v : v \in S_1^\perp, 0 \le v \le u\}.$$

Theorem 6.23 (Hewitt-Yosida decomposition) *Let $D(\mathbf{L}, \mathbf{S})$ be a lattice. For every $m \in D(\mathbf{L}, \mathbf{S})$, there exists a unique $m_1 \in D_\downarrow(\mathbf{L}, \mathbf{S})$ and $m_2 \in D_\downarrow(\mathbf{L}, \mathbf{S})^\perp$ such that*

$$m = m_1 \,\square\, m_2\,.$$

Proof. By Theorem 6.21 $D_\downarrow(\mathbf{L}, \mathbf{S})$ is a band in $D(\mathbf{L}, \mathbf{S})$, then by the preceding Theorem 6.22 the desired decomposition follows. □

6.7 Notes and References

We have adapted the proof of the Hahn decomposition Theorem 6.2 to null-additive signed fuzzy measures using [Liu 1993]. The Jordan decomposition theorem in the present form is, to our knowledge, new.

Open problem: Does there exist for any null-additive signed fuzzy measure m a representation $m = v_1 - v_2$, where v_1 and v_2 are null-additive fuzzy measures?

The Lebesgue decomposition theorem for null-additive set functions is new, and for σ−null-additive set functions has been taken from [de Lucia and Pap 1994b].

Atoms are presented using [Suzuki 1991] and Saks decomposition theorem is from [Pap 1994a].

Generalized measures were introduced in [Klement and S. Weber 1991] and the Hewitt-Yosida decomposition theorem was prepared by [Pap 1993b] and [Pap 1994b].

Chapter 7

Choquet and Sugeno Integrals

7.1 Measurable Functions

7.1.1 Measurability

Let $\mathcal{B}$ be the Borel σ-algebra of the real line, which is the smallest σ-algebra containing all intervals. More generally the Borel σ-algebra of a topological space is the smallest σ-algebra containing all open sets.

Usually the measurability (Σ−measurability) of a real valued function f with respect to a σ-algebra Σ is given by the condition

$$f^{-1}(B) = \{x : f(x) \in B\} \in \Sigma$$

for every Borel subset B of the real line. Then this is equivalent that every set $\{x : f(x) \geq a\}$ for any real a or every set $\{x : f(x) \leq a\}$ belongs to Σ.

If we start with an arbitrary family $\mathcal{D}$ of subsets of the set X with the property $\emptyset \in \mathcal{D}$ (paving) we can define the $\mathcal{D}$-measurability of an extended real valued function $f \colon X \to [-\infty, +\infty]$ such that the sets $\{x : f(x) \geq a\}$ and $\{x : f(x) \leq -a\}$ belong to $\mathcal{D}$ for every positive element a in $[0, +\infty]$. We denote by $\mathcal{L}(\mathcal{D})$ the family of all $\mathcal{D}$-measurable extended real valued functions on X. If $\mathcal{D}$ is a σ-ring, then this notion of measurability coincides with the usual measurability.

It is easy to check that if $f \in \mathcal{L}(\mathcal{D})$, then $f \wedge a$, $f - f \wedge a$, $f \vee (-a)$, $f - f \vee (-a)$, $f^+ = \max(f, 0)$, $f^- = \max(-f, 0)$, a $\cdot f$ belong to $\mathcal{L}(\mathcal{D})$ for every positive real number a.

Let $\mathcal{F}$ be the family of all finite subsets of $[-\infty, +\infty]$ which contain zero. For $A \subset [-\infty, +\infty]$ we denote by $A^+ = A \cap [0, +\infty]$ and $A^- = A \cap [-\infty, 0]$ and $A + b = \{x + b : x \in A\}$.

If $f: X \to [-\infty, +\infty]$ and $F \in \mathcal{F}$ we define the corresponding

$$f_F = \sum_{i=1}^{n} (a_i - a_{i-1})\chi_{A_i} + \sum_{j=1}^{k} (b_j - b_{j-1})\chi_{B_j}$$

where $F = \{b_k, \ldots, b_1, 0, a_1, \ldots, a_n\}$ with

$$b_k < \cdots < b_1 < b_0 = 0 = a_0 < a_1 < \cdots < a_n ,$$

$A_i = \{x : f(x) \geq a_i\}$ and $B_j = \{x : f(x) \leq b_j\}$. If $f_F \in \mathcal{L}(\mathcal{D})$ then f_F is called a simple function. We remark, that this notion coincides with the usual notion of simple function in classical measure theory for non-negative functions.

It is easy to see that if $f \in \mathcal{L}(\mathcal{D})$ and $F \in \mathcal{F}$, then $f_F \in \mathcal{L}(\mathcal{D})$ and that for each $f \in \mathcal{L}(\mathcal{D})$ there exists a sequence of simple functions $\{f_n\}$ (take for example $f_n = f_{F_n}$ for

$$F_n = \left\{\frac{i}{2^n} : i = 0, \pm 1, \ldots, \pm n2^n\right\}$$

which converges pointwise to f on X.

Further, if $\mathcal{D}$ is closed with respect to intersection and countable union, then $f, g \in \mathcal{L}(\mathcal{D})$ and $f, g \geq 0$ implies $f + g$, $f \wedge g \in \mathcal{L}(\mathcal{D})$ and $\sup_n f_n \in \mathcal{L}(\mathcal{D})$ for every sequence from $\mathcal{L}(\mathcal{D})$.

If $\mathcal{D} = \Sigma$ is a σ-algebra, then the $\Sigma-$measurability of f can be equivalently defined by families

$$\{\{x : f(x) \geq a\} : a \in (-\infty, +\infty)\} \text{ or } \{\{x : f(x) > a\} : a \in (-\infty, +\infty)\}.$$

We have two kinds of $\mathcal{D}-$measurability of a non-negative function.

Definition 7.1 A c-measurable (respectively o-measurable) function f on X is a function $f: X \to [0, \infty]$ such that $\{x : f(x) \geq r\} \in \mathcal{D}$ (respectively $\{x : f(x) > r\} \in \mathcal{D}$) for every $r \in [0, \infty)$. A function f is measurable if it is c-measurable or o-measurable.

Propositon 7.1 *Let m be a positive monotone continuous set function defined on $\mathcal{D}$ and f a function on X which is both o-measurable and c-measurable. Then*

$$m(\{x : f(x) > r\}) = m(\{x : f(x) \geq r\})$$

except at most countable values of r.

Proof. The monotonicity of m implies that the functions $m(\{f > r\})$ and $m(\{f \geq r\})$ are non-increasing. In addition, we have

$$m(\{f > r\}) \leq m(\{f \geq r\}) \leq \lim_{s \uparrow r} m(\{f > s\}) \qquad (r \in [0, \infty]).$$

Therefore the assertion follows. □

7.1.2 s-Null-Sets

The notion of 'almost everywhere' is one of the key concepts in measure theory. It is defined by the notion of 'null set'. A null set is defined as a measurable set of measure zero. But this definition of null sets is not convenient for non-additive set functions, or even monotone set functions. If m is a monotone set function, then it is possible that there exist measurable sets A and B such that $m(A) = m(B) = 0$ but $m(A \cup B) > 0$.

Definition 7.2 Let m be a positive monotone set function. A subset N of X is called an s-null-set if, for every pair (A, B) of measurable sets,

$$A \subset B \cup N \quad \text{implies} \quad m(A) \leq m(B).$$

A proposition $P(x)$ concerning the points of X is said to be true s-almost everywhere (or s-a.e. for short) if there exists a null set N such that $x \in X \setminus N$ implies that $P(x)$ is true.

Note that an s-null set is not always measurable. The following simple properties of s-null-sets easily follow from the definition.

Propositon 7.2 *Let $(X, \mathcal{D}, m)$ be a monotone measure space, i.e., m is a positive monotone set function on $\mathcal{D}$. Then we have*

(i) The empty set is an s-null set.

(ii) If $N \in \mathcal{D}$ is an s-null set, then $m(N) = 0$.

(iii) A subset of an s-null set is an s-null set.

Propositon 7.3 *Let $(X, \mathcal{D}, m)$ be a monotone measure space, $\mathcal{D}$ be closed under finite union and $N \in \mathcal{D}$. Then the following conditions (i) and (ii) are equivalent.*

(i) N is an s-null set.

(ii) $m(A \cup N) = m(A) \quad (A \in \mathcal{D})$.

Proof. (i) $\Rightarrow$ (ii). Since $(A \cup N) \subset A \cup N$, by the definition of s-null sets we have $m(A \cup N) \leq m(A)$. Hence it follows from the monotonicity of $m(A \cup N) = m(A)$.

(ii) $\Rightarrow$ (i). Let $A, B \in \mathcal{D}$. If $A \subset B \subset N$, then $m(A) \leq m(B \cup N) = m(B)$. Hence N is an s-null set. □

The importance of null-additive set functions is stressed by the following consequence.

Corollary 7.1 *Let m be a null-additive monotone set function and $N \in \mathcal{D}$. Then N is an s-null set if and only $m(N) = 0$.*

Corollary 7.2 *Let $(X, \mathcal{D}, m)$ be a monotone measure space. If $\mathcal{D}$ is closed under finite union, then a finite union of measurable s-null sets is an s-null set. Moreover, if $\mathcal{D}$ is closed under countable union and if m is continuous from below, then a countable union of measurable s-null sets is an s-null set.*

Propositon 7.4 *Let $(X, \mathcal{D}, m)$ be a monotone measure space and $N \subset X$. N is a s-null set if and only if, for every pair (A, B) of measurable sets*

$$A \cap N^c \subset B \quad \textit{implies} \quad m(A) \leq m(B).$$

Proof. Since $A \subset B \cup N$ is equivalent with $A \cap N^c \subset B$, our assertion is immediate from the definition of s-null-sets. □

Propositon 7.5 *Let $(X, \mathcal{D}, m)$ be a monotone measure space, $\mathcal{D}$ be closed under finite intersection and $N^c \in \mathcal{D}$. Then the following conditions are equivalent*

(i) N is a null set.

(ii) $m(A \cap N^c) = m(A) \quad (A \in \mathcal{D})$.

Corollary 7.3 *Let $(X, \mathcal{D}, m)$ be a monotone measure space.*

(i) Let $\mathcal{D}$ be closed under finite intersection. If $\{N_i : i = 1, \ldots, n\}$ is a finite sequence of null sets such that $N_i^c \in \mathcal{D}$ for $i = 1, \ldots, n$, then $\bigcup_{i=1}^{n} N_i$ is a null set.

(ii) Let $\mathcal{D}$ be closed under countable intersection and m be continuous from above. If $\{N_n\}$ is a sequence of s-null sets such that $N_n^c \in \mathcal{D}$ $(n \in \mathbf{N})$, then $\bigcup_{n=1}^{n} N_n$ is an s-null set.

7.1.3 Egoroff's and Riesz' Theorems

Let m be a positive monotone set function defined on $\mathcal{D}$.

Definition 7.3 We say that the sequence $\{f_n\}$ of measurable functions converges almost uniformly on $A \in \mathcal{D}$ if there exists $\{E_k\} \subset \mathcal{D}$ such that $\lim_{k\to\infty} m(E_k) = 0$ then $f_n \to f$ on $A \setminus E_k$ uniformly for every $k \in \mathbf{N}$, in notation $f_n \overset{a.u.}{\to} f$.

Definition 7.4 We say that the sequence $\{f_n\}$ of measurable functions converges to measurable function f in m (or, converges in measure) if for every $\epsilon > 0$

$$\lim_{n\to\infty} m(\{x : |f_n(x) - f(x)| \geq \epsilon\} \cap A) = 0 .$$

In notation $f_n \overset{m}{\to} f$.

We have two generalizations of classical theorems on the convergences of measurable functions.

First we give a version of Egoroff's theorem.

Theorem 7.1 *Let m be a positive monotone set function autocontinuous from above. If a sequence of measurable functions $\{f_n\}$ converges a.e. to an a.e. finite-valued measurable function f on $A \in \Sigma$, $m(A) < \infty$, then for every $\epsilon > 0$, there exists $E \in \Sigma$ with $m(E) < \epsilon$, such that $\{f_n\}$ converges to f uniformly on $A \setminus E$.*

Proof. By autocontinuity from above m is null-additive. So we can suppose that $f_n \to f$ everywhere and f is finite on A. We introduce

$$E_n^j = \bigcap_{i=n}^{\infty} \left\{x : |f_i - f| < \frac{1}{j}\right\} \qquad (j \in \mathbf{N}),$$

then $E_1^j \subset E_2^j \subset \dots$. Since $f_n \to f$ everywhere on A, we have $\lim_{n\to\infty} E_n^j \supset A$, therefore $\lim_{n\to\infty}(A \setminus E_n^j) = 0$. It follows, by the continuity of m, that $\lim_{n\to\infty} m(A \setminus E_n^j) = 0$. Then, for every $j \in \mathbf{N}$, there exists n_j, such that

$$m(A \setminus E_{n_j}^j) < \frac{1}{j}\,.$$

If $F_j = A \setminus E_{n_j}^j$, then we have $m(F_j) \to 0$ as $j \to \infty$. We choose a subsequence $\{F_{j_i}\}$ of $\{F_j\}$ in the following way. For a given $\epsilon > 0$, we take F_{j_1}, such that $m(F_{j_1}) < \frac{\epsilon}{2}$. Autocontinuity from above of m implies

$$\lim_{j\to\infty} m(F_{j_1} \cup F_j) = m(F_{j_1})\,.$$

Therefore we can choose F_{j_2}, such that

$$m(F_{j_1} \cup F_{j_2}) < \frac{\epsilon}{2} + \frac{\epsilon}{2^2} = \frac{3}{4}\epsilon\,.$$

Continuing this procedure we obtain $F_{j_1}, \dots, F_{j_k}$ such that

$$m\Big(\bigcup_{i=1}^{k} F_{j_i}\Big) < \sum_{i=1}^{k} \frac{\epsilon}{2^i} < \epsilon \qquad (k \in \mathbf{N})\,.$$

Finally, using the continuity of m we obtain a subsequence $\{F_{j_i}\}$ of $\{F_j\}$ such that

$$m\Big(\bigcup_{i=1}^{\infty} F_{j_i}\Big) < \epsilon\,.$$

Put $E = \cup_{i=1}^{\infty} F_{j_i} \in \Sigma$, then $m(E) < \epsilon$. We will prove that $\{f_n\}$ converges to f uniformly on $A \setminus E$. For $\epsilon' > 0$, take k such that $k = j_{i_0} > 1/\epsilon'$. If $x \in A \setminus E$, then $x \in A$, but $x \notin E$, therefore $x \in A$ and $x \notin F_k$. Then $x \in E_{n_k}^k$. Hence

$$x \in \bigcap_{i=n_k}^{\infty} \Big\{x : |f_i - f| < \frac{1}{k}\Big\},$$

i.e., $|f_i - f| < \frac{1}{k} < \epsilon'$ for $i \geq n_k$. □

We have in the converse direction (with the classical proof):

Theorem 7.2 *If $\{f_n\}$ is a sequence of measurable functions which converges to f almost uniformly, then $\{f_n\}$ converges to f a.e.*

Theorem 7.3 *Almost uniform convergence implies convergence in measure.*

Definition 7.5 A monotone set function has the property (S) if for every sequence $\{A_n\}$ from Σ such that $\lim_{n\to\infty} m(A_n) = 0$ there exists a subsequence $\{A_{n_k}\}$ of $\{A_n\}$ such that

$$m(\limsup A_{n_k}) = 0\,,$$

where

$$\limsup A_{n_k} = \bigcap_{i=1}^{\infty} \bigcup_{k=1}^{\infty} A_{n_k}\,.$$

The following theorem is a generalization of the Borel-Cantelli lemma.

Theorem 7.4 *Let $m\colon \Sigma \to [0,\infty]$ be a monotone and continuous from below set function. If m is autocontinuous from above then it has the property (S). If m is a finite null-additive monotone set function with the property (S) then it is autocontinuous from above.*

Proof. Let m be autocontinuous from above and $\{A_n\} \subset \Sigma$ such that $\lim_{n\to\infty} m(A_n) = 0$. Since m is also continuous from below there exists a subsequence $\{A_{n_i(1)}\}$ of $\{A_n\}$ such that

$$m\Big(\bigcup_{i=1}^{\infty} A_{n_i(1)}\Big) < 1\,.$$

In the same way there exists a subsequence $\{A_{n_i(2)}\}$ of $\{A_{n_i(1)}\}$ such that

$$m\Big(\bigcup_{i=1}^{\infty} A_{n_i(2)}\Big) < \frac{1}{2}\,.$$

Continuing this procedure we obtain a subsequence $\{A_{n_i(k)}\}$ of $\{A_{n_i(k-1)}\}$ such that

$$m\Big(\bigcup_{i=1}^{\infty} A_{n_i(k)}\Big) < \frac{1}{k}\,.$$

Taking $n_i = n_{i(i)}$ for $i \in \mathbf{N}$, we obtain a subsequence $\{A_{n_i}\}$ of $\{A_n\}$ which satisfies

$$m(\limsup A_{n_i}) = 0\,.$$

Suppose now that m is null-additive and has the property (S). If $\{B_n\}$ is a sequence from Σ such that $\lim_{n\to\infty} m(B_n) = 0$, then there exists a subsequence $\{B_{n_i}\}$ of $\{B_n\}$ such that

$$\overline{\lim_{n\to\infty}}\, m(A \cup B_n) = \lim_{i\to\infty} m(A \cup B_{n_i})\,.$$

By the property (S) there exists a subsequence $\{B_{n_{i_k}}\}$ of $\{B_{n_i}\}$ such that

$$m(\limsup B_{n_{i_k}}) = 0\,. \tag{7.1}$$

The Proposition 2.1 implies

$$\begin{aligned}\overline{\lim_{n\to\infty}}\, m(A \cup B_n) &= \lim_{k\to\infty} m(A \cup B_{n_{i_k}}) \le m(\limsup(A \cup B_{n_{i_k}}))\\ &= m(A \cup \limsup B_{n_{i_k}})\,.\end{aligned}$$

Hence by null-additivity of m and (7.1) we obtain

$$\overline{\lim_{n\to\infty}}\, m(A \cup B_n) = m(A)\,.$$

Since by monotonicity of m we have $m(A) \le \underline{\lim}_{n\to\infty} m(A \cup B_n)$, we finally obtain $\lim_{n\to\infty} m(A \cup B_n) = m(A)$, i.e., m is autocontinuous from below.□

Second generalization of convergence theorems is a Riesz type theorem.

Theorem 7.5 *If $m\colon \Sigma \to [0, +\infty]$ is a monotone and continuous from below set function which is continuous from above then it has the Riesz property* (R)*: for every sequence $\{f_n\}$ of measurable functions such that $f_n \xrightarrow{m} f$ there exists a subsequence $\{f_{n_k}\}$ of $\{f_n\}$ such that $f_{n_k} \to f$ almost everywhere.*

Proof. Since $f_n \xrightarrow{m} f$ we have for every $i \in \mathbf{N}$

$$\lim_{n\to\infty} m\Big(\Big\{x : |f_n(x) - f(x)| \ge \frac{1}{i}\Big\}\Big) = 0\,.$$

Hence there exists an increasing sequence $\{n_i\}$ of natural numbers such that

$$m(A_i) < \frac{1}{i} \qquad (i \in \mathbf{N}),$$

where

$$A_i = \left\{x : |f_{n_i}(x) - f(x)| \geq \frac{1}{i}\right\}.$$

Since $\lim_{i\to\infty} m(A_i) = 0$ and m is autocontinuous from above, we have by Theorem 7.4

$$m(\limsup A_{i_k}) = 0$$

for a subsequence $\{A_{i_k}\}$ of $\{A_n\}$.

Now it is easy to show that the sequence $\{f_{n_{i_k}}\}$ converges to f on the set $X \setminus \limsup A_{i_k}$, i.e., $f_{n_{i_k}} \overset{a.e.}{\to} f$. □

We can describe more closely now the relation between autocontinuity and null-additivity.

Theorem 7.6 *Let $m\colon \Sigma \to [0, +\infty)$ be a monotone set function continuous from above and below. Then the following statements are equivalent:*
(1) m is null additive and has the property (R);
(2) m is null additive and has the property (S);
(3) m is autocontinuous from above;
(4) m is autocontinuous.

Proof. (2) $\Rightarrow$ (1) Analogous to the proof of Theorem 7.5

(1) $\Rightarrow$ (2) Suppose that $\{A_n\}$ is a sequence from Σ such that $\lim_{n\to\infty} m(A_n) = 0$. For a special sequence $f_n(x) = \chi_{A_n}(x)$ $(x \in X, n \in \mathbf{N})$ we have $f_n \overset{m}{\to} 0$. By the property (R) there exists a subsequence $\{f_{n_i}\}$ of $\{f_n\}$ such that $f_{n_i} \overset{a.e.}{\to} 0$. Hence

$$m\Big(\bigcap_{k=1}^{\infty}\bigcup_{i=k}^{\infty}(\{x : |f_{n_i}(x) - 0| > \epsilon\})\Big) = 0$$

for every $\epsilon > 0$.

Therefore

$$m\Big(\bigcap_{k=1}^{\infty}\bigcup_{i=k}^{\infty} A_{n_i}\Big) = 0.$$

(2) $\iff$ (3) Follows from Theorem 7.4.

(3) $\iff$ (4) Follows from the proved property of autocontinuity.

7.2 Choquet Integral for Non-Negative Case

Let m be a positive monotone set function defined on $\mathcal{D}$, i.e., $(X, \mathcal{D}, m)$ is a monotone measure space, and f non-negative function on X. By Proposition 7.1 we can define Choquet's integral as follows:

Definition 7.6 Let $(X, \mathcal{D}, m)$ be a monotone measure space. Choquet's integral of a measurable function f with respect to m, denoted by $(C) \int f \, dm$, is defined by

$$(C) \int f \, dm = \begin{cases} \int_0^\infty m(\{x : f(x) > r\}) \, dr & \text{if } f \text{ is } o\text{-measurable,} \\ \int_0^\infty m(\{x : f(x) \geq r\}) \, dr & \text{if } f \text{ is } c\text{-measurable.} \end{cases}$$

Most of the theorems and propositions in this section are special cases of the statements in the forthcoming sections, so they will be not proved in this section.

Propositon 7.6 *Let $(X, \mathcal{D}, m)$ be a monotone measure space and f a measurable function. If, for every $r \in [0, \infty)$*

$$G(r) = \begin{cases} -m(\{f > r\}) & \text{if } f \text{ is o-measurable,} \\ -m(\{f \geq r\}) & \text{if } f \text{ is c-measurable,} \end{cases}$$

then

$$(C) \int f \, dm = \int_0^\infty r \, dG(r) .$$

The integral on the right side is a Stieltjes integral.

Propositon 7.7 *Let $(X, \mathcal{D}, m)$ be a monotone measure space. Let f and g measurable functions and $\{f_n\}$ a sequence of measurable functions.*

(i) $f \leq g$ implies $(C) \int f \, dm \leq (C) \int g \, dm$.

(ii) $(C) \int af \, dm = a(C) \int f \, dm$ for every $a \geq 0$.

(iii) If m is a classical measure, then Choquet's integral coincides with the Lebesgue integral.

(iv) $(C) \int \chi_A \, dm = m(A) \quad (A \in \mathcal{D})$.

(v) If m is continuous from below, then $f_n \uparrow f$ implies that

$$(C) \int f_n dm \uparrow (C) \int f \, dm .$$

(vi) Let m be continuous from above. If $f_n \uparrow f$ and if for at least one value of n the function f_n is bounded and $m(\{f_n > 0\}) < \infty$, then

$$(C)\int f_n dm \uparrow (C)\int f\,dm\,.$$

The next proposition gives an expression of Choquet's integral for a simple function.

Propositon 7.8 *Let $(X, \mathcal{D}, m)$ be a monotone measurable space. Every simple function f on X can be represented as*

$$f = \sum_{i=1}^{n}(a_i - a_{i-1})\chi_{A_i}\,, \tag{7.2}$$

where $0 = a_0 \leq a_1 \leq \cdots \leq a_n$ and $A_1 \supset A_2 \supset \cdots \supset A_n$. Choquet's integral of a simple function f written as (7.2) with respect to m can be represented as

$$(C)\int f\,dm = \sum_{i=1}^{n}(a_i - a_{i-1})m(A_i). \tag{7.3}$$

If a monotone set function m is continuous from below, then we can define the integral like a Lebesgue integral: for a simple function f defined as (7.2), we define the integral of f as (7.3), and for a measurable function f, we define the integral of f as

$$(C)\int f\,dm = \lim_{n\to\infty}(C)\int f_n\,dm\,,$$

where $\{f_n\}$ is a sequence of simple functions such that for every $x \in X$, $f_n(x) \uparrow f(x)$.

Since a monotone set function generally is non-additive, Choquet's integral also is generally non-additive. Concerning the additivity of Choquet's integral, the next theorem holds.

Theorem 7.7 *Let $(X, \mathcal{D}, m)$ be a monotone measure space. We have*

$$(C)\int (f+g)\,dm \leq (C)\int f\,dm + (C)\int g\,dm$$

for every pair (f,g) of measurable functions if and only if

$$m(A \cup B) + m(A \cap B) \leq m(A) + m(B) \qquad (A \in \mathcal{D},\ B \in \mathcal{D}).$$

We have

$$(C)\int (f+g)\,dm \geq (C)\int f\,dm + (C)\int g\,dm$$

for every pair (f,g) of measurable functions if and only if

$$m(A \cup B) + m(A \cap B) \geq m(A) + m(B) \qquad (A \in \mathcal{D},\ B \in \mathcal{D}).$$

There are functions f and g such that $f = g$ a.e. in the sense of ordinary null sets (i.e., a measurable set of measure zero) but $(C)\int f\,dm \neq (C)\int g\,dm$.

Example 7.1 Let $X = \{x_1, x_2\}$ and m a monotone set function on $\mathcal{P}(X)$ given by $m(\{x_1\}) = 0$, $m(\{x_2\}) = 0$, and $m(X) = 1$. Let $f(x_1) = 0$, $f(x_2) = 1$, $g(x_1) = 2$, and $g(x_2) = 1$. Then $f = g$ a.e. in the ordinary sense, i.e., $m(\{f \neq g\}) = m(\{x_1\}) = 0$. But we have $(C)\int f\,dm = 0$ and $(C)\int g\,dm = 1$.

Lemma 7.1 *Let $(X, \mathcal{D}, m)$ be a monotone measure space, N an s-null set, and $\mathcal{D}' = \{A \cap N^c : A \in \mathcal{D}\}$. Then $(N^c, \mathcal{D}')$ is a monotone measurable space and there exists a monotone set function m' on $\mathcal{D}$ such that*

$$m'(A \cap N^c) = m(A) \qquad (A \in \mathcal{D}').$$

Proof. It is obvious that $(N^c, \mathcal{D}')$ is a monotone measurable space. It follows that

$$m(A) = m(B) \quad \text{whenever} \quad A \cap N^c = B \cap N^c.$$

Therefore we can define a set function m' on $\mathcal{D}'$ by

$$m'(A \cap N^c) = m(A) \qquad (A \in \mathcal{D}),$$

and obviously m' is a monotone set function. □

Definition 7.7 Let $(X, \mathcal{D}, m)$ be a monotone measure space, N an s-null set, and $\mathcal{D}' = \{A \cap N^c : A \in \mathcal{D}\}$. The monotone set function m' on $(N^c, \mathcal{D}')$ defined by

$$m'(A \cap N^c) = m(A) \qquad (A \in \mathcal{D})$$

is called the restriction of m to $(N^c, \mathcal{D}')$ (or to N^c).

Lemma 7.2 *Let $(X, \mathcal{D}, m)$ be a monotone measure space, N an s-null set, and m' the restriction of m to N^c. For any measurable function f,*

$$(C)\int f\,dm' = (C)\int f|_{N^c}\,dm\,,$$

where $f|_{N^c}$ is the restriction of f to N^c.

Proof. Suppose that f is an o-measurable function. The analogous proof holds for the case where f is c-measurable.

$$\begin{aligned}(C)\int f\,dm &= \int_0^\infty m(\{f > r\})\,dr = \int_0^\infty m'(\{f > r\} \cap N^c)\,dr \\ &= \int_0^\infty m'(\{f|_{N^c}) > r\})\,dr = (C)\int f|_{N^c}\,dm'\,. \qquad \square\end{aligned}$$

Theorem 7.8 *Let $(X, \mathcal{D}, m)$ be a monotone measure space. If f and g are measurable functions such that $f = g$ s-a.e. (a.e. for null-additive set function m), then*

$$(C)\int f\,dm = (C)\int g\,dm\,.$$

Proof. Let N be a null set such that $f(x) = g(x)$ $(x \in N^c)$. Let m' be the restriction of m to N^c. Lemma 7.2 implies

$$(C)\int f\,dm = (C)\int f|_{N^c}\,dm' = (C)\int g|_{N^c}\,dm' = (C)\int g\,dm\,. \qquad \square$$

Example 7.2 Let $\mathcal{D}$ be a paving. If $B \subset X$ and $B \neq \emptyset$, then the 0–1 possibility measure focused on B, denoted by π_B, is a set function on $\mathcal{D}$ for every $A \in \mathcal{D}$ defined by

$$\pi_B(A) = \begin{cases} 1 & \text{if } A \cap B \neq \emptyset\,, \\ 0 & \text{otherwise.} \end{cases}$$

The 0–1 necessity measure focused on B, denoted by N_B, is a set function on $\mathcal{D}$ for every $A \in \mathcal{D}$ defined by

$$N_B(A) = \begin{cases} 1 & \text{if } B \subset A, \\ 0 & \text{otherwise.} \end{cases}$$

We have for measurable function f

(i) $(C) \int f \, d\pi_B = \sup_{x \in B} f(x)$;

(ii) $(C) \int f \, dN_B = \inf_{x \in B} f(x)$;

(iii) if m is a positive monotone set function such that $m(X) = 1$, then

$$\inf_{x \in X} f(x) \le (C) \int f \, dm \le \sup_{x \in X} f(x)\,.$$

7.3 Symmetric Choquet (Šipoš) Integral

The integral in this section (Šipoš integral) will be considered for positive monotone extended real-valued set functions defined on families of subsets which do not have to be σ-algebras, (they are pavings (pre-spaces)) and measurable functions. For the definition of this integral we do not need the notion of the Lebesgue integral, as was the case in the previous section in the definition of the Choquet integral. Since the Šipoš integral coincides with the Choquet integral for non-negative integrable functions, we can consider the Šipoš integral as a way of direct definition of Choquet's integral.

We repeat for the sake of completeness that a real net is a triple $(s, \ge D)$ where D is a directed set and s is a real function with the domain D.

By a paving (pre-space) we mean a pair $(X, \mathcal{D})$, where $\emptyset \in \mathcal{D}$ and $\mathcal{D} \subset \mathcal{P}(X)$. An extended real valued monotone set function m defined on $\mathcal{D}$ with $m(\emptyset) = 0$.

A function $f\colon X \to [-\infty,\infty]$ is $\mathcal{D}$-measurable (or only measurable) if the sets $\{x : f(x) \geq a\}$ and $\{x : f(x) \leq -a\}$ are in $\mathcal{D}$ for every $a > 0$.

We denote by $\mathcal{L}(\mathcal{D})$ the set of all $\mathcal{D}$-measurable functions. A function $f \in \mathcal{L}(\mathcal{D})$ is called a simple function if its range is finite.

Let $\mathcal{F}$ be the family of all finite subsets of $[-\infty,\infty]$ which contain zero. Let $F \in \mathcal{F}$ with $F = \{b_k, b_{k-1}, \dots, b_0, a_1, \dots, a_n\}$, where $b_k < b_{k-1} < \cdots < b_0 = 0 = a_0 < a_1 < \cdots < a_n$, and let f be a $\mathcal{D}$-measurable function. We introduce by

$$f_F = \sum_{i=1}^{n}(a_i - a_{i-1})\chi_{A_i} + \sum_{j=1}^{k}(b_j - b_{j-1})\chi_{B_j}$$

where $A_i = \{f(x) \geq a_i\}$ and $B_j = \{x : f(x) \leq b_j\}$, the integral sum for f

$$S_F(f) = \sum_{i=1}^{n}(a_i - a_{i-1})m(\{x : f(x) \geq a_i\} + \sum_{j=1}^{k}(b_j - b_{j-1})m(\{x : f(x) \leq b_j\})$$

if the right-hand side expression contains no expression of the type $\infty - \infty$. Since $\mathcal{F}$ is directed by inclusion, the triple $(S_F(f), \supset, \mathcal{F})$ is a net. We define the Symmetric Choquet (Šipoš) integral of f as

$$(\check{S})\int f\,dm = \lim_{F\in\mathcal{F}} S_F(f)$$

if the limit exists. A function f is called Šipoš integrable if and only if $(\check{S})\int f\,dm$ is finite. We denote by $\mathcal{L}_1(X,\mathcal{D},m)$ the set of all integrable functions.

The following theorem proves that the Šipoš integral for non-negative functions coincides with the usual Choquet integral.

Theorem 7.9 *If $f \in \mathcal{L}_1^+(\mathcal{D},m)$ then*

$$(\check{S})\int f\,dm = \int_0^\infty m(\{f \geq t\})\,dt = (C)\int f\,dm$$

where in the middle there is an ordinary Lebesgue integral.

Proof. Let f be bounded by a positive constant K, $f < K$. Then the function $g(y) = m(\{f \geq t\})$ vanishes on $[K, +\infty]$. Let $\epsilon > 0$. Choose $F_0 = \{0 = a_0 < a_1 < \cdots < a_n\}$ with $a_n \leq K$ and such that for $F \supset F_0$ it holds:

$$(\check{S}) \int f \, dm - S_F(f) < \epsilon .$$

Choose such a $\delta > 0$ that for every partition Δ of the interval $[0, K]$ with the norm less then δ there holds $|(\check{S}) \int_0^K g(t)\, dt - \Sigma(g, \Delta)| < \epsilon$, where $\Sigma(g, \Delta)$ is a Riemann integral sum of g with respect to the partition Δ. We take $F \supset F_0$ such that $F \cap [0, K]$ is a partition of $[0, K]$ with the norm less then δ. Then

$$\begin{aligned} &\left|(\check{S}) \int f \, dm - (\check{S}) \int_0^K g(t)\, dt\right| \\ &= \left|(\check{S}) \int f \, dm - S_F(f) + S_F(f) - (\check{S}) \int_0^K g(t)\, dt\right| \\ &= \left|(\check{S}) \int f \, dm - S_F(f) + \Sigma(g, F \cap [0, K]) - (\check{S}) \int_0^K g(y)\, dt\right| < 2\epsilon , \end{aligned}$$

because $S_F(f) = \Sigma(g, F \cap [0, K])$. Hence we get

$$(\check{S}) \int f \, dm = \int_0^K g(y)\, dt = (\check{S}) \int g \, d\mu .$$

If f is not bounded, then we have

$$\begin{aligned} (\check{S}) \int f \, dm &= \lim_{K \to \infty} (\check{S}) \int (f \wedge K)\, dm = \lim_{K \to \infty} \int_0^K m(\{f \geq t\})\, dt \\ &= \int_0^{+\infty} g d\mu , \end{aligned}$$

where μ is the Lebesgue measure. □

We are now able to define on a σ-algebra a general (symmetric) Choquet integral for any measurable function using the properties of the Šipoš integral:

$$\begin{aligned} (C) \int f \, dm = (\check{S}) \int f d m &= (\check{S}) \int f^+ \, dm - (\check{S}) \int f^- \, dm \\ &= (C) \int f^+ \, dm - (C) \int f^- \, dm \\ &= \int_0^\infty m(\{f \geq t\})\, dt - \int_0^\infty m(\{-f \geq t\})\, dt . \end{aligned}$$

The basic properties of Šipoš integral are contained in the following theorem.

Theorem 7.10 *Let f be a $\mathcal{D}$-measurable function on X.*

(i) If $f \geq 0$, then $(\check{S}) \int f\,dm$ exists and $(\check{S}) \int f\,dm \geq 0$. Moreover in this case we have

$$(\check{S}) \int f\,dm = \sup_{F \in \mathcal{F}} S_F(f)$$

(ii) $(\check{S}) \int f\,dm$ is a monotone functional.

(iii) If $(\check{S}) \int f\,dm$ exists, then for every real c

$$(\check{S}) \int (c \cdot f)\,dm = c \cdot (\check{S}) \int f\,dm$$

(iv) If $a \geq 0$, then the integral is additive in the horizontal sense

$$(\check{S}) \int f\,dm = (\check{S}) \int (f \wedge a)\,dm + (\check{S}) \int (f - f \wedge a)\,dm$$

if one of the right-hand side expression is finite.

(v) If $(\check{S}) \int f^+\,dm$ or $(\check{S}) \int f^-\,dm$ is finite, then $(\check{S}) \int f\,dm$ exists and

$$(\check{S}) \int f\,dm = (\check{S}) \int f^+\,dm - (\check{S}) \int f^-\,dm\,.$$

Moreover if f is integrable, then the last equality holds too, and f^+, f^- are integrable too.

(vi) Let $(\check{S}) \int f^+\,dm$ or $(\check{S}) \int f^-\,dm$ be finite. Then f is a limit of a sequence $\{f_n\}_{n=1}^{\infty}$ of simple functions and

$$(\check{S}) \int f\,dm = \lim_{n \to \infty} (\check{S}) \int f_n\,dm\,.$$

If f is non-negative, then f_n may be taken to be non-negative and the sequence $\{f_n\}$ may be assumed to be increasing.

(vii) If f is a non negative $\mathcal{D}$-measurable function, then

$$(\check{S}) \int f\,dm = \sup\Big\{(\check{S}) \int g\,dm : g \leq f,\ g \text{ is a simple function}\Big\}.$$

For proving Theorem 7.10 we need two lemmas.

Lemma 7.3 *Let f be an integrable function and let $F \in \mathcal{F}$, then we have*

(i) $S_F(f) = S_{F^+}(f) + S_{F_-}(f)$.

(ii) If $F_1, F_2 \in \mathcal{F}$ with $F_1 \subset F_2$, then $S_{F_1^+}(f) \leq S_{F_2^+}(f)$ and $S_{F_1^-}(f) \geq S_{F_2}^-(f)$.

Proof. (i) Consequence of the definition of the integral sum.

(ii) Let $F = \{0, a_1, a_2, \ldots, a_n\}$ and let $F^* = F \cup \{a\}$, where

$$0 = a_0 < a_1 < \cdots < a_{i-1} < a < a_j < \cdots < a_n .$$

We prove first that

$$S_F(f) = S_{F^*}(f) .$$

Denote

$$y = \sum_{\substack{j=1 \\ j \neq i}}^{n} (a_j - a_{j-1}) m(\{x : f(x) \geq a_i\}) ;$$

then

$$\begin{aligned} S_F(f) &= y + (a_i - a_{i-1}) m(\{x : f(x) \geq a_i\}) \\ &\leq y + (a_i - a) m(\{x : f(x) \geq a_i\}) + (a - a_{i-1}) m(\{x : f(x) \geq a\}) \\ &= S_{F^*}(f) , \end{aligned}$$

where we have used the monotonicity of m and the fact that

$$\{x : f(x) \geq a_i\} \supset \{x : f(x) \geq a\} .$$

If now $F_1 \subset F_2$, then $F_1^+ \subset F_2^+$. Let $c_1, c_2, \ldots, c_n$ be such real numbers that $F_2^+ = F_1^+ \cup \{c_1, c_2, \ldots, c_n\}$. Then from the first part of this proof we obtain

$$S_{F_1^+}(f) \leq S_{F_1^+ \cup \{c_1\}}(f) \leq \cdots \leq S_{F_2^+}(f) .$$

The proof of $S_{F_1^-}(f) \geq S_{F_2^-}(f)$ is analogous. □

Lemma 7.4 *(i) $S_F(cf) = S_{\frac{1}{c}F}(f)$ for $c \neq 0$.*

(ii) $S_F(f) = S_F(f^+) - S_{-F}(f^-)$.

(iii) $S_F(f) = S_F(f \wedge a) + S_{F-a}(f - f \wedge a)$ if $a \in F$ and $a \geq 0$.

Proof. (i) Follows from the equalities

$$\{x : c \cdot f(x) \geq a\} = \{x : f(x) \geq a/c\}$$

and

$$\{x : c \cdot f(x) \leq b\} = \{x : f(x) \leq b/c\}$$

for $c > 0$ and from the analogous equalities for $c < 0$.

(ii) If $f \geq 0$ ($f \leq 0$), then $S_{F^-}(f) = 0$ (resp. $S_{F^+}(f) = 0$) for every $F \in \mathcal{F}$. Hence by the definition of f^+ and f^- and from (i) of this lemma it follows that

$$S_F(f) = S_{F^+}(f^+) - S_{F^-}(f^-).$$

(iii) By Lemma 7.3(i)

$$S_F(f) = S_{F^+}(f) + S_{F^-}(f).$$

Therefore it is enough to prove the assertion if $F \in \mathcal{F}^+$. Let $F=\{a_0, a_1, \ldots, a_n\}$ where $0 = a_0 < a_1 < \cdots < a_n$. Then

$$\begin{aligned} S_F(f) &= \sum_{i=1}^{n} (a_i - a_{i-1}) m(\{x : f(x) \geq a_i\}) \\ &= \sum_{a_i \leq a} (a_i - a_{i-1}) m(\{x : f(x) \geq a_i\}) \\ &\quad + \sum_{a_i > a} (a_i - a_{i-1}) m(\{x : f(x) \geq a_i\}). \end{aligned}$$

However,

$$\begin{aligned} &\sum_{a_i \leq a} (a_i - a_{i-1}) m(\{x : f(x) \geq a_i\}) \\ &\qquad = \sum_{a_i \leq a} (a_i - a_{i-1}) m(\{x : f(x) \geq a_i\}) \\ &\qquad\quad + \sum_{a_i > a} (a_i - a_{i-1}) m(\{x : (f \wedge a)(x) \geq a_i\}) \\ &\qquad = \sum_{i=1}^{n} (a_i - a_{i-1}) m(\{x : (f \wedge a)(x) \geq a_i\}) \\ &\qquad = S_F(f \wedge a). \end{aligned}$$

On the other hand

$$\sum_{a_i > a} (a_i - a_{i-1}) m(\{x : f(x) \geq a_i\}) = \sum_{a_i > a} (a_i - a - (a_{i-1} - a))$$
$$\times\, m(\{x : (f - f \wedge a)(x) \geq a_i - a\})$$
$$= S_{F-a}(f - f \wedge a)\,.$$

Hence

$$S_F(f) = \sum_{a_i \leq a} + \sum_{a_i > a} = S(f \wedge a, f) + S_{F-a}(f - f \wedge a). \qquad \square$$

Proof of Theorem 7.10. (i) Is a simple conclusion of Lemma 7.3(ii) and the fact that monotone real valued nets always have limits (finite or infinite).

(ii) Follows from the inequalities

$$m(\{x : f(x) \geq a\}) \leq m(\{x : g(x) \geq a\})$$

and

$$-m(\{x : f(x) \leq -a\} \leq m(\{x : g(x) \leq -a\})$$

if $f \leq g$ and $a > 0$.

(iii) If $c = 0$, then it trivially holds. If $c \neq 0$, then the assertion follows from (i) of Lemma 7.4.

(iv) Let $(\check{S}) \int (f \wedge a)\, dm$ be finite. Choose $F_1 \in \mathcal{F}$ with $a \in F_1$ and such that for $F \supset F_1$ we have

$$\left| S_F(f \wedge a) - (\check{S}) \int (f \wedge a)\, dm \right| < \epsilon$$

and $f - f \wedge a$ is a non-negative measurable function, hence by (i) of this theorem $(\check{S}) \int (f - f \wedge a)\, dm$ exists. Take $F_2 \in \mathcal{F}^+$ such that for every $F \supset F_2$ we have

$$S_F(f - f \wedge a) \geq n \quad \text{if} \quad (\check{S}) \int (f - f \wedge a) = \infty$$

and

$$\left| (\check{S}) \int (f - f \wedge a)\, dm - S_F(f - f \wedge a) \right| < \epsilon \quad \text{if} \quad (\check{S}) \int (f - f \wedge a)\, dm < \infty\,.$$

Let $F_0 = F_1 \cup (F_2 + a)$ and let $F \supset F_0$. Therefore

$$S_F(f) = S_F(f \wedge a) + S_F(f - f \wedge a) \geq n + (\check{S}) \int (f \wedge a)\, dm - \epsilon\,,$$

if $(\check{S}) \int (f - f \wedge a)\, dm = \infty$. In the other case

$$\begin{aligned} &\left| S_F(f) - (\check{S}) \int (f - f \wedge a)\, dm - (\check{S}) \int (f \wedge a)\, dm \right| \\ &\qquad \leq \left| S_F(f - f \wedge a)\, dm - (\check{S}) \int (f - f \wedge a)\, dm \right| \\ &\qquad\quad + \left| S_{F-a}(f \wedge a)\, dm - (\check{S}) \int (f \wedge a)\, dm \right| \leq 2\epsilon\,, \end{aligned}$$

since $F - a \supset F_2$. Hence $(\check{S}) \int f\, dm$ exists and

$$(\check{S}) \int f\, dm = (\check{S}) \int (f \wedge a)\, dm + (\check{S}) \int (f - f \wedge a)\, dm\,.$$

The proof for the infinite $(\check{S}) \int (f \wedge a)\, dm$ is analogous.

(v) Take $a = 0$ in (iv). Then

$$\begin{aligned} (\check{S}) \int f\, dm &= (\check{S}) \int (f \wedge 0)\, dm + (\check{S}) \int (f - f \wedge 0)\, dm \\ &= (\check{S}) \int (-f^-)\, dm + (\check{S}) \int f^+\, dm \\ &= (\check{S}) \int f^+\, dm - (\check{S}) \int f^-\, dm. \end{aligned}$$

For the second part of (v) it is sufficient to prove that $(\check{S}) \int f^+\, dm$ and $(\check{S}) \int f^-\, dm$ are finite. Let $F_0 \in \mathcal{F}$ be such that

$$|S_F(f) - (\check{S}) \int f\, dm| < \epsilon \quad \text{for} \quad F \supset F_0\,.$$

We choose F such that $F^- = F_0^-$.

Then by (ii) of Lemma 7.4 we obtain

$$S_F(f) = S_F(f^+) - S_F(f^-) = S_F(f^+) - S_{-F_0}(f^-)\,.$$

Hence

$$S_F(f^+) = S_F(f^+) \le S_{-F_0}(f^-) + (\check{S})\int f\,dm + \epsilon$$

for every $F \in \mathcal{F}$, $F \supset F_0$, $F^- = F_0^-$. Since $(\check{S})\int f^+\,dm = \sup_{F\in\mathcal{F}} S_{F^+}(f^+)$, we get $(\check{S})\int f^+\,dm < \infty$.

Since $(\check{S})\int(-f)\,dm = -(\check{S})\int f\,dm$ and $f^- = (-f)^+$, we get that $(\check{S})\int f^-\,dm$ is finite too.

(vii) Suppose first that $f \ge 0$. It follows from (i) that $(\check{S})\int f\,dm$ exists. Let $\{G_n\}$ be a sequence of sets from $\mathcal{F}$ with

$$\lim_{n\to\infty} S_{G_n}(f) = (\check{S})\int f\,dm.$$

We denote

$$F_n = \bigcup_{i=1}^{n} G_i \cup \{j/2^n : j = 1, 2, \ldots, n\cdot 2^n\}$$

and

$$f_n = f_{F_n}\,.$$

f_n is a non-negative simple function and the sequence $\{f_n\}$ is increasing. For $f(x) < n$ we have

$$0 \le f(x) - f_n(x) \le 1/2^n.$$

For $f(x) = \infty$ we have $f_n(x) \ge n$ and so

$$\lim_n f_n(x) = f(x) \quad (x \in X)\,.$$

Moreover

$$(\check{S})\int f\,dm = \lim_n S_{G_n}(f) = \lim_n S_{F_n}(f) = \lim_n (\check{S})\int f_n\,dm \le (\check{S})\int f\,dm\,.$$

Hence

$$\lim_n (\check{S})\int f_n = (\check{S})\int f\,dm\,.$$

The proof for a not necessarily positive function f follows if we apply the result just proved separately to f^+, and f^- and from (v). □

Theorem 7.11 *If f is a simple function with the range F and $(\check{S})\int f\,dm$ exists, then*

$$(\check{S})\int f\,dm = S_F(f)\,.$$

If $\mathcal{D}$ is a σ-ring and μ is a σ-additive measure on $\mathcal{D}$ and f is a simple function, then

$$(\check{S})\int f\,d\mu = \int f\,d\mu\,,$$

where $\int f\,d\mu$ is the Lebesgue integral of the function f.

Proof. Let $f \geq 0$. Let $F_1 \in \mathcal{F}$ with $F_1 \supset F$.

Since $\{x : f(x) \geq c\} = \{x : f(x) \geq a\}$ if $a = \min\{x \in F : x \geq c\}$, in the case $\{x \in F : x \geq c\} \neq \emptyset$ and $a = \max F$ in the case $\{x \in F : x \geq c\} = \emptyset$, we have $S_{F_1}(f) = S_F(f)$ for $F_1 \supset F$ and so

$$(\check{S})\int f = S_F(f)\,.$$

For a not necessarily positive function we apply the result just proved separately to f^+ and f^- and (v) of Theorem 7.10. The second part is obvious.□

Corollary 7.4 *Let $A_1 \supset A_2 \supset \cdots \supset A_n$ be measurable sets. Let c_i be positive real numbers and let $f_i = c_i \cdot \chi_{A_i}$ $(i = 1, 2, \ldots, n)$. Then we have*

$$(\check{S})\int \Big(\sum_{i=1}^{n} f_i\Big)\,dm = \sum_{i=1}^{n} (\check{S})\int f_i\,dm\,.$$

Theorem 7.12 *(i) If $|f| \in \mathcal{L}(\mathcal{D})$ and $|f| \in \mathcal{L}_1(\mathcal{D})$, then $f \in \mathcal{L}_1(\mathcal{D})$ and*

$$\Big|(\check{S})\int f\,dm\Big| \leq (\check{S})\int |f|\,dm\,.$$

(ii) If $f, |f| \in \mathcal{L}(\mathcal{D})$, $g \in \mathcal{L}_1(\mathcal{D})$ and $|f| \leq g$, then $f \in \mathcal{L}_1(\mathcal{D})$.

It is interesting that integrability of f does not imply the integrability of $|f|$.

Example 7.3 Take $X = (-1, 0) \cup (0, 1)$ and $\mathcal{D} = \mathcal{P}(X)$. Define m by

$$m(A) = \begin{cases} 1 & \text{if } A \cap (-1, 0) \neq \emptyset, \\ 1 & \text{if } A \cap (0, 1) \neq \emptyset, \\ 0 & \text{otherwise.} \end{cases}$$

If we take $f(x) = \frac{1}{x}$, then $(\check{S})\int f\,dm = 0$ and $(\check{S})\int |f|\,dm = \infty$.

7.4 Convergence Theorems for Symmetric Choquet (Šipoš) Integral

Let m be a positive continuous and monotone set function defined on $\mathcal{D}$.

Theorem 7.13 *Let $\{f_n\}$ be the sequence of non-negative integrable functions with $(\check{S})\int f_n\,dm \le c$ and let $f_n \nearrow f$. Then f is integrable $(\check{S})\int f\,dm \le c$ and*

$$(\check{S})\int f\,dm = \lim_n (\check{S})\int f_n\,dm\,.$$

Proof. Let $\epsilon > 0$ and $F = \{a_0, a_1, \ldots, a_n\} \in \mathcal{F}$ where $0 = a_0 < a_1 < \cdots < a_n$. We choose $\delta > 0$ such that

$$\delta c/(a_1 - \delta) < \epsilon$$

and

$$0 < 2\delta < \min\{a_j - a_{j-1} : j = 1, 2, \ldots, k\}\,.$$

Take $b_0 = 0$, $b_i = a_i - \delta$ $(i = 1, 2, \ldots, k)$. From

$$\bigcup_n \{x : f_n(x) \ge a_i - \delta\} \supset \{x : f(x) \ge a_i\}$$

and by the continuity of m we obtain

$$\begin{aligned}
S_F(f) &= \sum_{i=1}^{k} (a_i - a_{i-1})\, m(\{x : f(x) \ge a_i\}) \\
&\le \lim_n \sum_{i=1}^{k} (a_i - a_{i-1})\, m(\{x : f_n(x) \ge a_i - \delta\}) \\
&= \lim_n \Big[\sum_{i=1}^{k} (b_i - b_{i-1}) m(\{x; f_n(x) \ge b_i\}) + \delta \cdot m(\{x; f_n(x) \ge a_1 - \delta\})\Big] \\
&\le \lim_n (\check{S})\int f_n\,dm + \frac{\delta}{a_1 - \delta}\lim_n (\check{S})\int f_n\,dm \le \lim_n (\check{S})\int f_n\,dm + \epsilon\,.
\end{aligned}$$

Hence

$$(\check{S})\int f\,dm = \sup_{F\in\mathcal{F}} S_F(f) \le \lim_{n\to\infty} (\check{S})\int f_n\,dm + \epsilon\,.$$

Since ϵ was arbitrary, we get

$$(\check{S})\int f\,dm \leq \lim_{n\to\infty}(\check{S})\int f_n\,dm\,.$$

The opposite inequality follows form the monotonicity of the integral $(\check{S})\int$. □

Lemma 7.5 *Let f be a non-negative integrable function, then*

(i) $\lim_{A\to\infty}(\check{S})\int(f - f\wedge A)\,dm = 0$.

(ii) $\lim_{a\to 0}(\check{S})\int(f\wedge a)\,dm = 0$.

Proof. (i) Let $F\in\mathcal{F}$, $\epsilon > 0$ and let $(\check{S})\int f\,dm - S_F(f) < \epsilon$. Clearly $S_F(f) < \infty$, then $f_F \leq f\wedge A \leq f$. By the monotonicity of the integral we obtain

$$S_F(f) = (\check{S})\int f\,dm \leq (\check{S})\int(f\wedge A)\,dm \leq (\check{S})\int f\,dm\,.$$

Hence

$$(\check{S})\int(f - f\wedge A)\,dm = (\check{S})\int f\,dm - (\check{S})\int(f\wedge A)\,dm$$
$$\leq (\check{S})\int f\,dm - S_F(f) < \epsilon\,.$$

(ii) It is enough to consider the case $a_n \searrow 0$. Then the sequence $\{f_n\}$, $0\leq f_n = f - f\wedge a_n$ is increasing, $f_n \nearrow f$ and $(\check{S})\int f_n\,dm \leq (\check{S})\int f\,dm < \infty$, by the integrability of f.

Theorem 7.13 implies

$$\lim_{n\to\infty}(\check{S})\int f_n\,dm = (\check{S})\int f\,dm\,.$$

Since

$$(\check{S})\int(f\wedge a_n)\,dm = (\check{S})\int f\,dm - (\check{S})\int f_n\,dm\,,$$

we obtain

$$\lim_{n\to\infty}(\check{S})\int(f\wedge a_n)\,dm = 0\,.$$ □

Theorem 7.14 *Let $\{f_n\}$ be a sequence of non-negative integrable functions. Let f be a measurable function and let $f_n \searrow f$. Then f is integrable and*

$$\lim_{n\to\infty} (\check{S}) \int f_n \, dm = (\check{S}) \int f \, dm \, .$$

Proof. Since $0 \le f \le f_1$, we have that f is integrable. We have

$$\begin{aligned}
&\left|(\check{S}) \int f_n \, dm - (\check{S}) \int f \, dm\right| \\
&\quad = \left|(\check{S}) \int (f_n \wedge A) \, dm + (\check{S}) \int (f_n - f_n \wedge A) \, dm \right. \\
&\qquad \left. - (\check{S}) \int (f \wedge A) \, dm - (\check{S}) \int (f - f \wedge A) \, dm\right| \\
&\quad \le \left|(\check{S}) \int (f_n - A) \, dm - (\check{S}) \int (f \wedge A) \, dm \right. \\
&\qquad \left. + (\check{S}) \int (f_n - f_n \wedge A) \, dm + (\check{S}) \int (f - f \wedge A) \, dm\right| \\
&\quad = \left|(\check{S}) \int (f_n \wedge A) \, dm - (\check{S}) \int (f \wedge a) \, dm\right| + 2 \cdot (\check{S}) \int (f_1 - f_1 \wedge A) \, dm.
\end{aligned}$$

Therefore by $\lim_{A\to\infty} (\check{S}) \int (f_1 - f_1 \wedge A) \, dm = 0$ it follows that we may assume that the functions f_n and f are bounded with a number A.

Further, by

$$\begin{aligned}
&\left|(\check{S}) \int f_n \, dm - (\check{S}) \int f \, dm\right| \\
&\le \left|(\check{S}) \int (f_n - f_n \wedge a) \, dm - (\check{S}) \int (f - f \wedge a) \, dm\right| + 2(\check{S}) \int (f_1 \wedge a) \, dm,
\end{aligned}$$

$$\lim_{a\to 0} (\check{S}) \int (f_1 \wedge a) \, dm = 0$$

and

$$\begin{aligned}
m(\{x : (f_1 - f_1 \wedge a)(x) \ge t\}) &= m(\{x : f_1(x) \ge a + t\}) \\
&= (a+t)^{-1}(a+t) m(\{x : f_1(x) \ge a + t\}) \\
&\le a^{-1} (\check{S}) \int f_1 \, dm
\end{aligned}$$

for every $t > 0$, we may assume that there exists a real constant M such that

$$m(\{x : f_1(x) \geq t\}) \leq M < \infty \quad \text{for} \quad t > 0\,.$$

The functions

$$g_n(t) = m(\{x : f_n(x) \geq t\})$$

and

$$g(t) = m(\{x : f(x) \geq t\}$$

are real non-negative monotone functions defined on $[0, \infty)$, which vanish on $[A, \infty)$ and they are bounded with M. By the continuity of m and by the fact that $f_n \searrow f$ we get $g_n \searrow g$. For an arbitrary but fixed $\epsilon > 0$ we choose $F_0 = \{a_0, a_1, \ldots, a_k\}$ where $0 = a_0 < a_1 < \cdots < a_k$ with $a_k \leq A$ and such that for $F \supset F_0$ there holds: $(\check{S}) \int f\, dm - S_F(f) < \epsilon/2$. Choose a $\delta > 0$ such that for every partition Δ of the interval $[0, A]$ with the norm less then δ there holds $| \int_0^A g(t)\, dt - \Sigma(g, \Delta)| < \epsilon/2$, where $\Sigma(g, \Delta)$ is a Riemann integral sum of g with respect to the partition Δ. Let $F \supset F_0$ be such that $F \cap [0, A]$ is a partition of $[0, A]$ with the norm less then δ. Then

$$\begin{aligned}\left|(\check{S}) \int f\, dm - \int_0^A g(t)\, dt\right| &= \left|(\check{S}) \int f\, dm - S_F(f) + S_F(f) - \int_0^A g(t)\, dt\right| \\ &= \left|(\check{S}) \int f\, dm - S_F(f) + \Sigma(g, F \cap [0, A]) - \int_0^A g(t)\, dt\right| < \epsilon\,,\end{aligned}$$

because $S_F(f) = \Sigma(g, F \cap [0, A])$. Thus we get

$$(\check{S}) \int f\, dm = \int_0^A g(t)\, dt\,.$$

Analogously

$$(\check{S}) \int f_n\, dm = \int_0^A g_n(t)\, dt\,.$$

Since $g_n \searrow g$, we obtain

$$(\check{S}) \int f_n\, dm = \int_0^A g_n(t)\, dt \searrow \int_0^A g(t)\, dt = (\check{S}) \int f\, dm\,.$$

Hence

$$\lim_{n \to \infty} (\check{S}) \int f_n\, dm = (\check{S}) \int f\, dm\,. \qquad \square$$

Theorem 7.15 *Let $\{f_n\}$ be a sequence of integrable functions. Let $f_n \nearrow f$, where f is a measurable function and $(\check{S})\int f_n\,dm \le c < \infty$ for every $n \in \mathbf{N}$. Then f is integrable and*

$$(\check{S})\int f_n\,dm \nearrow (\check{S})\int f\,dm\,.$$

Proof. We have that $f_n \nearrow f$ implies $f_n^+ \nearrow f^+$ and $f_n^- \nearrow f^-$. Since

$$\begin{aligned}(\check{S})\int f_n^+\,dm &= (\check{S})\int f_n\,dm + (\check{S})\int f_n^-\,dm \le (\check{S})\int f_n\,dm + (\check{S})\int f_1^-\,dm\\ &\le c + (\check{S})\int f_1^-\,dm < \infty\,,\end{aligned}$$

Theorem 7.13 implies $(\check{S})\int f_n^+\,dm \nearrow (\check{S})\int f^+\,dm$. Analogously by Theorem 7.14 we have $(\check{S})\int f_n^-\,dm \nearrow (\check{S})\int f^-\,dm$. Hence

$$\begin{aligned}\lim_{a\to\infty}(\check{S})\int f_n\,dm &= \lim_{n\to\infty}(\check{S})\int f_n^+\,dm - \lim_{n\to\infty}(\check{S})\int f_n^-\,dm\\ &= (\check{S})\int f^+\,dm - (\check{S})\int f^-\,dm = (\check{S})\int f\,dm.\end{aligned}$$ □

Now we have the Lebesgue convergence type theorem.

Theorem 7.16 *Let $\mathcal{D} = \Sigma$ be a σ-algebra of subsets. If $\{f_n\}$ is a sequence of measurable functions which converges pointwise to a measurable function f and if g is an integrable function with $|f_n| \le g$ for $n \in \mathbf{N}$ then f is integrable and*

$$\lim_{n\to\infty}(\check{S})\int f_n = (\check{S})\int f\,dm\,.$$

Proof. It follows from $|f_n| \le g$ and $|f| \le g$ that the functions f_n and f are integrable. Let

$$h_n = \inf_{i\ge n} f_i \qquad \text{and} \qquad g_n = \sup_{i\ge n} f_i\,.$$

h_n and g_n are measurable functions. We have $h_n \le f_n \le g_n$, and

$$\lim_{n\to\infty} h_n = \lim_{n\to\infty} f_n = f$$

and analogously

$$\lim_{n\to\infty} g_n = f\,.$$

The functions h_n and g_n are integrable and $h_n \nearrow f$ and $g_n \to f$. By Theorem 7.15 we have

$$(\check{S})\int h_n\,dm \nearrow (\check{S})\int f\,dm \quad\text{and}\quad (\check{S})\int g_n\,dm \to (\check{S})\int f\,dm\,.$$

Therefore by

$$(\check{S})\int h_n\,dm \le (\check{S})\int f_n\,dm \le (\check{S})\int g_n\,dm,$$

we obtain

$$\lim_{n\to\infty}(\check{S})\int f_n\,dm = (\check{S})\int f\,dm\,.$$

□

7.5 The Submodular Theorem

In this section the family $\mathcal{D}$ is assumed to be a σ-algebra Σ and m is a positive monotone and continuous set function defined on $\mathcal{D}$.

Theorem 7.17 *Let f and g be non-negative measurable functions. Then*

$$(\check{S})\int (f\wedge g)\,dm + (\check{S})\int (f\vee g)\,dm \le (\check{S})\int f\,dm + (\check{S})\int g\,dm$$

if m is a submodular. We have

$$(\check{S})\int (f\wedge g)\,dm + (\check{S})\int (f\vee g)\,dm \ge (\check{S})\int f\,dm + (\check{S})\int g\,dm$$

if m is supermodular.

Proof. Let $F \in \mathcal{F}$ with $F = \{a_0, a_1, a_2, \ldots, a_n\}$ where $a_0 = 0 < a_1 < a_2 < \cdots < a_n$. Take

$$A_i = \{x : f(x) \geq a_i\} \quad \text{and} \quad B_i = \{x : g(x) \geq a_i\} \quad (i = 0, 1, 2, \ldots, n).$$

Since m is submodular we obtain

$$\begin{aligned} &m(\{x : (f \wedge g)(x) \geq a_i\}) + m(\{x : \ (f \vee g)(x) \geq a_i\}) \\ &\quad = m(A_i \cap B_i) + m(A_i \cup B_i) \leq m(A_i) + m(B_i). \end{aligned}$$

Multiplying by $(a_i - a_{i-1})$ and summing over i we get

$$\begin{aligned} S_F(f \wedge g) + S_F(f \vee g) &\leq \sum_{i=1}^{n} (a_i - a_{i-1})(m(A_i) + m(B_i)) \\ &= S_F(f) + S_F(g). \end{aligned}$$

For $(\check{S}) \int f\, dm + (\check{S}) \int g\, dm = \infty$ the assertion is trivial. Let $(\check{S}) \int f\, dm + (\check{S}) \int g\, dm < \infty$. Let $\epsilon > 0$. Choose $F_0 \in \mathcal{F}$ with $|S_F(f) - (\check{S}) \int f\, dm| < \epsilon$ and $|S_F(g) - (\check{S}) \int g\, dm| < \epsilon$ for $F \supset F_0$. Then

$$\begin{aligned} (\check{S}) \int f\, dm + (\check{S}) \int g\, dm &> S_f(f) + S_F(g) - 2\epsilon \\ &\geq S_F(f \wedge g)\, dm + S_F(f \vee g)\, dm - 2\epsilon. \end{aligned}$$

Since ϵ was arbitrary, we obtain

$$(\check{S}) \int f\, dm + (\check{S}) \int g\, dm \geq (\check{S}) \int (f \wedge g)\, dm + (\check{S}) \int (f \vee g)\, dm.$$

The proof for the supermodular case is analogous. □

Corollary 7.5 *If f and g are non-negative measurable functions and m is a subadditive, then*

$$(\check{S}) \int (f \vee g)\, dm \leq (\check{S}) \int f\, dm + (\check{S}) \int g\, dm.$$

We will show that $(\check{S}) \int$ has similar properties on non-negative Σ-measurable functions as m has on Σ.

Lemma 7.6 *Let $A_1 \supset A_2 \supset \cdots \supset A_n$ and A be measurable sets and let m be a submodular set function on Σ then*

$$\sum_{i=1}^{n+1} m((A \cap A_{i-1}) \cup A_i) \leq \sum_{i=1}^{n+1} m(A_i) + m(A),$$

where $A_0 = A$ and $A_{n+1} = \emptyset$. If m is a submodular set function on Σ then the opposite inequality holds.

Proof. For $m(A) = \infty$ the assertion is trivial. Suppose that $m(A)$ is finite. For $n = 2$, using the fact that m is submodular, we obtain

$$\begin{aligned} m(A \cup A_1) &+ m((A \cap A_1) \cup A_2) + m(A \cap A_2) \\ &\leq m(A) + m(A_1) - m(A \cap A_1) + m(A \cap A_1) + m(A_2) \\ &\quad - m(A \cap A_1 \cap A_2) + m(A \cap A_2) \\ &\leq m(A) + m(A_1) + m(A_2). \end{aligned}$$

For $n > 2$ the assertion can be similarly proved by induction. The proof of the inequality for a supermodular set function is analogous. □

Lemma 7.7 *Let f and g be non-negative simple functions $g = c\chi_A$ and m be a submodular set function. Then we have*

$$(\check{S}) \int (f + g)\, dm \leq (\check{S}) \int f\, dm + (\check{S}) \int g\, dm.$$

If m is a supermodular set function, then

$$(\check{S}) \int (f + g)\, dm \geq (\check{S}) \int f\, dm + (\check{S}) \int g\, dm.$$

Proof. Let m be a submodular set function. If $m(A) = \infty$, then the proof is trivial. Let $m(A) < \infty$. Let $\{a_1, \ldots, a_n\}$, $0 = a_0 < a_2 < \cdots < a_n$, be the range of f.

Denote $A_i = \{x : f(x) \leq a_i\}$ for $i = 1, 2, \ldots, n$, and $A_0 = A$, $A_{i+1} = \emptyset$, $c_{n+1} = 0$, and $c_i = a_i - a_{i-1}$. Suppose first that $c \leq c_i$, $i = 1, 2, \ldots, n$. Then

$$f + g = \sum_{i=1}^{n+1} (c \cdot \chi_{(A \cap A_{i-1}) \cup A_i} + (c_i - c)\chi_{A_i})$$

and

$$A \cup A_1 \supset A_1(A \cap A_1) \cup A_2 \supset A_2 \supset \cdots \supset (A \cap A_{n-1}) \cup A_n \supset A_n \supset A \cap A_n .$$

By Corollary 7.4 we have

$$\begin{aligned}
(\check{S}) \int (f+g)\, dm &= \sum_{i=1}^{n+1} (cm((A \cap A_{i-1} \cup A_i) + (c_i - c)m(A_i)) \\
&= \sum_{i=1}^{n+1} c_i m(A_i) + c\Big(\sum_{i=1}^{n+1} m((A \cap A_{i-1}) \cup A_i) - \sum_{i=1}^{n+1} m(A_i)\Big) \\
&\leq (\check{S}) \int f\, dm + (\check{S}) \int g\, dm,
\end{aligned}$$

where the last inequality follows from Lemma 7.6. We have proved the lemma for c with $0 < c \leq c_i$, $i = 1, 2, \ldots, n$. Let now c be an arbitrary positive number and let $c = k \cdot b$, where k is a natural number and

$$b \leq \min\{c_i : i = 1, 2, \ldots, n\} .$$

Denote

$$g_j = b \cdot \chi_A, \quad \text{for} \quad j = 1, 2, \ldots, k .$$

Then

$$\begin{aligned}
(\check{S}) \int (f+g)\, dm &= (\check{S}) \int (f + g_1 + g_2 + \cdots + g_k)\, dm \\
&\leq (\check{S}) \int (f + g_1 + \cdots + g_{k-1})\, dm + (\check{S}) \int g_k\, dm \\
&\leq (\check{S}) \int f\, dm + (\check{S}) \int g_1\, dm + \cdots + (\check{S}) \int g_k\, dm \\
&= (\check{S}) \int f\, dm + (\check{S}) \int g\, dm .
\end{aligned}$$

Here we have proved the first part of this proof. In the case of a supermodular set function the proof is analogous. □

Propositon 7.9 *Let f and g be non-negative simple functions and m be a submodular set function. Then*

$$(\check{S})\int (f+g)\,dm \le (S)\int f\,dm + \int g\,dm\,.$$

If m is a supermodular set function then

$$(\check{S})\int (f+g)\,dm \ge (\check{S})\int f\,dm + (\check{S})\int g\,dm\,.$$

Proof. Let $\{b_1,\ldots,b_n\}$ with $0=b_0<b_1<\cdots<b_n$ be the range of g. Denote $g_i=(b_i-b_{i-1})\cdot\chi_{\{x:g(x)\ge b_i\}}$ for $i=1,2,\ldots,n$. Then

$$g=\sum_{i=1}^{n} g_i \qquad \text{and} \qquad (\check{S})\int g\,dm = \sum_{i=1}^{n}(\check{S})\int g_i\,dm\,.$$

Further

$$\begin{aligned}(\check{S})\int (f+g)\,dm &= (\check{S})\int (f+g_1+\cdots+g_n)\,dm\\ &\le (\check{S})\int (f+g_1+\cdots+g_{n-1})\,dm + (\check{S})\int g_n\,dm\\ &\le (\check{S})\int f\,dm + (\check{S})\int g_1\,dm+\cdots+(\check{S})\int g_n\,dm\\ &= (\check{S})\int f\,dm + (\check{S})\int g\,dm,\end{aligned}$$

using Lemma 7.7. The case of a supermodular set function is analogous.□

Now we can prove the main result of this section.

Theorem 7.18 *Let f and g be measurable non-negative functions and m a submodular set function. Then*

$$(\check{S})\int (f+g)\,dm \le (\check{S})\int f\,dm + (\check{S})\int g\,dm\,.$$

If m is a supermodular set function, then

$$(\check{S})\int (f+g)\,dm \ge (\check{S})\int f\,dm + (\check{S})\int g\,dm\,.$$

Proof. Let m be a submodular set function. If $(\check{S})\int f\,dm$ or $(\check{S})\int g\,dm$ is infinite, then the assertion is clear. Let $(\check{S})\int f\,dm$ and $(\check{S})\int g\,dm$ be finite. Then by

$$f+g\leq 2\cdot(f\vee g)$$

and the monotonicity of $(\check{S})\int$ we have

$$(\check{S})\int(f+g)\,dm\leq 2\cdot(\check{S})\int(f\vee g)\,dm\leq 2\cdot((\check{S})\int f\,dm+(\check{S})\int g)\,dm<\infty\,.$$

Take a simple function $h\leq f+g$. It is obvious that $m(S_h)<\infty$, where $S_h=\{x\in X:h(x)\neq 0\}$. Let $F=\{a_0,a_1,\ldots,a_n\}\in\mathcal{F}$, $a_0=0<a_1<\cdots<a_n$, be such that $\max\{a_i-a_{i-1}\}<\epsilon/2$,

$$(\check{S})\int f\,dm-S_F<\epsilon/2\quad\text{and}\quad(\check{S})\int g\,dm-S_F(g)<\epsilon/2\,.$$

Let $(f+g)_F(x)=a_i$, $f_F(x)=a_j$ and $g_F(x)=a_k$ then $a_i\leq f(x)+g(x)$, $a_j\leq f(x)<a_{j+1}$ and $a_k\leq g(x)<a_{k+1}$. Since $\max\{a_i-a_{i-1}\}<\epsilon/2$, we obtain

$$\begin{aligned}h(x)\leq h_F(x)\leq(f+g)_F(x)=a_i\leq f(x)+g(x)\leq a_{j+1}+a_{k+1}\\ \leq a_j+a_k+\epsilon=f_F(x)+g_F(x)+\epsilon\,,\end{aligned}$$

and so

$$h(x)\leq f_F(x)+g_F(x)+\epsilon\chi_{S_h}(x)\,.$$

Hence

$$\begin{aligned}(\check{S})\int h\,dm&\leq(\check{S})\int(f_F+g_F)\,dm+\epsilon m(S_h)\\ &\leq(\check{S})\int f_F\,dm+(\check{S})\int g_F\,dm+\epsilon\cdot m(S_h)\\ &\leq(\check{S})\int f\,dm+(\check{S})\int g\,dm+\epsilon\cdot m(S_h)\,.\end{aligned}$$

Since $m(S_h)<\infty$, we have

$$(\check{S})\int h\,dm\leq(\check{S})\int f\,dm+(\check{S})\int g\,dm\,.$$

The proof now follows from (vii) in Theorem 7.10.

If m is a supermodular set function, then the case $(\check{S})\int(f+g)\,dm = \infty$ is clear. Let $(\check{S})\int(f+g)\,dm < \infty$. Then by the monotonicity of integral $(\check{S})\int$ we have that $(\check{S})\int f\,dm$ and $(\check{S})\int g\,dm$ are finite. Let $\tilde{f}$, $\tilde{g}$ be measurable simple functions with $\tilde{f} \le f$ and $\tilde{g} \le g$; then $(\check{S})\int \tilde{f}\,dm + (\check{S})\int \tilde{g}\,dm \le (\check{S})\int(\tilde{f}+\tilde{g})\,dm \le (\check{S})\int(f+g)\,dm$ and by Theorem 7.10

$$(\check{S})\int f\,dm + (\check{S})\int g\,dm \le (\check{S})\int (f+g)\,dm\,. \qquad \square$$

Corollary 7.6 *Let f be a measurable function. Let m be a submodular set function. Then f is integrable if and only if $|f|$ is integrable.*

Proof. Let f be integrable. Then

$$(\check{S})\int |f|\,dm = (\check{S})\int (f^+ + f^-)\,dm \le (\check{S})\int f^+\,dm + (\check{S})\int f^-\,dm < \infty\,.$$

And so $|f|$ is also integrable. The opposite implication follows from (i) in Theorem 7.12.

As a consequence of Theorem 7.18 we obtain for measures.

Theorem 7.19 *Let m be a measure on Σ. Let f and g be non-negative Σ-measurable functions, then*

$$\begin{aligned}(\check{S})\int (f+g)\,dm &= (\check{S})\int (f\wedge g)\,dm + (\check{S})\int (f\vee g)\,dm \\ &= (\check{S})\int f\,dm + (\check{S})\int g\,dm\,.\end{aligned}$$

The last theorem can be generalized using a similar approach in the following sense.

Theorem 7.20 *Let m be a monotone set function on a paving $\mathcal{D}$ and let $\mathcal{M} \subset \mathcal{D}$ be a subsystem closed under finite unions and denumerable intersection. Then the Šipos integral (and hence the symmetric Choquet integral) is additive on $\mathcal{L}^+(\mathcal{M})$ if and only if m is a valuation on $\mathcal{M}$.*

Proof. If m is not a valuation, then there are $A, B \in \mathcal{M}$ such that $m(A \cup B) + m(A \cap B) \neq m(A) + m(B)$. Put $f = \chi_A$, $g = \chi_B$. Then $f, g \in \mathcal{L}^+(\mathcal{M})$, $(\check{S}) \int f\, dm = m(A)$, $(\check{S}) \int g\, dm = m(B)$ and

$$(\check{S}) \int (f+g)\, dm = m(A \cup B) + m(A \cap B).$$

It follows

$$(\check{S}) \int (f+g)\, dm \neq (\check{S}) \int f\, dm + (\check{S}) \int g\, dm.$$

Sufficiency goes in an analogous way as the proof of Theorem 7.19. □

Corollary 7.7 *Let Σ be a σ-algebra and let m be a positive monotone set function on Σ. Then the Šipoš integral (general symmetric Choquet integral) is linear on the system $\mathcal{L}_1(X, \Sigma, m)$ of all integrable measurable functions if and only if m is a valuation on Σ, i.e., m is an additive set function on Σ.*

Proof. If $0 \leq g \leq f$, $f, g \in \mathcal{L}_1$, then by Theorem 7.20

$$(\check{S}) \int f\, dm = (\check{S}) \int g\, dm + (\check{S}) \int (f-g)\, dm$$

and hence

$$(\check{S}) \int (f-g)\, dm = (\check{S}) \int f\, dm - (\check{S}) \int g\, dm.$$

For every $A \in \Sigma$ and $f \in \mathcal{L}_1$ we have

$$(\check{S}) \int f\, dm = (\check{S}) \int f \cdot \chi_A\, dm + (\check{S}) \int f \cdot \chi_A\, dm \qquad (f \in \mathcal{L}_1).$$

If $f, g \in \mathcal{L}_1$, then

$$\begin{aligned} f + g &= (f^+ - f^-) + (g^+ - g^-) \\ &= (f^+ + g^+ - f^- - g^-) \cdot \chi_{\{f^+ + g^+ \geq f^- + g^-\}} \\ &\quad - (f^- + g^- - f^+ - g^+) \cdot \chi_{\{f^- + g^- > f^+ + g^+\}}. \end{aligned}$$

The first term in the last expression is equal to $(f+g)^+$ and the second term is equal to $(f+g)^-$. Denote $A = \{f^+ + g^+ \geq f^- + g^-\}$. Hence

$$
\begin{aligned}
(\check{S})\int (f+g)\,dm &= (\check{S})\int (f+g)^+\,dm - (\check{S})\int (f+g)^-\,dm \\
&= (\check{S})\int f^+\cdot\chi_A\,dm + (\check{S})\int g^+\cdot\chi_A\,dm - (\check{S})\int f^-\cdot\chi_A\,dm \\
&\quad - (\check{S})\int g^-\cdot\chi_A\,dm - ((\check{S})\int f^-\cdot\chi_A\,dm + (\check{S})\int g^-\cdot\chi_A\,dm \\
&\quad - (\check{S})\int f^+\cdot\chi_A\,dm - (\check{S})\int g^+\cdot\chi_A\,dm) \\
&= (\check{S})\int f^+\,dm - (\check{S})\int f^-\,dm + (\check{S})\int g^+\,dm - (\check{S})\int g^-\,dm \\
&= (\check{S})\int f\,dm + (\check{S})\int g\,dm\,.
\end{aligned}
$$

Hence by the homogeneity of the Šipoš integral we obtain the linearity of the Šipoš integral on $\mathcal{L}_1$. □

Definition 7.8 Two measurable non-negative functions f and g are comonotonic (compatible) if and only if for every pair (x, x_1) of elements of X

$$f(x) < f(x_1) \quad \text{implies} \quad g(x) \leq g(x_1)\,.$$

Remark 7.1 An equivalent definition is the following one: f and g are compatible iff the system $\{\{f \geq t\}, \{g \geq s\} : t, s \in [0, \infty)\} = \mathcal{D}_{f,g}$ forms a chain of subset of X.

Corollary 7.8 *Let $\mathcal{D}$ be a paving and $f, g \in \mathcal{L}^+(\mathcal{D})$. Then the following conditions are equivalent:*
(1) f and g are compatible;
(2) For every positive monotone set function m on $\mathcal{D}$ it holds

$$(\check{S})\int (f+g)\,dm = (\check{S})\int f\,dm + (\check{S})\int g\,dm\,.$$

Proof. By Theorem 7.9 we have the equality of the Choquet a(nd the Šipoš integral for all $f \in \mathcal{L}^+(\mathcal{D})$.

(2) $\Rightarrow$ (1) If f and g are non-compatible, then there are $t, s \in (0,\infty)$ such that there exist $x \in \{f \geq t\} \setminus \{g \geq s\}$ and $y \in \{g \geq s\} \setminus \{f \geq t\}$. We define a monotone set function m on $\mathcal{D}$ by

$$m(A) = \begin{cases} 1 & \text{if } \{x,y\} \subset A \\ 0 & \text{otherwise.} \end{cases}$$

Then $(\check{S}) \int f\,dm < t$, $(\check{S}) \int g\,dm < s$ and $(\check{S}) \int (f+g)\,dm \geq t+s$. Hence

$$(\check{S}) \int (f+g)\,dm > (\check{S}) \int f\,dm + (\check{S}) \int g\,dm\,.$$

(1) $\Rightarrow$ (2) Let f and g be compatible, i.e., $\mathcal{D}_{f,g}$ forms a chain (by Remark 7.1). It is obvious that $\mathcal{D}_{f,g}$ is closed under finite unions and denumerable intersection and that both f and g are elements of $\mathcal{L}^+(\mathcal{D}_{f,g})$. For any A, B of $\mathcal{D}_{f,g}$ we have $\{A, B\} = \{A \cap B, A \cup B\}$ and hence for monotone set functions with $m(\emptyset) = 0$ the following holds

$$m(A) + m(B) = m(A \cap B) + m(A \cup B)\,.$$

Hence by Theorem 7.20

$$(\check{S}) \int (f+g)\,dm = (\check{S}) \int f\,dm + (\check{S}) \int g\,dm\,. \qquad \square$$

7.6 Asymmetric Choquet Integral

We introduce another extension of the Choquet integral in this section, which is only positively homogenous, i.e., it is asymmetric with respect to multiplication by -1. This integral is defined also for non-monotone set functions.

Definition 7.9 The (asymmetric) Choquet integral of a measurable function $f\colon X \to \mathbf{R}$ with respect to a real-valued set function m is given by

$$(C)\int f\,dm = \int_{-\infty}^{0} (m(\{x : f(x) > r\}) - m(X))\,dr + \int_{0}^{+\infty} m(\{x : f(x) > r\})\,dr,$$

whenever the right-hand side exists.

It is obvious that for a monotone positive set function and a non-negative measurable function the preceding definition of the Choquet integral coincides with Definition 7.6. The preceding integral is well defined also if Σ is finite.

Theorem 7.21 *If $m \in BV$ then the Choquet integral with respect to m has the Lebesgue-Stieltjes integral representation*

$$(C)\int f\,dm = \int_{[-\infty,+\infty]} r\,dG(r), \qquad \textit{where} \tag{7.4}$$

$$G(r) = \begin{cases} -m(X) & r = -\infty\,, \\ -m(\{x : f(x) > r\} & -\infty < r < +\infty, \\ 0 & r = +\infty\,. \end{cases}$$

Moreover if either f is bounded or m is continuous both at X and at $\emptyset$, then

$$(C)\int f\,dm = \int_{-\infty}^{+\infty} r\,dG(r)\,.$$

If f is bounded, then the Choquet integral always exists.

Proof. Since $m \in BV$ implies that G is a function of bounded variation there exist two nondecreasing functions G^+ and G^- such that $G = G^+ - G^-$. We divide $(0, +\infty]$ on subintervals $I(n,i) = \left(\frac{i}{2^n}, \frac{i+1}{2^n}\right]$ for $i = 0, 1, \ldots, n2^n - 1$ and $I(n, n2^n) = (n, +\infty]$. Now we define for every $n \in \mathbf{N}$

$$g_n = \sum_{i=1}^{n \cdot 2^n} \left(\frac{i}{2^n}\right) \chi_{I(n,i)}\,,$$

$$h_n^+ = \sum_{i=0}^{n2^n - 1} \left[G^+(+\infty) - G^+\left(\frac{i+2}{2^n} + 0\right)\right] \chi_{I(n,i)}\,.$$

Hence

$$\int_{[0,+\infty]} g_n(r)\,dG^+(r) = \int_0^{+\infty} h_n^+(r)\,dr. \tag{7.5}$$

Since $\{g_n\}$ and $\{h_n^+\}$ are nondecreasing sequences of non-negative measurable functions and

$$\lim_{n\to\infty} g_n(r) = r \qquad (r \in [0,+\infty]),$$
$$\lim_{n\to\infty} h_n^+(r) = G^+(+\infty) - G^+(r+0) \qquad (r \in [0,+\infty)),$$

we obtain by the classical monotone convergence theorem that

$$\lim_{n\to\infty} \int_{[0,+\infty]} g_n(r)\,dG(r) = \int_{[0,+\infty]} r\,dG^+(r)$$
$$\lim_{n\to\infty} \int_0^{+\infty} h_n^+(r)\,dr = \int_0^{+\infty} [G^+(+\infty) - G^+(r+0)]\,dr\,.$$

Hence by (7.5)

$$\int_{[0,+\infty]} r dG^+(r) = \int_0^{+\infty} [G^+(+\infty) - G^+(r+0)]\,dr\,.$$

In a quite analogous way we have

$$\int_{[0,+\infty]} r\,dG^-(r) = \int_0^{+\infty} [G^-(+\infty) - G^-(r+0)]\,dr\,,$$
$$\int_{[-\infty,0]} r\,dG^\pm(r) = \int_{-\infty}^{0} [G^\pm(r+0) - G^\pm(-\infty)]\,dr\,,$$
$$\int_{[-\infty,0]} r\,dG^-(r) = \int_{-\infty}^{0} [G^-(r+0) - G^-(-\infty)]\,dr\,.$$

Since $G(+\infty) - G(r+0) = m(\{x : f(x) > r\})$ for all but at most countable many r in $[0,+\infty)$ and

$$G(r+0) - G(-\infty) = m(\{x:\ f(x) > r\}) - m(X)$$

for all but at most countable many r in $(-\infty, 0)$ we obtain (7.4).

Let the function f be bounded. Then the function

$$m_f(r) = \begin{cases} m(\{x : f(x) > r\}) - m(X) & r < 0, \\ m(\{x : f(x) > r\}) & r \geq 0 \end{cases}$$

is zero outside the interval

$$\left[\inf_{x \in X} f(x),\ \sup_{x \in X} f(x)\right].$$

Since $m \in BV$, the function m_f is of bounded variation and so measurable. Hence the Choquet integral of f exists. □

We remark, that the defined Choquet integral is linear with respect to linear combination of set functions.

Theorem 7.22 *For all $m \in BV(\Sigma)$, the asymmetric Choquet integral is positively homogenous and comonotonicaly additive. If additionally m is continuous from above and below then the corresponding asymmetric Choquet integral is also continuous, i.e., continuous with respect to the norm $\|f\| = \sup_{x \in X} |f(x)|$ on the space $B(X, \Sigma)$ of all real-bounded measurable functions.*

Proof. We will prove only the comonotonicity of the integral. Since $m \in BV$, then there exist two positive monotone set functions m^+ and m^- such that $m = m^+ - m^-$. Then we have

$$(C)\int_X f\,dm = (C)\int_X f\,dm^+ - (C)\int_X f\,dm^-$$

and we can apply Corollary 7.8 on both integrals on the right side of the preceding equality. □

7.7 Sugeno Integral

Let Σ be a σ-algebra of subsets of X and m be a positive monotone set function on Σ.

Definition 7.10 Let f be a non-negative extended real-valued measurable function on (X, Σ, m) and $A \in \Sigma$, the Sugeno (fuzzy) integral of f on A with respect to m is defined by

$$(S)\int_A f\,dm = \sup_{\alpha\in[0,\infty]} [\alpha \wedge m(A \cap F_\alpha)],$$

where $F_\alpha = \{x : f(x) \geq \alpha\}$, $\alpha \in [0,\infty]$. F_α is called α-cut.

Propositon 7.10 *The Sugeno integral has the following representations*

$$(S)\int_A f\,dm = \sup_{\alpha\in[0,\infty)} [\alpha \wedge m(A \cap F_\alpha)]$$

$$= \sup_{E\in\Sigma}\left[(\inf_{x\in E} f(x)) \wedge m(A \cap E)\right] = \sup_{E\in\Sigma(f)}\left[(\inf_{x\in E} f(x)) \wedge m(A \cap E)\right],$$

where $\Sigma(f)$ is the σ-algebra generated by f, i.e., the smallest σ-algebra in which f is measurable.

Propositon 7.11 *The Sugeno integral has the following properties:*

(1) $(S)\int_A f\,dm = (S)\int f \cdot \chi_A\,dm$.

(2) *For any constant $a \in [0,\infty]$, $(S)\int_A a\,dm = a \wedge m(A)$.*

(3) *If $f_1 \leq f_2$ on A, then $(S)\int_A f_1\,dm \leq (S)\int_A f_2\,dm$.*

(4) *If m is continuous from below and $(S)\int_A f\,dm = 0$, then*

$$m(A \cap \{f > 0\}) = 0.$$

(5) *For any constant $a \in [0,\infty]$,*

$$(S)\int_A (f + a)\,dm \leq (S)\int_A dm + a \wedge m(A)$$

$$= (S)\int_A f\,dm + (S)\int_A a\,dm.$$

Proof. The proofs of (1)–(3) are obtained directly by Definition 7.10, whereas the proof of (4) is obtained by contradiction. The proof of (5) follows from Proposition 7.10. □

We suppose in the rest of this section that m is monotone and continuous.

Theorem 7.23 *Let $a \in [0, \infty]$. If $|f_1 - f_2| \le a$ on A, then*

$$\left|(S)\int_A f_1\, dm - (S)\int_A f_2\, dm\right| \le a \wedge m(A)\,.$$

Proof. From $f_1 \le f_2 + a$ on A, we have

$$(S)\int_A f_1\, dm \le (S)\int_A (f_2 + a)\, dm \le (S)\int_A f_2\, dm + a \wedge m(A)$$

by using (3) and (5) in Proposition 7.11. Analogously, from $f_2 \le f_1 + a$ on A, we have

$$(S)\int_A f_2\, dm \le (S)\int f_1\, dm + a \wedge m(A).$$

Hence

$$\left|(S)\int_A f_1\, dm - (S)\int_A f_2\, dm\right| \le a \wedge m(A).$$ □

Lemma 7.8 *Let h_α, $\alpha \in [0, \infty]$, be a non-negative decreasing extended real-valued function of α. If $\alpha_0 = \sup_{\alpha \in [0,\infty)}[\alpha \wedge h_\alpha]$, then $h_{\alpha_0 - 0} \ge \alpha_0 \ge h_{\alpha_0 + 0}$, where $h_{\alpha_0-0} = \lim_{\alpha \to \alpha_0 - 0} h_\alpha$, $h_{\alpha_0+0} = \lim_{\alpha \to \alpha_0+0} h_\alpha$.*

Proof. If $\alpha_0 = 0$, then $h_\alpha = 0$ for every $\alpha > 0$. If $\alpha_0 = \infty$, then $h_\alpha = \infty$ for every $\alpha \in [0, \infty)$. In these two cases, the conclusion of the lemma is obvious. Suppose $\alpha_0 \in (0, \infty)$. For every $\alpha > \alpha_0$, since $\alpha \wedge h_\alpha \le \alpha_0$, we have $h_\alpha \le \alpha_0$. Therefore $h_{\alpha_0+0} \le \alpha_0$. On the other hand, we assume that $h_{\alpha_0-0} < \alpha_0$, then there exists $\alpha' < \alpha_0$, such that $h_{\alpha'} < \alpha_0$. Using the monotonicity of h_α, then $h_\alpha \le h_{\alpha'} < \alpha_0$ as $\alpha \ge \alpha'$ holds there. Consequently,

$$\begin{aligned}\sup_{\alpha \in [0,\infty)} [\alpha \wedge h_\alpha] &= \sup_{\alpha \in [0,\alpha')} [\alpha \wedge h_\alpha] \vee \sup_{\alpha \in [\alpha',\infty)} [\alpha \wedge h_\alpha] \\ &\le \sup_{\alpha \in [0,\alpha')} \alpha \vee \sup_{\alpha \in [\alpha',\infty)} h_\alpha < \alpha_0\,,\end{aligned}$$

we obtain a contradiction. Therefore, $h_{\alpha_0-0} \ge \alpha - 0$. □

If we denote $\{x : f \ge \alpha\}$ by F_α and $\lim_{\alpha' \to \alpha+0} F_{\alpha'}$ by $F_{\alpha+0}$, then we have

Theorem 7.24 *We have*
(1) $(S)\int_A f\,dm = \alpha$ *iff* $m(A \cap F_\alpha) \geq \alpha \geq m(A \cap F_{\alpha+0})$,
(2) $(S)\int_A f\,dm > \alpha$ *implies* $m(A \cap F_\alpha) > \alpha$,
(3) $(S)\int_A f\,dm \geq \alpha$ *iff* $m(A \cap F_\alpha) \geq \alpha$.

Proof. (1) From the definition of $(S)\int_A f\,dm$ directly, it is obvious that $m(A \cap F_\alpha) \geq \alpha \geq m(A \cap F_{\alpha+0})$ implies $(S)\int_A f\,dm = \alpha$. For the converse proposition, we write $h_\alpha = m(A \cap F_\alpha)$, then h_α is decreasing with respect to α, and $(S)\int_A f\,dm = \sup_{\alpha\in[0,\infty)}[\alpha \wedge h_\alpha]$. Using that $h_{\alpha-0} = h_\alpha$, the desired conclusion is obtained by Lemma 7.8.

(2) Let $(S)\int_A f\,dm = \alpha_0$. Then by $\alpha_0 > \alpha$, we have

$$m(A \cap F_\alpha) \geq m(A \cap F_{\alpha_0}) \geq \alpha_0 > \alpha\,.$$

(3) From the definition of $(S)\int_A f\,dm$, we have

$$m(A \cap F_\alpha) \geq \alpha \Rightarrow (S)\int_A f\,dm \geq \alpha\,.$$

The converse proposition is true by (1) and (2). □

In classical measure theory, for two measurable functions f_1 and f_2, if $f_1 = f_2$ a.e., then their integrals are equal. Is this still true for the Sugeno integral? Again, null-additivity will play a crucial role.

Theorem 7.25 *Whenever,* $f_1 = f_2$ *a.e.,* $(S)\int f_1\,dm = (S)\int f_2\,dm$, *if and only if* m *is null-additive.*

Proof. If m is null-additive, then for any $\alpha \geq 0$, from $m(\{f_1 \neq f_2\}) = 0$, we have

$$\begin{aligned} m(\{f_2 \geq \alpha\}) &\leq m(\{f_1 \geq \alpha\} \cup \{f_1 \neq f_2\}) = m(\{f_1 \geq \alpha\}) \\ &\leq m(\{f_2 \geq \alpha\} \cup \{f_1 \neq f_2\}) = m(\{f_2 \geq \alpha\})\,. \end{aligned}$$

It follows that $(S)\int f_1\,dm = (S)\int f_2\,dm$.

Conversely, for any $E \in \Sigma$ and $F \in \Sigma$ with $m(F) = 0$, take $f_1 = \infty$, $x \in E$, and $f_1(x) = 0$ for $x \notin E$, $f_2(x) = \infty$ for $x \in E \cup F$, and $f_2(x) = 0$ for $x \notin E \cup F$. Then $f_1 = f_2$ a.e. It follows, from $(S)\int f_1\,dm = (S)\int f_2\,dm$, that

$$m(E) = m(E \cup F)\,.$$

That is, m is null-additive. □

Corollary 7.9 *If m is null-additive, then for every $E \in \Sigma$, $F \in \Sigma$, with $m(F) = 0$, we have*

$$(S)\int_{E\cup F} f\,dm = (S)\int_E f\,dm\,.$$

Now we will give some convergence results.

Definition 7.11 $\{f_n\}$ is said to mean converge to an a.e. finite measurable function f, if

$$\lim_{n\to\infty}(S)\int |f_n - f|\,dm = 0\,.$$

We have the following obvious proposition.

Propositon 7.12 *Mean convergence is equivalent to convergence in measure.*

Theorem 7.26 (Monotone convergence theorem) *Let $\{f_n\}$ be increasing, and $A \in \Sigma$, then*

$$\lim_{n\to\infty}(S)\int_A f_n\,dm = (S)\int_A \lim_{n\to\infty} f_n\,dm\,.$$

Proof. By (1) in Proposition 7.11 we can take $A = X$ without loosing generality. If $f = \lim_{n\to\infty} f_n$, then f is a nonnegative extended real-valued measurable function. Write $c = (S)\int f\,dm$. If $c = 0$, then, by Proposition 7.11(3)

$$0 \le (S)\int f_n\,dm \le (S)\int f\,dm = 0 \qquad (n \in \mathbf{N})\,.$$

Let $0 < c < \infty$. For $\epsilon \in (0, c/2)$, there exists α_0, such that $c \ge \alpha_0 \wedge m(\{f \ge \alpha_0\}) > c - \epsilon$; of course, $\alpha_0 > \epsilon$. We write $E_n = \{f_n \ge \alpha_0 - \epsilon\}$, then $\{E_n\}$ is an increasing sequence of sets, and $E_n \nearrow E = \bigcup_{n=1}^{\infty} E_n$. It follows, by monotonicity and continuity of m, that $m(E_n) \nearrow m(E)$. Since $f_n \nearrow f$, we have $E \supset \{f \ge \alpha_0\}$ (for any $x \in \{f \ge \alpha_0\}$, there exists n, such that $f_n(x) \ge \alpha_0 - \epsilon$, i.e., $x \in E_n \subset E$), it follows that $m(E) \ge m(\{f \ge \alpha_0\})$. Therefore, there exists n_0, such that $m(E_n) \ge m(\{f \ge \alpha_0\}) - \epsilon$ as $n \ge n_0$. Thus, we have

$$\begin{aligned}(S)\int f_n\,dm &= \sup_{\alpha\in[0,\infty)} [\alpha \wedge m(\{f_n \ge \alpha\})] \ge (\alpha_0 - \epsilon) \wedge m(\{f_n \ge \alpha_0 - \epsilon\}) \\ &\ge (\alpha_0 - \epsilon) \wedge [m(\{f \ge \alpha_0\}) - \epsilon] = [\alpha_0 \wedge m(\{f \ge \alpha_0\})] - \epsilon \\ &> c - 2\epsilon\end{aligned}$$

for all $n \geq n_0$. Using Proposition 7.11(3), we have

$$(S) \int f_n \, dm \leq c.$$

Hence

$$\lim_{n \to \infty} (S) \int f_n \, dm = c.$$

For $c = \infty$ we have $m(\{f \geq \alpha\}) = \infty$ for every $\alpha \in [0, \infty)$. For arbitrarily given $N > 0$, write $F_N^n = \{f_n \geq N\}$, then $F_N^n \nearrow F_N^\infty = \bigcup_{n=1}^\infty F_N^n$, and therefore $m(F_N^n) \nearrow m(F_n^\infty)$. On the other hand, since $f_n \nearrow f$, we have $F_N^\infty \supset \{f \geq N+1\}$, and therefore $m(F_N^\infty) \geq m(\{f \geq N+1\}) = \infty$. Hence there exists n_0, such that $m(F_N^n) \geq N$ as $n \geq n_0$. Theorem 7.24 implies $(S) \int f_n \, dm \geq N$ for $n \geq n_0$. □

For a decreasing sequence, we have

Theorem 7.27 *If $\{f_n\}$ is decreasing and converges to f on $A \in \Sigma$, and if there exist n_0 and a constant $c' \leq (S) \int_A f \, dm$, such that $m(\{f_{n_0} > c'\} \cap A) < \infty$, then*

$$\lim_{n \to \infty} (S) \int_A f_n \, dm = (S) \int_A f \, dm.$$

Proof. We can take $A = X$ without loosing generality. Denoting $c = (S) \int f \, dm$, we have $\lim_{n \to \infty} (S) \int f_n \, dm \geq c$. For $c = \infty$ the conclusion of the theorem is true. If $c < \infty$, we assume that $\lim_{n \to \infty} (S) \int f_n \, dm > c$, then there exists $\delta > 0$, such that

$$\lim_{n \to \infty} (S) \int f_n \, dm \geq c + \delta.$$

Since $\{f_n\}$ is decreasing, we have $(S) \int f_n \, dm \geq c + \delta$, and therefore, by Theorem 7.24,

$$m(\{f_n \geq c + \delta\}) \geq c + \delta$$

for every n. From the hypothesis that $m(\{f_{n_0} > c'\}) < \infty$, we have

$$m(\{f_{n_0} \geq c + \delta\}) \leq m(\{f_{n_0} > c'\}) < \infty.$$

Therefore, since $f_n \searrow f$, $\{\{f_n \geq c+\delta\}\}$ is decreasing with respect to n, we have

$$\{f_n \geq c+\delta\} \searrow \bigcap_{n=1}^{\infty}\{f_n \geq c+\delta\} = \{f \geq c+\delta\}.$$

It follows, by the continuity of m, that

$$m(\{f \geq c+\delta\}) = \lim_{n\to\infty} m(\{f_n \geq c+\delta\}) \geq c+\delta.$$

Hence by Theorem 7.24, $(S)\int f\,dm \geq c+\delta$. This is a contradiction. □

Corollary 7.10 *If $f_n \searrow f$ and m is finite, then $(S)\int f_n\,dm \searrow (S)\int f\,dm$.*

In Theorem 7.27, the hypothesis of finiteness cannot be dropped. The following simple example will prove this.

Example 7.4 Let $X = (0,\infty), \Sigma$ be a Borel field on X and m be the Lebesgue measure. We take $f_n(x) = x/n$, $(n \in \mathbf{N})$. Then $f_n \searrow f \equiv 0$. Obviously, this does not satisfy the hypotheses of Theorem 7.27. Consequently, $(S)\int f_n\,dm = \infty$, $(n \in \mathbf{N})$ and $(S)\int f\,dm = 0$. Hence $\lim_{n\to\infty}(S)\int f_n\,dm \neq (S)\int f\,dm$.

Corollary 7.11 *Let m be null-additive, then*

(1) $f_n \nearrow f$ a.e. implies $(S)\int f_n\,dm \nearrow (S)\int f\,dm$;

(2) $f_n \searrow f$ a.e. and there exist n_0 and $c' \leq (S)\int f\,dm$, such that $m(\{f_{n_0} > c'\}) < \infty$, imply $(S)\int f_n\,dm \searrow (S)\int f\,dm$.

Now, we have a general convergence theorem.

Theorem 7.28 *If $\{f_n\}$ converges to f on $A \in \Sigma$, and there exist n_0 and a constant $c' \leq (S)\int_A f\,dm$, such that*

$$m\left(\left\{\sup_{n\geq n_0} f_n > c'\right\} \cap A\right) < \infty,$$

then $\lim_{n\to\infty}(S)\int_A f\,dm$ exists, and $\lim_{n\to\infty}(S)\int_A f_n\,dm = (S)\int_A f\,dm$.

Proof. We can suppose that $A = X$. Put $\bar{f}_n = \sup_{i\geq n} f_i$ and $f_n^* = \inf_{i\geq n} f_i$, then $\bar{f}_n$, f_n^* $(n \in \mathbf{N})$ are measurable, and $\bar{f}_n \searrow f$, $f_n^* \nearrow f$. We have by $f_n^* \leq f_n \leq \bar{f}_n$ that

$$(S)\int f_n^*\,dm \leq (S)\int f_n\,dm \leq (S)\int \bar{f}_n\,dm.$$

Hence

$$\lim_{n\to\infty}(S)\int f_n^* \, dm \leq \underline{\lim}_{n\to\infty}(S)\int f_n \, dm \leq \overline{\lim}_{n\to\infty}(S)\int f_n \, dm$$
$$\leq \lim_{n\to\infty}(S)\int \bar{f}_n \, dm \, .$$

By Theorems 7.26 and 7.27, we have

$$\lim_{n\to\infty}(S)\int \overline{f}_n \, dm = \lim_{n\to\infty}(S)\int \bar{f}_n \, dm = (S)\int f \, dm \, .$$

Hence $\underline{\lim}_{n\to\infty}(S)\int f_n \, dm = \overline{\lim}_{n\to\infty}(S)\int f \, dm$. □

As a consequence of Theorem 7.25 we obtain:

Corollary 7.12 *If $f_n \to f$ a.e., and there exist n_0 and a constant $c' \leq (S)\int f \, dm$ such that $m(\{\sup_{n\geq n_0} f_n > c'\}) < \infty$, then $(S)\int f_n \, dm \to (S)\int f \, dm$, if and only if m is null-additive.*

Corollary 7.13 *If m is finite and subadditive and $f_n \to f$ a.e., then*

$$(S)\int f_n \, dm \to (S)\int f \, dm \, .$$

We give the next theorem, in which autocontinuity, and hence again null-additivity plays a crucial role, without proof.

Theorem 7.29 *If $\{f_n\}$ converges in measure to an a.e. finite measurable function f on $A \in \Sigma$, then $(S)\int_A f_n \, dm \to (S)\int_A f \, dm$, if and only if m is autocontinuous.*

Together with Theorem 7.26 we obtain the following statement.

Corollary 7.14 *Whenever $\{f_n\}$ mean converges to an a.e. finite measurable function f on $A \in \Sigma$, then $(S)\int_A f_n \, dm \to (S)\int_A f \, dm$, if and only if m is autocontinuous.*

Now, we construct an example in which $f_n \to f$ in measure, but $(S)\int f_n \, dm \not\to (S)\int f \, dm$.

Example 7.5 Let $X = \mathbf{N}$, $\Sigma = \mathcal{P}(X)$, $m(E) = k \cdot \sum_{i \in E}(1/2^i)$ for $E \in \Sigma$, where k is the number of points in E. m is autocontinuous from below, but not from above. Take $f(x) = \chi_{(1)}(x)$, $f_n(x) = \chi_{(1,n)}(x)$, then, for arbitrary $\epsilon \in (0, 1)$,

$$m(\{|f - f_n| \geq \epsilon\}) = m(\{n\}) = (1/2^n) \to 0\,,$$

i.e., $f_n \to f$ in measure. But, $(S) \int f\, dm = \frac{1}{2}$ and $(S) \int f_n\, dm = 1$, $(n \in \mathbf{N})$. Hence $(S) \int f_n\, dm \not\to (S) \int f\, dm$.

7.8 Set Functions Defined by Sugeno Integrals

Let m be a positive σ-finite monotone continuous set function defined on Σ. For a given and fixed function f, the Sugeno integral determines a set function v on Σ:

$$v_f(E) = (S) \int_E f\, dm \qquad (E \in \Sigma)\,. \tag{7.6}$$

v_f is called the set function induced by f (and m). Throughout this section v is used instead of v_f if there is no confusion.

Propositon 7.13 *Let v be the set function induced by f. Then*

(i) $\alpha \wedge m(\{f \geq \alpha\} \cap E) \leq v(E) \leq \alpha \wedge m(\{f \geq \alpha\} \cap E)$ for any $\alpha \geq 0$ and $E \in \Sigma$,

(ii) $v(E) \leq m(E)$ for any $E \in \Sigma$, and in particular $v(\emptyset) = 0$,

(iii) $E, F \in \Sigma$ and $E \subset F$ implies $v(E) \leq v(F)$.

The inequality in (ii) above implies that the induced set function v is absolutely continuous with respect to the original fuzzy measure m, i.e., $v \ll m$. The following proposition gives an equivalent condition of the inverse relation $m \ll v$.

Propositon 7.14 *Let m satisfy the condition*

$$E, F \in \Sigma \quad \text{and} \quad m(E) = m(F) = 0 \quad \text{implies } m(E \cup F) = 0\,. \tag{7.7}$$

Then, $m \ll v$ if and only if $m(\{f = 0\}) = 0$.

Proof. The 'only if' part is trivial because $v(\{f = 0\}) = 0$. Now let $m(\{f = 0\}) = 0$ and $m(E) > 0$. Since

$$\lim_{n\to\infty} m(\{f \geq 1/n\} \cap E) = m(\{f > 0\} \cap E) > 0$$

by the condition (7.7), there exists $\alpha_0 > 0$ such that $m(\{f \geq \alpha_0\} \cap E) > 0$. Hence

$$v(E) \geq \alpha_0 \wedge m(\{f \geq \alpha_0\} \cap E) > 0 .$$

□

Remark 7.2 Without the condition (7.7), the 'if' part of the proposition is not true. Let $X = \{1,2\}$, $m(X) = 1$, and $m(\{1\}) = m(\{2\}) = 0$. Then a function f which takes values $f(1) = 0$ and $f(2) = 1$, for example, induces a set function v with $v(X) = 0$.

We are interested in the case where v also becomes a monotone continuous measure by choosing an appropriate function f when a monotone continuous measure m is given. The two properties: monotonicity and $v(\emptyset) = 0$, are always satisfied as stated in Proposition 7.13(ii) and (iii). Continuity from below also holds because of Theorem 7.26. Continuity from above does not hold for an induced set function v unless f satisfies some conditions.

To investigate those conditions, Theorem 7.27 is very helpful.

For a given m, we denote by $FM(m)$ the family of non-negative measurable functions f which satisfy $m(\{\alpha \leq f < \beta\}) < \infty$ for any α and β with $0 < \alpha < \beta$.

$FM(m)$ contains a wide class of functions.

Example 7.6 If $X = (-\infty, +\infty)$ and m is the Lebesgue measure on X, then the function $f \vee 0$ is contained in $FM(m)$ whenever f is m-integrable in the ordinary sense. If m is finite, every measurable non-negative function belongs to $FM(m)$.

We give the following theorem without the proof.

Theorem 7.30 *Let (X, Σ, m) be a σ-finite monotone continuous measure space satisfying the condition*

$$m(E) < \infty \quad \textit{and} \quad m(F) < \infty \Rightarrow m(E \cup F) < \infty . \tag{7.8}$$

Then the following three statement are equivalent:

(i) $f \in FM(m)$,

(ii) if there exists $E \in \Sigma$ *such that*

$$m(\{f \geq \alpha\} \cap E) < \infty \tag{7.9}$$

for some $\alpha > 0$, *then* (7.9) *is true for any* $\alpha > 0$,

(iii) the set function v *induced by* f *is a monotone continuous measure.*

7.9 Comparison of Choquet and Sugeno Integrals

We restrict in this section to monotone set functions with values in $[0, 1]$.

Theorem 7.31 *If* $f\colon X \to [0,1]$ *is a measurable function and* $m\colon \Sigma \to [0,1]$ *a monotone continuous from above and below set function such that* $m(X) = 1$, *then*

$$\left|(S)\int_X f\,dm - (C)\int_X f\,dm\right| < \frac{1}{4}.$$

Proof. We denote by $S = (S)\int_X f\,dm$. Obviously $0 \leq S \leq 1$. Since $m(\{x : f(x) \geq r\}) \geq S$ for all r such that $0 \leq r \leq S$ and $m(\{x : f(x) \geq r\}) \leq S$ for all r such that $S < r \leq 1$, we have

$$\begin{aligned}
(C)\int_X f\,dm - (S)\int_X f\,dm &= \int_0^1 m(\{x : f(x) \geq r\})\,dm - S \\
&= \int_0^S m(\{x : f(x) \geq r\})\,dr + \int_S^1 m(\{x : f(x) \geq r\})\,dr - S \\
&\leq \int_0^S dr + \int_S^1 S\,dr - S = S(1-S) \leq \frac{1}{4}.
\end{aligned}$$

Analogously

$$\begin{aligned}
&(S)\int_X f\,dm - (C)\int_X f\,dm \\
&\quad = S - \int_0^S m(\{x : f(x) \geq r\})\,dr - \int_S^1 m(\{x : f(x) \geq r\})\,dr \\
&\quad \leq S - \int_0^S S\,dr = S(1-S) \leq \frac{1}{4}.
\end{aligned}$$

□

In a special case we have

Theorem 7.32 *If m is a 0–1 monotone set function, i.e., whose range is $\{0,1\}$ then for every measurable function $f: X \to [0,1]$*

$$(C)\int_X f\,dm = (S)\int_X f\,dm\,.$$

Proof. First we will prove that

$$(C)\int_X f\,dm = \sup_{A:m(A)=1} \inf_{x\in A} f(x)\,.$$

We denote by c the right side of the last equality. It is enough to prove that

$$m(\{x : f(x) > r\}) = \begin{cases} 1 & \text{if } r < c \\ 0 & \text{if } r > c. \end{cases}$$

Case $r < c$. Then there exists a set $A \in \Sigma$ such that $m(A) = 1$ and $r < \inf_{x\in A} f(x)$. Hence $A \subset \{x : f(x) > r\}$, which implies $m(\{x : f(x) > r\}) = 1$.

Case $r > c$. Suppose that $m(\{x : f(x) > r\}) = 1$. This implies $r > c \geq \inf_{x\in\{x:f(x)>r\}} f(x) \geq r$, which is impossible.

Then the opposite statement is true, i.e.,

$$m(\{x : f(x) > r\}) = 0\,.$$

Now we will prove that also $(S)\int_X f\,dm = c$. This fact follows by Proposition 7.10 and since m is 0–1 valued set function

$$(S)\int_X f\,dm = \sup_{A\in\Sigma}\{[\inf_{x\in A} f(x)] \wedge m(A)\} = c\,. \qquad \square$$

7.10 Notes and References

The notion of 's-null-sets' was introduced in [Murofushi and Sugeno 1991a]. Egoroff's type theorem was proved in [Z. Wang 1984] (see also [Z. Wang and Klir 1992]). The relation between Riesz property and (S) property was discussed in [Sun 1994]. The Choquet integral was introduced in [Choquet

1953–1954], and it is elaborated in many papers and books, for example [Murofushi and Sugeno 1991a], [Murofushi and Sugeno 1989], [Murofushi and Sugeno 1993], [Murofushi and Sugeno 1994], [Z. Wang and Klir 1992], [Denneberg 1994].

For many applications see the references at the end of this book. The symmetric Choquet integral is presented by [Šipoš 1979a] and [Šipoš 1979b] (see also [Denneberg 1994]). The asymmetric integral was investigated in [Schmeidler 1989], [Denneberg 1994], [Murofushi and Sugeno 1994]. We have used the approach in [Murofushi and Sugeno 1994]. The Sugeno (fuzzy) integral was introduced in [Sugeno 1974] and then elaborated in many papers and books, for example in [Ralexu and Adams 1980], [Z. Wang 1984], [Z. Wang 1985], [Z. Wang and Klir 1992], [Suzuki 1988]. For many applications see the references at the end of the book. Set functions induced by the Sugeno integral are investigated in [Suzuki 1988]. The presented relation between the Choquet and Sugeno integrals was obtained in [Murofushi and Sugeno 1993].

Choquet and Sugeno integrals are very useful in the multicriteria decision theory in modeling the utility function. Namely, on a set X of consequences (implied by the choice of alternatives (criterions)) is considered the preference relation, i.e., a binary relation $\geq$ on X which is transitive and comparable (always is $x \geq y$ or $y \geq x$). The main goal is the existence of a utility function $u \colon X \to \mathbf{R}$ with the property

$$x > y \iff u(x) > u(y) .$$

This utility function can be represented by Choquet or Sugano integral with respect to a non-additive set function which models the dependence between criterions. The main difficulty in practical applications is the indentification of the non-additive set function. Today, this procedure is managed by heuristic methods or by variuos optimization algorithms.

Chapter 8

Integrals Based on Decomposable Measures

8.1 Pseudo-Integral

Let $[a, b]$ be a closed (in some cases we will take semiclosed) subinterval of $[-\infty, +\infty]$. We shall consider a partial order $\leq$ on $[a, b]$, which can be the usual order of the real line, but it can also be another order. All future considerations will be with respect to the order $\leq$.

The operation $\oplus$ (pseudo-addition) is a function $\oplus\colon [a, b] \times [a, b] \to [a, b]$ which is commutative, nondecreasing (with respect to $\leq$) associative and either a or b is a zero element, denoted by $\mathbf{0}$, i.e. for each $x \in [a, b]$

$$\mathbf{0} \oplus x = x \quad \text{holds.}$$

Let $[a, b]_+ = \{x : x \in [a, b],\ x \geq \mathbf{0}\}$. The operation $\otimes$ (pseudo-multiplication) is a function $\otimes\colon [a, b] \times [a, b] \to [a, b]$ which is commutative, positively nondecreasing, i.e. $x \leq y$ implies $x \otimes z \leq y \otimes z$, $z \in [a, b]_+$, associative and for which there exist a unit element $\mathbf{1} \in [a, b]$, i.e., for each $x \in [a, b]$

$$\mathbf{1} \otimes x = x\,.$$

We suppose, further, $\mathbf{0} \otimes x = \mathbf{0}$ and that $\otimes$ is a distributive pseudo-multiplication with respect to $\oplus$, i.e.,

$$x \otimes (y \oplus z) = (x \otimes y) \oplus (x \otimes z)\,.$$

Then usually the structure $([a, b], \oplus, \otimes)$ will be called a semiring.

Some examples of these are:

$$\begin{aligned} x \oplus y &= (x^p + y^p)^{\frac{1}{p}},\ p > 0 \quad \text{and} \quad x \otimes y = x \cdot y \text{ on } [0, +\infty] \\ \text{or } x \oplus y &= \max\{x, y\} \qquad\qquad \text{and} \quad x \otimes y = x + y \text{ on } [-\infty, +\infty). \end{aligned}$$

Pseudo-addition $\oplus$ is idempotent if for any $x \in [a, b]$

$$x \oplus x = x \quad \text{holds}.$$

Let X be a non-empty set. Let Σ be a σ-algebra of subsets of X. A set function $m\colon \Sigma \to [a, b]$ (or semiclosed interval) is a $\bigoplus$-decomposable measure if there hold $m(\emptyset) = \mathbf{0}$ (if $\oplus$ is not idempotent), $m(A \cup B) = m(A) \oplus m(B)$ for $A, B \in \Sigma$ such that $A \cap B = \emptyset$.

In the case when $\oplus$ is idempotent, it is possible that m is not defined on an empty set.

A $\bigoplus$-decomposable measure m is σ-$\bigoplus$-decomposable if

$$m\Big(\bigcup_{i=1}^{\infty} A_i\Big) = \bigoplus_{i=1}^{\infty} m(A_i)$$

holds for any sequence $\{A_i\}$ of pairwise disjoint sets from Σ.

Propositon 8.1 *If $m\colon \Sigma \to [a, b]$ is a $\bigoplus$-decomposable measure with respect to the idempotent pseudo-addition $\oplus$, then we have*

$$m(A \cup B) = m(A) \oplus m(B)$$

for any $A, B \in \Sigma$.

Let m be a σ-$\bigoplus$-decomposable measure. A function $f\colon X \to [a, b]$ is measurable from below if for any $c \in [a, b]$ the sets $\{x : f(x) \leq c\}$ and $\{x : f(x) < c\}$ belong to Σ. f is measurable, if it is measurable from below and the sets $\{x : f(x) \geq c\}$ and $\{x : f(x) > c\}$ belong to Σ.

Let f and g be two functions defined on X and with values in $[a, b]$. Then, we define for any $x \in X$

$$\begin{aligned} (f \oplus g)(x) &= f(x) \oplus g(x), \\ (f \otimes g)(x) &= f(x) \otimes g(x) \end{aligned}$$

and for any $c \in [a,b]$

$$(c \otimes f)(x) = c \otimes f(x).$$

We suppose further that $([a,b],\oplus)$ and $([a,b],\otimes)$ are complete lattice ordered semigroups. A complete lattice means that for each set $A \subset [a,b]$ bounded from above (below) there exists $\sup A$ ($\inf A$). Further, we suppose that $[a,b]$ is endowed with a metric d compatible with sup and inf, i.e., $\sup x_n = x$ and $\inf x_n = x$ imply $\lim_{n\to\infty} d(x_n,x) = 0$, and which satisfies at least one of the following conditions:

(a) $d(x \oplus y,\ x' \oplus y') \leq d(x,x') + d(y,y')$

(b) $d(x \oplus y,\ x' \oplus y') \leq \max\{d(x,x'),\ d(y,y')\}$.

Both conditions (a) and (b) imply that:

$$d(x_n,y_n) \to 0 \quad \text{implies} \quad d(x_n \oplus z,\ y_n \oplus z) \to 0.$$

Condition (b) implies

$$d\Big(\bigoplus_{i=1}^{n} x_i,\ \bigoplus_{j=1}^{n} y_j\Big) \leq \min_j \max_i d(x_i,y_j).$$

We suppose further the monotonicity of the metric d, i.e.,

$$x \leq z \leq y \quad \text{implies} \quad d(x,y) \geq \max\{d(y,z), d(x,z)\}.$$

Example 8.1 The following metrics on the corresponding semirings satisfy the preceding conditions

(a) $([-\infty,0),\max,+)$ with $d(x,y) = e^{\max(x,y)} - e^{\min(x,y)}$;

(b) $([-\infty,+\infty],\max,\min)$ with
$d(x,y) = \arctan(\max(x,y)) - \arctan(\min(x,y))$;

(c) $((-\infty,+\infty],\min,+)$ with $d(x,y) = e^{-\max(x,y)} - e^{-\min(x,y)}$.

We note that in the case (b) all sets are bounded. The preceding examples can be extended for n-dimensional case, taking the lexicographic order of n-tuples.

Let ε be a positive real number, and $B \subset [a,b]$. A subset $\{l_i^\varepsilon\}$ is a ε-net if for each $x \in B$ there exists l_i^ε such that $d(l_i^\varepsilon,x) \leq \varepsilon$. If we have $l_i^\varepsilon \leq x$, then we shall call $\{l_i^\varepsilon\}$ a lower ε-net. If $l_i^\varepsilon \leq l_{i+1}^\varepsilon$ holds, then $\{l_i^\varepsilon\}$ is monotone.

We define the characteristic function

$$\chi_A(x) = \begin{cases} 0 & x \notin A, \\ 1 & x \in A. \end{cases}$$

A mapping $e \colon X \to [a,b]$ is an elementary (measurable) function if it has the following representation

$$e = \bigoplus_{i=1}^{\infty} a_i \otimes \chi_{A_i} \quad \text{for} \quad a_i \in [a,b]$$

and $A_i \in \Sigma$ disjoint if $\oplus$ is not idempotent.

Theorem 8.1 *Let $f \colon X \to [a,b]$ be a measurable from below function. If the pseudo-addition is idempotent, or f is measurable and for each positive real number ε there exists a monotone ε-net in $f(X)$. Then there exists a sequence $\{\varphi_n\}$ of elementary functions such that, for each $x \in X$,*

$$d(\varphi_n(x), f(x)) \to 0 \quad \textit{uniformly as} \quad n \to \infty .$$

Proof. Suppose first that $\oplus$ is not idempotent. We take a lower monotone ε-net $\{f_i^\varepsilon\}$ on $f(X)$.

Let

$$e^\varepsilon = \bigoplus_{i=1}^{\infty} f_i^\varepsilon \otimes \chi_{X_i^\varepsilon},$$

where $X_i^\varepsilon = \{x : f_{i+1}^\varepsilon > f(x) \geq f_i^\varepsilon\}$.

For each point x from X there exists f_i^ε such that

$$d(f_i^\varepsilon(x), f(x)) \leq \varepsilon ,$$

where $f_i^2(x) = f_i^\varepsilon \otimes \chi_{X_i^\varepsilon}(x)$. Hence, by the monotonicity of d and $e^\varepsilon(x) \geq f_i^\varepsilon(x)$, we obtain

$$d(e^\varepsilon(x), f(x)) < \varepsilon \qquad (x \in X).$$

Taking $\varepsilon = \frac{1}{n}$ we define the desired sequence (φ_n) as $\varphi_n = e^{\frac{1}{n}}$.

If the pseudo-addition $\oplus$ is idempotent, then we take a lower ε-net $\{f_i^\varepsilon\}$ in this procedure and

$$X_i^\varepsilon = \{x : f(x) \geq f_i^\varepsilon\}.$$

We have used that $e^{\varepsilon}(x) \le f(x)$ holds, which follows by

$$e^{\varepsilon}(x) \le e^{\varepsilon}(x) \oplus f(x) \le \bigoplus_{i=1}^{\infty}((f_i^{\varepsilon} \otimes \chi_{X_i^{\varepsilon}}(x)) \oplus f(x)) \le f(x). \qquad \square$$

The pseudo-integral of a simple function $s = \bigoplus_{i=1}^{n} a_i \otimes \chi_{A_i}$ for $a_i \in [a,b]$ with disjoint $A_1, A_2, \ldots, A_n$, if $\oplus$ is not idempotent, is defined by

$$\int_X^{\oplus} s \otimes dm = \bigoplus_{i=1}^{n} a_i \otimes m(A_i).$$

The pseudo-integral of an elementary function

$$e = \bigoplus_{i=1}^{\infty} a_i \otimes \chi_{A_i} \quad \text{for} \quad a_i \in [a,b] \qquad (i \in \mathbf{N})$$

with $\{A_i\}$ disjoint if $\oplus$ is not idempotent, is defined by

$$\int_X^{\oplus} e \otimes dm = \bigoplus_{i=1}^{\infty} a_i \otimes m(A_i). \tag{8.1}$$

The pseudo-integral of a bounded measurable (from below for $\oplus$ idempotent) function $f\colon X \to [a,b]$, for which, if $\oplus$ is not idempotent for each $\varepsilon > 0$, there exists a monotone ε-net in $f(X)$, is defined by

$$\int_X^{\oplus} f \otimes dm = \lim_{n\to\infty} \int_X^{\oplus} \varphi_n(x) \otimes dm, \tag{8.2}$$

where $\{\varphi_n\}$ is the sequence of elementary functions constructed in Theorem 8.1.

Theorem 8.2 *The integral defined in* (8.2) *is independent of the choice of sequence* $\{\varphi_n\}$.

Theorem 8.3 *Let $\oplus$ and $\otimes$ be continuous and $\oplus$ infinitely commutative and associative. Then the integrals defined by* (8.1) *and* (8.2) *have the following properties:*

(i) $\int_X^{\oplus}(f \oplus g) \otimes dm = \int_X^{\oplus} f \otimes dm \oplus \int_X^{\oplus} g \otimes dm$,

(ii) $\int_X^{\oplus}(c \otimes f) \otimes dm = c \otimes \int_X^{\oplus} f \otimes dm$ *for any* $c \in [a,b]$.

Proof. (i) Let f and g be elementary functions, i.e.,

$$f = \bigoplus_{i=1}^{\infty}(a_i \otimes \chi_{A_i}), \quad g = \bigoplus_{i=1}^{\infty} b_i \otimes \chi_{B_i},$$

where $\{A_i\}$ and $\{B_i\}$ are (if $\oplus$ is not idempotent, disjoint) partitions of X in Σ. Hence, $f \oplus g$ is also an elementary function and

$$f \oplus g = \bigoplus_{i=1}^{\infty}\bigoplus_{j=1}^{\infty}(a_i \oplus b_j) \otimes \chi_{A_i \cap B_j} \,.$$

By (8.1) we obtain

$$\begin{aligned}
\int_X^{\oplus} (f \oplus g) \otimes dm &= \bigoplus_{i=1}^{\infty}\bigoplus_{j=1}^{\infty}(a_i \oplus b_j) \otimes m(A_i \cap B_j) \\
&= \bigoplus_{i=1}^{\infty}\bigoplus_{j=1}^{\infty}(a_i \otimes m(A_i \cap B_j)) \oplus \bigoplus_{i=1}^{\infty}\bigoplus_{j=1}^{\infty} b_j \otimes m(A_i \cap B_j) \\
&= \bigoplus_{i=1}^{\infty}\Big(a_i \otimes m\Big(A_i \cap \bigcup_{j=1}^{\infty} B_j\Big)\Big) \oplus \bigoplus_{j=1}^{\infty}(b_j \otimes m\Big(\bigcup_{i=1}^{\infty} A_i \cap B_j\Big) \\
&= \bigoplus_{i=1}^{\infty}(a_i \otimes m(A_i)) \oplus \bigoplus_{j=1}^{\infty}(b_j \otimes m(B_j)) \\
&= \int_X^{\oplus} f \otimes dm \oplus \int_X^{\oplus} g \otimes dm \,.
\end{aligned}$$

Now, let f and g be measurable (from below if $\oplus$ is idempotent). Let $\{\varphi_n\}$ and $\{\psi_n\}$ be the corresponding sequences from Theorem 8.1 to f and g, respectively. The integral

$$\int_X^{\oplus} (f \oplus g) \otimes dm$$

exists, since it can be defined by the sequence $\{\varphi_n(x) \oplus \psi_n(x)\}$

$$\int_X^{\oplus} (f \oplus g) \otimes dm = \lim_{n \to \infty} \int_X^{\oplus} (\varphi_n(x) \oplus \psi_n(x)) \otimes dm$$

and $\{\varphi_n(x)\}$ and $\{\psi_n(x)\}$ satisfy, for any x,

$$d(\varphi_n(x), f(x)) \to 0 \qquad \text{and} \qquad d(\psi_n(x), g(x)) \to 0\,.$$

Hence, since d satisfies (a) or (b)

$$d(\varphi_n(x) \oplus \psi_n(x),\ f(x) \oplus g(x)) \to 0\,.$$

Now, we have

$$\begin{aligned}
\int_X^{\oplus} (f \oplus g) \otimes dm &= \lim_{n\to\infty} \int_X^{\oplus} (\varphi_n \oplus \psi_n) \otimes dm \\
&= \lim_{n\to\infty} \Big(\int_X^{\oplus} \varphi_n \otimes dm \oplus \int_X^{\oplus} \psi_n \otimes dm \Big) \\
&= \lim_{n\to\infty} \int_X^{\oplus} \varphi_n \otimes dm \oplus \lim_{n\to\infty} \int_X^{\oplus} \psi_n \otimes dm \\
&= \int_X^{\oplus} f \otimes dm \oplus \int_X^{\oplus} g \otimes dm\,.
\end{aligned}$$

Property (ii) easily follows by the continuity of $\otimes$. □

Example 8.2 For any function h we can define

$$m(A) = \sup_{x \in A} h(x) \qquad (A \in \mathcal{B})\,,$$

where $\mathcal{B}$ is the Borel σ-algebra on $[-\infty, \infty)$.

Taking $\oplus = \max = \sup$, $\otimes = +$, we obtain

$$\int_{\mathbf{R}}^{\oplus} f \otimes dm = \sup_{x \in \mathbf{R}} (f(x) + h(x))\,,$$

f bounded above.

Example 8.3 If $\oplus$ is a strict pseudo-addition with a monotone generator g, $g \circ m \colon \Sigma \to [0, g(c)]$ and $c \in [a, b]$ is a measure and we have for the simple function

$$\int_X^{\oplus} s \otimes dm = g^{-1}\Big(\sum_{i=1}^{n} g(a_i) \cdot (g \circ m)(A_i) \Big)$$

and for the measurable function f

$$\int_X^{\oplus} f \otimes dm = g^{-1}\Big(\int_X (g \circ f) \cdot dx \Big),$$

where $dx = d(g \circ m)$ is the Lebesgue measure and $u \otimes v = g^{-1}(g(u) \cdot g(v))$.

Example 8.4 If $c > 0$, then we define $u \oplus v = -c \ln(e^{-\frac{u}{c}} + e^{-\frac{v}{c}})$ and $u \otimes v = u + v$. The corresponding integral is

$$\int_{\mathbf{R}}^{\oplus} f \otimes dm = -c \ln \int_{\mathbf{R}} e^{(-\frac{f(x)}{c})} dx .$$

8.2 Some Other Types of Integrals

In this section we list some important integrals which are also constructed by decomposable measures. They are all based (also the pseudo-integral from the previous section) on the Lebesgue integral or on the similar procedure which gives the Lebesgue integral.

8.2.1 Weber Integral

The Weber integral is based on Archimedean t-conorm decomposable measures.

Definition 8.1 Let m be a σ-S-decomposable measure on Σ, S being an Archimedean t-conorm on $[0, M]$ with some generator g, $B \in \Sigma$ and $m_B(A) = m(B \cap A)$ the restriction of m to $B \cap \Sigma = \{B \cap A : A \in \Sigma\}$, and $\mathbf{F}$ the collection of all measurable functions $f \colon X \to [0, 1]$.

Case (A). If m_B is of type (A) then the integral of $f \in \mathbf{F}$ over B with respect to the S-decomposable measure m is defined by

$$(W) \int_B f \; S \; m = g^{-1}\Big(\int_B f \, d(g \circ m)\Big).$$

Case (P). If m_B is of type (P) and if B is m_B-achievable, i.e., there exists a disjoint sequence $\{B_k\}$ in $B \cap \Sigma$ with $m(B_k) < M$ and $B = \bigcup_{k \in \mathbf{N}} B_k$, then the integral of $f \in \mathbf{F}$ over B with respect to the σ-S-decomposable measure m is defined by

$$(W) \int_B f \; S \; m = S_{k \in \mathbf{N}}\Big((W) \int_{B_k} f \; S \; m\Big).$$

For case (A) the integral $\int_B f\, d(g\circ m)$ on the right-hand side is a classical Lebesgue integral, and the definition does not depend on the special choice of the generator g. In case (P), the additional condition $m(B_k) < M$ ensures that m_{B_k} is of type (A) whence the integral $(W)\int_{B_k} f\ S\ m$ exists according to case (A). Moreover, the definition is independent of the choice of the B_ks. Using an arbitrary generator g of S, the definition can be rewritten as

$$(W)\int_B f\ S\ m = g^{(-1)}\Big(\sum_{k=1}^{\infty}\int_{B_k} f\, d(g\circ m)\Big).$$

This means that any integral in case (A) can be written as in case (P) with an arbitrary measurable partition $\{B_k\}_{k\in\mathbf{N}}$ of B. In the classical situation, when m itself is additive, i.e., $+$-decomposable, the new integral reduces to the Lebesgue integral.

Example 8.5 Let $X = \{x_1, \ldots, x_N\}$ be a discrete (finite of countable infinite) set and $m_{\lambda,\alpha}$ the σ-$S^{\mathbf{S}}_{\lambda}$-decomposable measure ($\lambda > -1$, $\lambda \neq 0$, $\alpha > 0$). For each $f \in \mathbf{F}$ we get

$$(W)\int_X f S^{\mathbf{S}}_{\lambda} m_{\lambda,\alpha} = \begin{cases} \frac{1}{\lambda}\Big[\exp\big(\frac{\lambda}{\alpha}\cdot\sum_{i=1}^{N} f(x_i)\big) - 1\Big] & \text{if } \sum_{i=1}^{N} f(x_i) \leq \frac{\alpha}{\lambda} ln(1+\lambda), \\ 1 & \text{if } \sum_{i=1}^{N} f(x_i) > \frac{\alpha}{\lambda} ln(1+\lambda). \end{cases}$$

The integral naturally induces a measure in two different ways: For any fixed $B \in \Sigma$ the mapping

$$\mathbf{I}_B : \mathbf{F} \to [0, M], \quad f \mapsto (W)\int_B f\ S\ m,$$

is a $[0, M]$-valued $\square$-measure on $\mathbf{F}$ with $\square = S$ (see Section 6.6). For any fixed $f \in \mathbf{F}$ the mapping

$$\mathbf{N}_f : \Sigma \to [0, M], \quad B \mapsto (W)\int_B f\ S\ m,$$

is a S-decomposable measure on Σ with values in $[0, M]$.

The Lebesgue integral preserves the linear structure, interpreted as mapping from the vector space of all real valued measurable functions on X into the vector space of real numbers, i.e., in both the domain and the range of the integral the linear structures are given by or derived from sum and product on $\mathbf{R}$.

The present situation is quite different in two respects. First, the domain $\mathbf{F}$ of the integrals of Definition 8.1 has a structure which is (pointwise) derived from $([0,1], S_{\mathbf{L}}, \cdot)$ where $S_{\mathbf{L}}$ is the bounded sum on $[0,1]$. This space is called a semi-vectorspace, where 'semi' indicates that there is no inverse with respect to $S_{\mathbf{L}}$, and that the two axioms of distributivity are valid only in the subset where $S_{\mathbf{L}}$ coincides with the sum $+$.

The range of the integrals is contained in $[0, M]$ where a binary operation S is defined. With the product $*$ defined by $a * x = g^{-1}(a \cdot g(x))$ for all 'vectors' $x \in [0, M]$ and 'scalars' $a \in [0,1]$ we obtain a semi-vectorspace $([0, M], S, *)$, where in analogy to the particular case $([0,1], S_{\mathbf{L}}, \cdot)$ 'semi' indicates that there is no inverse with respect to S and the two axioms of distributivity are valid only in the following sense:

$$\begin{array}{ll} a * (x\ S\ y) = (a * x)\ S\ (a * y) & \text{for all } (x, y) \text{ with } g(x\ S\ y) = g(x) + g(y), \\ (a + b) * x = (a * x)\ S\ (b * x) & \text{for all } (a, b) \text{ with } a + b \leq 1\,. \end{array}$$

These two restrictions give the following results.

Theorem 8.4 *For any fixed $B \in \Sigma$ with $m(B) > 0$ the mapping $\mathbf{I}_B$ satisfies $\mathbf{I}_B(f + h) = \mathbf{I}_B(f)\ S\ \mathbf{I}_B(h)$ for all $f, h \in \mathbf{F}$ with $f + h \leq 1$ (strong S-decomposability), $\mathbf{I}_B(a \cdot f) = a * \mathbf{I}_B(f)$ for all $a \in [0,1]$, $f \in \mathbf{F}$ ($*$-homogeneity); for measures of type* (P) *we need the additional assumption $\mathbf{I}_B(f) < M$.*

8.2.2 Murofushi-Sugeno Integral

Let $\oplus$ a be pseudo-addition on $[0, +\infty]$, i.e., a binary operation on $[0, \infty]$ such that (see Subsection 2.2.2)

(P1) $x \oplus 0 = 0 \oplus x = x$ (neutral element);

(P2) $(x \oplus y) \oplus z = x \oplus (y \oplus z)$ (associativity);

(P3) $x \leq x', y \leq y' \Rightarrow x \oplus y \leq x' \oplus y'$ (monotonicity in both components);

(P4) $x_n \rightarrow x,\ y_n \rightarrow y \Rightarrow x_n \oplus y_n \rightarrow x \oplus y$ (continuity).

Remark 8.1 If $(0, M)$ is contained in the representation of $\oplus$ and the range m is contained in $[0, M]$, then the definition of a $\oplus$-measure (see Subsection 2.2.2) coincides with the Weber definition of $\oplus$-decomposable measures.

It is natural to assume that an integral of a function

$$f(x) = \begin{cases} a & x \in A, \\ 0 & x \notin A \end{cases}$$

with respect to a $\oplus$-measure m depends only on a and $m(A)$, i.e., $\int f\, dm = J(a, m(A))$. The binary operation J is called a pseudo-multiplication (corresponding to $\oplus$), $J(a, x) = a \odot x$. Further, it is natural to require that the indefinite integral with respect to a $\oplus$-measure is also a $\oplus$-measure. Hence we need the following conditions for $\odot$:

(M1) $a \odot (x \oplus y) = (a \odot x) \oplus (a \odot y)$,

(M2) $a \leq b \Rightarrow a \odot x \leq b \odot x$,

(M3) $a \odot x = 0 \Longleftrightarrow a = 0$ or $x = 0$,

(M4) $\exists\, e \in (0, +\infty], \forall x \in [0, \infty] : e \odot x = x$,

(M5) $0 < a < +\infty$, $a_n \to a$ and $x_n \to x \Rightarrow a_n \odot x_n \to a \odot x$, and $(+\infty) \odot x = \lim_{a \to +\infty} a \odot x$.

We have that $\odot$ is a pseudo-multiplication corresponding to $\oplus \approx \{\langle(\alpha_k, \beta_k), g_k\rangle : k \in K\}$ iff there exists a family of nondecreasing continuous functions $\{h_k : k \in K\}$ such that

$$\begin{aligned} a \odot x &= g_k^*(h_k(a) g_k(x)) \quad \text{for } a > 0,\ x \in [\alpha_k, \beta_k], \\ h_k(e) &= 1,\ 0 < h_k(a) < +\infty \quad \text{for } 0 < a < +\infty, \end{aligned}$$

and if $(\alpha_k, \beta_k]$ is nilpotent (i.e., $g_k(\beta_k) < \infty$; $g_k^* \neq g_k^{-1}$), then $h_k(a) = 1$ for $a \leq e$.

On the set of all idempotent elements, the structure of $\odot$ cannot characterized.

Let

$$f(x) = \begin{cases} a_j & \text{if } x \in A_j,\ j = 1, \ldots, n, \\ 0 & \text{otherwise} \end{cases}$$

be a simple function, where $A_j \in \Sigma$, $0 \leq a_j < +\infty$ for $j = 1, \ldots, n$ and $A_i \cap A_j = \emptyset$ for $i \neq j$. The integral of f over $B \in \Sigma$ is defined by

$$(MS) \int_B^{\oplus} f\,dm = \bigoplus_{j=1}^{n} a_j \odot m(A_j \cap B).$$

Let f be a non-negative measurable function on X and let $\{f_n\}$ be a sequence of non-negative measurable simple functions such that $f_n(x) \uparrow f(x)$ for every $x \in B$. We define the integral of f over B to be

$$(MS) \int_B^{\oplus} f\,dm = \lim_{n \to \infty} (MS) \int_B^{\oplus} f_n\,dm.$$

It is obvious that the integral is well defined. Note that the definition depends on the choice of multiplication.

This integral has similar properties as Lebesgue's integral. We define the characteristic function of $A \in \Sigma$ by χ_A:

$$\chi_A(x) = \begin{cases} e & \text{if } x \in A \\ 0 & \text{if } x \notin A. \end{cases}$$

Then we have

(i) $0 \leq f \leq g \ m - a \cdot e \Rightarrow \forall B \in \Sigma : (MS) \int_B^{\oplus} f\,dm \leq (MS) \int_B^{\oplus} g\,dm$;

(ii) $A \cap B = \emptyset$, $A, B \in \Sigma \Rightarrow (MS) \int_{A \cup B}^{\oplus} f\,dm = (MS) \int_A^{\oplus} f\,dm \oplus (MS) \int_B^{\oplus} f\,dm$;

(iii) $(MS) \int_X^{\oplus} \chi_A\,dm = m(A)$ for any $A \in \Sigma$;

(iv) $(MS) \int_X^{\oplus} f\,dm = 0 \Rightarrow f = 0 \ m - a \cdot e$;

(v) $(MS) \int_A^{\oplus} f\,dm = (MS) \int_X^{\oplus} \chi_A \odot f\,dm$;

(vi) If $v(A) = (MS) \int_A^{\oplus} f\,dm$ for every $A \in \Sigma$, then v is a $\oplus$-measure on Σ.

8.3 g-Calculus

We will consider in this section a special case of the pseudo-integral, introduced in Section 8.1, although all the considerations in this section can be taken independent of Section 8.1.

We shall work in this section with the extended real number system, i.e., the ordinary finite real numbers together with $-\infty$ and ∞. Let $[a,b]$ be a closed (in some cases semiclosed, see Remark 8.2) real interval. The operation $\oplus$ (from Section 8.1) is a function $\oplus\colon [a,b]\times[a,b] \to [a,b]$ which is commutative, nondecreasing, associative and has a zero element, denoted by $\mathbf{0}$ which is either a or b. In this section we will consider only strict pseudo-addition, i.e., such that the function $\oplus$ is continuous and strictly increasing in $(a,b)\times(a,b)$.

Remark 8.2 The requirement that the interval $[a,b]$ be closed does not restrict us in any way. If we begin with $(a,b]$ or (a,b) we can replace the condition $\mathbf{0}\oplus x = x$ by the condition: for all y in $(a,b]$ or in (a,b)

$$\lim_{x\to a} \oplus(x,y) = y, \quad \lim_{x\to a} \oplus(x,x) = a\,.$$

Adjoining the endpoint a to $(a,b]$, or both endpoints a and b to (a,b), we reduce, by continuity of $\oplus$, to the previous case.

By Aczel's representation theorem for each strict pseudo-addition $\oplus$ there exists a monotone function g (generator for $\oplus$), $g\colon [a,b] \to [0,\infty]$ such that either $g(a)=0$ or $g(b)=0$ and

$$u\oplus v = g^{-1}(g(u)+g(v))\,.$$

Hence g is an isomorphism of semigroup $([a,b],\oplus)$ with the semigroup $([0,\infty],+)$.

The operation $\otimes$ (pseudo-multiplication) is a function $\otimes\colon [a,b]\times[a,b]\to [a,b]$ which is commutative, nondecreasing, associative and has a unit element $\mathbf{1}$. We suppose further that the pseudo-multiplication $\otimes$ is defined as

$$u\otimes v = g^{-1}(g(u)g(v))\,,$$

where g is a generator of strict pseudo-addition $\oplus$. $\otimes$ is distributive with respect to $\oplus$.

Let g be an additive generator of the strict pseudo-addition $\oplus$ on $[a,b]$ such that g is continuously differentiable on (a,b). The corresponding pseudo-multiplication $\otimes$ will be always defined as

$$u \otimes v = g^{-1}(g(u)g(v)).$$

Let $f\colon [c,d] \to [a,b]$. If the function f is differentiable on (c,d) and has the same monotonicity as the function g, then we define the g-derivative of f at the point $x \in (c,d)$ as

$$\frac{d^{\oplus} f(x)}{dx} = g^{-1}\left(\frac{d}{dx} g(f(x))\right),$$

when the right part is meaningful.

Using the representations of the operations $\oplus$ and $\otimes$ by g and the preceding definition we have

Theorem 8.5 *Let f_1 and f_2 be two g-differentiable functions on (c,d) and with the values in $[a,b]$. Then we have for $\lambda \in [a,b]$*

(i) $\frac{d^{\oplus}(f_1 \oplus f_2)}{dx} = \frac{d^{\oplus} f_1}{dx} \oplus \frac{d^{\oplus} f_2}{dx}$

(ii) $\frac{d^{\oplus}(\lambda \otimes f_1)}{dx} = \lambda \otimes \frac{d^{\oplus} f}{dx}$;

(iii) $\frac{d^{\oplus} \lambda}{dx} = \mathbf{0}$.

Example 8.6 Let $g(u) = e^{-u}$. Then we have $u \oplus v = -\ln(e^{-u} + e^{-v})$ and $u \otimes v = u + v$. g-derivative is given as

$$\frac{d^{\oplus} f(x)}{dx} = -\ln\left(\frac{d}{dx} e^{-f(x)}\right) = f(x) - \ln(-f'(x))$$

for $f'(x) < 0$.

Example 8.7 Let

$$g(u) = ln\frac{1+u}{1-u}.$$

Then we have

$$g^{-1}(u) = \frac{e^u - 1}{e^u + 1}$$

and the g-derivative is given by

$$\frac{d^{\oplus} f(x)}{dx} = \frac{\exp\frac{2f'(x)}{1-f^2(x)} - 1}{\exp\frac{2f'(x)}{1-f^2(x)} + 1}.$$

The usual derivative could have an unusual behavior with respect to the pseudo-product. For example, if $\otimes$ is from Example 8.6, then we have

$$(f_1 \otimes f_2)' = f_1' \otimes f_2' .$$

For g-derivative case we have

Theorem 8.6 *Let f_1 and f_2 be two functions defined on $[c, d]$ and with values in $[a, b]$. If both functions are differentiable, then we have*

$$\frac{d^{\oplus}(f_1 \otimes f_2)}{dx} = \left(\frac{d^{\oplus} f_1}{dx} \otimes f_2\right) \oplus \left(f_1 \otimes \frac{d^{\oplus} f_2}{dx}\right).$$

Proof. We have

$$\begin{aligned}\frac{d^{\oplus}(f_1 \otimes f_2)}{dx} &= g^{-1}\left(\frac{d}{dx} g(f_1 \otimes f_2)\right) = g^{-1}\left(\frac{d}{dx} g(g^{-1}(g(f_1)g(f_2)))\right) \\ &= g^{-1}(g'(f_1)f_1'g(f_2) + g(f_1)g'(f_2)f_2') .\end{aligned}$$

On the other side we have

$$\begin{aligned}\left(\left(\frac{d^{\oplus} f_1}{dx}\right) \otimes f_2\right) \oplus \left(f_1 \otimes \frac{d^{\oplus} f_2}{dx}\right) &= g^{-1}\left(g\left(\frac{d^{\oplus} f_1}{dx} \otimes f_2\right) + g\left(f_1 \otimes \frac{d^{\oplus} f_2}{dx}\right)\right) \\ &= g^{-1}\left(g\left(g^{-1}\left(g\left(\frac{d^{\oplus} f_1}{dx}\right) g(f_2)\right)\right) + g\left(g^{-1}\left(g(f_1) g\left(\frac{d^{\oplus} f_2}{dx}\right)\right)\right)\right) \\ &= g^{-1}(g(g^{-1}(\frac{dg(f_1)}{dx}))g(f_2) + g(f_1)g(g^{-1}(\frac{dg(f_2)}{dx}))) \\ &= g^{-1}(g'(f_1)f_1'g(f_2) + g(f_1)g'(f_2)f_2') .\end{aligned}$$

□

We define recursively: if there exists the $(n-1)$-g-derivative of a function $f\colon [c, d] \to [a, b]$, then we define the n-g-derivative of f (if it exists)

$$\frac{d^{(n)\oplus} f}{dx^n} = \frac{d^{\oplus}}{dx}\left(\frac{d^{(n-1)\oplus} f}{dx^{n-1}}\right).$$

Theorem 8.7 *If there exists an n-g-derivative of f, then we have*

$$\frac{d^{(n)\oplus} f}{dx^n} = g^{-1}\left(\frac{d^n}{dx^n} g(f)\right).$$

In the next theorem the following notation we shall use: The value of function f at the point x_0 will be denoted by $f[x_0]$. So $\frac{d^{\oplus} f[x_0]}{dx}$ means the value of the g-derivative of f at the point x_0. By this convention we have

$$\frac{d^{\oplus}[x_0]}{dx} = \frac{d^{\oplus} i[x_0]}{dx},$$

where $i(x) = x$.

Theorem 8.8 *Suppose that h is a function on $[a,b] \subset [0,\infty]$ and $f\colon [c,d] \to [a,b]$ and*

$$F(x) = h(f(x)).$$

Suppose that f has a derivative at $x_0 \in (c,d)$ and that h has a derivative at $f(x_0)$. Then $\frac{d^{\oplus} F[x_0]}{dx}$ exists and

$$\frac{d^{\oplus} F[x_0]}{dx} = \frac{d^{\oplus}[h(f(x_o))]}{dx} \otimes \frac{d^{\oplus}}{dx} g^{-1}(h[f(x_0)]) \otimes \frac{d^{\oplus} g^{-1}(f[x_0])}{dx}.$$

Proof. We have

$$\begin{aligned}
\frac{d^{\oplus} F[x_0]}{dx} &= g^{-1}\Big(\frac{d}{dx} g(F[x_0])\Big) \\
&= g^{-1}\Big(\frac{d}{dx} g[h(f(x_0))] \frac{d}{dx} h[f(x_0)] \frac{d}{dx} f[x_0]\Big) \\
&= \frac{d^{\oplus}[h(f(x_0))]}{dx} \otimes \frac{d^{\oplus}}{dx} g^{-1}(h[f(x_0)]) \otimes \frac{d^{\oplus}}{dx} g^{-1}(f[x_0]),
\end{aligned}$$

where we have used the equality

$$\frac{d}{dx} s[x] = \frac{d}{dx} g(g^{-1}(s[x])).$$

□

For $g(x) = x$, using the fact $g'(x) = 1$ and hence $g'[h(f(x_0))] = 1$, we obtain the usual Chain rule.

For any measurable function $f\colon [c,d] \to [a,b]$ we define the g-integral

$$\int_{[c,d]}^{\oplus} f\, dx = g^{-1}\Big(\int_c^d g(f)\, dx\Big).$$

Theorem 8.9 *We have*

(i) $\int_{[c,d]}^{\oplus} (f_1 \oplus f_2)\, dx = \int_{[c,d]}^{\oplus} f_1\, dx \oplus \int_{[c,d]}^{\oplus} f_2\, dx;$

(ii) $\int_{[c,d]}^{\oplus} (\lambda \otimes f)\, dx = \lambda \otimes \int_{[c,d]}^{\oplus} f\, dx;$

(iii) $f_1 \leq f_2 \Longrightarrow \int_{[c,d]}^{\oplus} f_1\, dx \leq \int_{[c,d]}^{\oplus} f_2\, dx;$

(iv) $\int_{[c,d]\cup[e,f]}^{\oplus} f\, dx = \int_{[c,d]}^{\oplus} f\, dx \oplus \int_{[e,f]}^{\oplus} f\, dx.$

Example 8.8 For $\oplus$ and $\otimes$ as in Example 8.5 we have

$$\int_{[0,x]}^{\oplus} x\, dx = -\ln\Big(\int_0^x e^{-x}\, dx\Big) = x$$

and

$$\int_{[1,x]}^{\oplus} \ln x\, dx = -\ln \ln x\,.$$

Theorem 8.10 *Suppose that f is continuous on $[c,d]$. Then we have*

$$\frac{d^{\oplus}}{dx}\Big(\int_{[c,x]}^{\oplus} f\, dx\Big) = f(x)$$

for each $x \in (c,d)$.

Proof. We have for $x \in (c,d)$

$$\begin{aligned}\frac{d^{\oplus}}{dx}\Big(\int_{[c,x]}^{\oplus} f(x)\, dx\Big) &= \frac{d^{\oplus}}{dx}\Big(g^{-1}\Big(\int_c^x g((f(x))\, dx\Big)\Big) \\ &= g^{-1}\Big(\frac{d}{dx} g\Big(g^{-1}\Big(\int_c^x g(f(x))\, dx\Big)\Big)\Big) \\ &= g^{-1}\Big(\frac{d}{dx}\int_c^x g(f(x))\, dx\Big) = f(x)\,,\end{aligned}$$

where we have used the fundamental theorem of the usual calculus. □

Theorem 8.11 *Suppose that f has a continuous g-derivative on (c,d). Then we have*

$$\int_{[c,x]}^{\oplus} \frac{d^{\oplus} f}{dx}\, dx \oplus f(c) = f(x)$$

for $x \in (c,d)$.

Remark 8.3 As a generalization of g-derivative of a function f we can introduce a general g-derivative $f^{(h)}$ for a given integrable non-negative function h, defined by $f^{(h)} = g^{-1}((g \circ f)'/h)$, whenever the expression at the right side of the preceding equality exists.

8.4 Applications of g-Calculus to Nonlinear Differential Equations

8.4.1 Ordinary Differential equations

First we have a general result.

Theorem 8.12 *Suppose that $f\colon [c,d] \times [a,b] \to [a,b]$ is continuous, that ψ is defined and continuous on $J = \{x : x_0 - h < x < x_0 + h\} \subset [c,d]$ with values in $[a,b]$, and that $(x_0, y_0) \in [c,d] \times [a,b]$ with $\psi(x_0) = y_0$. Then the necessary and sufficient condition that ψ be a solution of*

$$\frac{d^{\oplus}\psi}{dx} = f(x, \psi(x))$$

on J is that ψ satisfies the g-integral equation

$$\psi(x) = y_0 \oplus \int_{[x_0,x]}^{\oplus} f(t, \psi(t))\, dt$$

for $x \in J$.

Now we will consider the following special ordinary differential equation of the second order for $p > 0$ and $n \in \mathbf{R}^+$

$$y'' + (p-1)y^{-1}(y')^2 + (2npx^{-1} - x^{-np}y^p)y' + n(np-1)x^{-2}y = 0\,.$$

We will show how we can apply the preceding g-calculus on this equation. Multiplying the preceding equation by $px^{np}y^{p-1}$ we obtain

$$py^{2p-1}y' = x^{np}py^{p-1}y'' + x^{np}p(p-1)y^{p-2}(y')^2 \\ + 2npx^{np-1}py^{p-1}y' + np(np-1)x^{np-2}y^p,$$

which implies

$$y \cdot py^{p-1}y' = x^{np}(py^{p-1}y'' + p(p-1)y^{p-2}(y')^2) \\ + 2npx^{np-1} \cdot py^{p-1}y' + np(np-1)x^{np-2} \cdot y^p.$$

We reduce the preceding equation on the form

$$y \cdot (y^p)' = (x^{np}y^p)'',$$

and finally

$$\left(y^p \cdot \left(\left(\frac{d}{dx}(y^p)\right)^{1/p}\right)^p\right)^{1/p} = \left(\frac{d^2}{dx^2}((x^n y)^p)\right)^{1/p}.$$

Introducing now the special generator g, $g(x) = x^p$ for $p > 0$, and the corresponding pseudo-addition $\oplus$ and pseudo-multiplication $\otimes$

$$u \oplus v = (u^p + v^p)^{1/p},$$
$$u \otimes v = u \cdot v,$$

we can write the preceding equation in the following form

$$y \otimes \frac{d^{\oplus}}{dx}(y) = \frac{d^{(2)\oplus}}{dx^2}(x^n \otimes y). \tag{8.3}$$

Now we shall simplify this equation using g-calculus and the following three lemmas

Lemma 8.1 *We mark by x^* the inverse element for element $x \in (a, b)$, with respect to the pseudo-multiplication $\otimes$. Then we have:*

$$x^* = g^{-1}\left(\frac{1}{g(x)}\right).$$

Lemma 8.2 *We have for $n \in \mathbf{N}$ and $x \in [a, b]$*

$$\underbrace{x \oplus x \oplus \ldots \oplus x}_{n} = g^{-1}(n) \otimes x\,.$$

Lemma 8.3 *We have for a g-derivable function $y = y(x)$*

$$y \otimes \frac{d^{\oplus} y}{dx} = g^{-1}\left(\frac{1}{2}\right) \otimes \frac{d^{\oplus}}{dx}(y \otimes y)\,.$$

Proof. We have by Lemma 8.1 and Lemma 8.2

$$\frac{d^{\oplus}}{dx}(y \otimes y) = \frac{d^{\oplus} y}{dx} \otimes y \oplus \frac{d^{\oplus} y}{dx} \otimes y = g^{-1}(2) \otimes y \otimes \frac{d^{\oplus} y}{dx}\,.$$

Hence

$$y \otimes \frac{d^{\oplus} y}{dx} = g^{-1}\left(\frac{1}{2}\right) \otimes \frac{d^{\oplus}}{dx}(y \otimes y).$$

□

Applying now the preceding lemma on (8.3) we obtain

$$g^{-1}\left(\frac{1}{2}\right) \otimes \frac{d^{\oplus}}{dx}(y \otimes y) = \frac{d^{(2)\oplus}}{dx^2}(x^n \otimes y)$$

and g-integrating this equation

$$g^{-1}\left(\frac{1}{2}\right) \otimes y \otimes y \oplus c_1 = \frac{d^{\oplus}}{dx}(x^n \otimes y),$$

where c_1 is an arbitrary constant.

By the definitions of the operations $\oplus$ and $\otimes$ and $\frac{d^{\oplus}}{dx}$ we can write

$$g^{-1}\left(g\left(g^{-1}\left(\frac{1}{2}\right) \cdot g(y) \cdot g(y) + g(c_1)\right)\right) = g^{-1}\left(\frac{d}{dx} g(x^n y)\right)$$

and

$$\frac{1}{2}(g(y))^2 + g(c_1) = \frac{d}{dx} g(x^n y).$$

Taking $g(y) = y^p$, we have for $c = g(c_1)$

$$\frac{1}{2} y^{2p} + c = (x^{np} y^p)' = npx^{np-1} y^p + x^{np} p y^{p-1} y'$$

and so for $t = y^p$

$$\frac{1}{2}t^2 + c = npx^{np-1}t + x^{np}t'.$$

This is a Riccati differential equation

$$t' = \frac{1}{2}x^{-np}t^2 - npx^{-1}t + cx^{-np}.$$

Specially, for $np = 1$ the preceding equation reduces on the equation

$$t' = x^{-1}\left(\frac{1}{2}t^2 - t + c\right),$$

which separates the variables. This case corresponds to the starting equation in the form

$$y'' + (p-1)y^{-1}(y')^2 + x^{-1}(2 - y^p)y' = 0.$$

This procedure can be applied also on the more general case

$$y \otimes \frac{d^{\oplus}y}{dx} = \frac{d^{(2)\oplus}}{dx^2}(h(x) \otimes y) \tag{8.4}$$

for a given function $h \in C^2$ such that h has the same monotonicity as the function g, for an arbitrary generator g. Namely, the equation (8.4) reduces in an analogous way as for special g on the equation

$$g^{-1}\left(\frac{1}{2}\right) \otimes \frac{d^{\oplus}}{dx}(y \otimes y) = \frac{d^{(2)\oplus}}{dx}(h(x) \otimes y)$$

and finally in an analogous way we obtain the Riccati equation

$$t' = \frac{1}{2\varphi(x)}t^2 - \frac{\varphi'(x)}{\varphi(x)}t + \frac{C}{\varphi(x)},$$

where $t = g(y)$ and $\varphi(x) = g(h(x))$.

8.4.2 The Burgers equation

We shall consider as an important nonlinear partial differential equation the Burgers equation

$$\frac{\partial u}{\partial t} + 1/2\left(\frac{\partial u}{\partial x}\right)^2 - c/2\frac{\partial^2 u}{\partial x^2} = 0$$

for $x \in \mathbf{R}$ and $t > 0$ with the initial condition $u(x,0) = u_0(x)$, where c is the given positive constant. We shall consider this equation by g-calculus with the generator $g(u) = e^{-u/c}$. The corresponding pseudo-addition is

$$u \oplus v = -c\ln(e^{-u/c} + e^{-v/c})$$

and the distributive pseudo-multiplication $u \otimes v = u + v$.

Then for solutions u_1 and u_2 the function $(\lambda_1 \otimes u_1) \oplus (\lambda_2 \otimes u_2)$ is also a solution of the Burgers equation.

The solution of the given initial problem is

$$u(x,t) = \frac{c}{2}\ln(2\pi ct) \otimes \int^{\oplus} \left[\frac{(x-s)^2}{2t}\right] \otimes u_0(s)\, ds,$$

where the integral is given by

$$\int^{\oplus} f(x)\, dx = -c\ln\Big(\int e^{-f(x)/c}\, dx\Big).$$

Taking $c \to 0$ in the Burgers equation we obtain Hamilton-Jacobi equation

$$\frac{\partial u}{\partial t} + \frac{1}{2}\left(\frac{\partial u}{\partial x}\right)^2 = 0\,.$$

In this case the pseudo-addition $\oplus$ reduces on min.

8.5 Applications of the Pseudo-Integral on Nonlinear Difference Equations

In this section we apply decomposable measures and corresponding integrals on a class of nonlinear difference equations important in the theory of the architecture of computers with parallel processing (neural nets).

The problem of the construction of the computers with parallel processing (neural nets) is an optimization problem in the following sense. If we denote by K the number of elementary processors (neurons) then the complexity of the algorithm grows with the number K, in the worst case as K^2. So it is important to examine the limit case $K \to \infty$. But for most of the considered problems the limit equation (obtained as limit of difference equations) has no differentiable solution, since the initial condition usually gives this.

Using the generalization of the theory of generalized functions the limit behavior of considered difference equation is characterized.

We shall consider the following difference equation

$$c_{m,n}^{k+1} = c_{m-1,n}^{k} \oplus c_{m,n-1}^{k} \tag{8.5}$$

where $k = 0, 1, 2, \ldots;\ m, n = 0, \pm 1, \pm 2, \ldots$, with the initial condition

$$c_{m,n}^{0} = \begin{cases} \mathbf{1} & \text{if } n = 0,\ m \geq 0, \\ \mathbf{1} & \text{if } m = 0,\ n \geq 0, \\ \mathbf{0} & \text{otherwise.} \end{cases} \tag{8.6}$$

Example 8.9 Let $\oplus = \min$ and $\otimes = +$ on $[-\infty, +\infty]$. Then we have $\mathbf{0} = +\infty$ and $\mathbf{1} = 0$. The equation (8.5) reduces on the Bellman equation

$$c_{m,n}^{k+1} = \min\{c_{m-1,n}^{k}, c_{m,n-1}^{k}\}$$

where $k = 0, 1, 2, \ldots;\ m, n = 0, \pm 1, \pm 2, \ldots$, with the initial conditions

$$c_{m,n}^{0} = \begin{cases} 0 & \text{if } n = 0,\ m \geq 0, \\ 0 & \text{if } m = 0,\ n \geq 0, \\ +\infty & \text{otherwise.} \end{cases}$$

Example 8.10 Let $\oplus = \max$ and $\otimes = +$ on $[-\infty, +\infty)$. Then we have $\mathbf{0} = -\infty$ and $\mathbf{1} = 0$. The equation (8.5) reduces to the equation

$$c_{m,n}^{k+1} = \max\{c_{m-1,n}^{k}, c_{m,n-1}^{k}\}$$

where $k = 0, 1, 2, \ldots;\ m, n = 0, \pm 1, \pm 2, \ldots$, with the initial conditions

$$c^0_{m,n} = \begin{cases} 0 & \text{if } n = 0,\ m \geq 0, \\ 0 & \text{if } m = 0,\ n \geq 0, \\ -\infty & \text{otherwise.} \end{cases}$$

Example 8.11 Let $\oplus = \max$ and $\otimes = \min$ on $[0, +\infty]$. Then we have $\mathbf{0} = 0$ and $\mathbf{1} = \infty$. The equation (8.5) reduces to the equation

$$c^{k+1}_{m,n} = \max\{c^k_{m-1,n}, c^k_{m,n-1})\}$$

where $k = 0, 1, 2, \ldots;\ m, n = 0, \pm 1, \pm 2, \ldots$, with the initial conditions

$$c^0_{m,n} = \begin{cases} \infty & \text{if } n = 0,\ m \geq 0, \\ +\infty & \text{if } m = 0,\ n \geq 0, \\ 0 & \text{otherwise.} \end{cases}$$

We shall introduce the operator $T\colon C^2 \to C$, where

$$C = \{c^k_{m,n} : k = 0, 1, 2 \ldots;\ m, n = 0, \pm 1, \pm 2, \ldots\}$$

and T acts in the following way

$$Tc^k_{m,n} = c^k_{m-1,n} \oplus c^k_{m,n-1}\,.$$

This operator is linear with respect to $\oplus$ and $\otimes$. Namely, we have by the commutativity and associativity of the operation $\oplus$

$$\begin{aligned} T(a^k_{m,n} \oplus b^k_{m,n}) &= ((a^k_{m-1,n} \oplus b^k_{m-1,n}) \oplus (a^k_{m,n-1} \oplus b^k_{m,n-1}) \\ &= T(a^k_{m,n}) \oplus T(b^k_{m,n})\,, \end{aligned}$$

and by the distributivity of the operation $\otimes$ with respect to the operation $\oplus$ we obtain

$$T(\gamma \otimes c^k_{m,n}) = (\gamma \otimes c^k_{m,n}) \oplus (\gamma \otimes c^k_{m,n-1}) = \gamma \otimes T(c^k_{m,n})\,.$$

Now we shall consider the corresponding continuous analog of the problem (8.5)–(8.6).

We consider functions defined on $[0, M] \times [-\infty, +\infty]^2$, where $M > 0$, and with values in $[a, b]$, i.e., $c\colon [0, M] \times [-\infty, +\infty]^2 \to [a, b]$. Then the corresponding equation to (8.5) of a continuous variable with mesh size $h > 0$ has the form

$$c_h(z + h, x, y) = c_h(z, x - h, y) \oplus c_h(z, x, y - h), \tag{8.7}$$

taking $z = kh$, $k = 0, 1, 2, \ldots$ and $c(kh, mh, nh) = c^k_{m,n}$.

The corresponding initial condition to (8.6) has the form

$$c_h(0, x, y) = c^0(x, y) = \begin{cases} \mathbf{1} & \text{if } y = 0,\ x \geq 0, \\ \mathbf{1} & \text{if } x = 0,\ y \geq 0, \\ \mathbf{0} & \text{otherwise.} \end{cases} \tag{8.8}$$

Introducing the operator $T_h\colon C_h^2 \to C_h$, where

$$\begin{aligned} C_h = \big\{c_h : c_h(kh, mh, nh) = c^k_{m,n} \quad & \text{for } k = 0, 1, 2, \ldots; \\ m, n = 0, \pm 1, \pm 2, \ldots \big\}, & \end{aligned}$$

such that

$$T_h c_h(z, x, y) = c_h(z, x - h, y) \oplus c_h(z, x, y - h)$$

for $z = kh$ ($k = 0, 1, 2, \ldots$), we obtain that the solution of the problem (8.7)–(8.8) is given by

$$c_h(z, x, y) = (T_h)^k c^0(x, y). \tag{8.9}$$

We introduce the pseudo-scalar product for functions with values in $[a, b]$ by

$$(f, g)_\oplus = \int^\oplus f(x, y) \otimes g(x, y)\, dx\, dy.$$

Example 8.12 For Min-Plus we have a pseudo scalar product

$$(f, g)_\oplus = \inf_{x,y} \{f(x, y) + g(x, y)\}.$$

Example 8.13 For Max-Plus we have a pseudo scalar product

$$(f,g)_\oplus = \sup_{x,y}\{f(x,y)+g(x,y)\}.$$

We take $k \to \infty$ or $h \to 0$ (since the interval $[0, M]$ is bounded it is the same) for a point $z_0 \in [0, M]$ that $kh \to z_0$. We have for a smooth function $f(x,y)$

$$\begin{aligned}(c_h(z_0,x,y), f(x,y))_\oplus &= ((T_h)^k c^0(x,y), f(x,y))_\oplus \\ &= (c^0(x,y), (T_h^*)^k f(x,y))_\oplus\,,\end{aligned}$$

where T_h^* is the adjoint operator of the operator T_h with respect to the pseudo-scalar product $(\cdot,\cdot)_\oplus$.

The weak-$\oplus$-limit of solutions $c_h(z_0,x,y)$ of the problem (8.7)–(8.8), since

$$\lim_{h\to 0}(c(z_0,x,y), f(x,y))_\oplus = F(f(x,y))$$

is a pseudo-linear functional, we shall denote by $c(z_0,x,y)$, i.e.,

$$F(f(x,y)) = (c(z_0,x,y), f(x,y))_\oplus\,.$$

On the other side we have

$$(c(z_0,x,y), f(x,y))_\oplus = \lim_{h\to 0}(c^0(x,y), (T_h^*)^k f(x,y))_\oplus\,.$$

Example 8.14 Take Min-Plus on the interval $[-\infty, +\infty]$. Then we have for the solution (8.9) of the problem (8.7)–(8.8)

$$\begin{aligned}c_h(z,x,y) &= (T_h)^k c^0(x,y) \\ &= \min_{r_1+r_2=k,\ r_i\in\{0,1,2,\dots\}}\{c^0(x-r_1h, y-r_2h)\}.\end{aligned}$$

The adjoint operator $(T_h^*)^k$ acts in the following way

$$(T_h^*)^k f(x,y) = \min_{r_1+r_2=k,\ r_i\in\{0,1,2,\dots\}}\{c^0(x+r_1h, y+r_2h)\}.$$

Taking the weak-$\oplus$-limit we obtain that

$$(c(z_0,x,y), f(x,y))_\oplus = (c^0(x,y), \min_{r_1+r_2=z_0,\ r_i\in[0,+\infty)}\{f(x+r_1, y+r_2)\}.$$

The limit equation to (8.7) on smooth function is

$$\frac{\partial c}{\partial z} = \min\left\{-\frac{\partial c}{\partial x}, -\frac{\partial c}{\partial y}\right\},$$

whose solution can be found by the Pontrjagin maximum principle.

8.6 Notes and References

The presented pseudo-integral is taken from [Maslov 1987], [Pap 1990d], [Pap 1992]. The Weber integral can be found in [S. Weber 1984], [S. Weber 1986], [Klement and S. Weber 1991]. Murofushi-Sugeno integral was introduced in [Sugeno and Murofushi 1987]. Pseudo-additivity of (MS) integral was investigated in [Mesiar and Pap 1992]. Other similar types of integral based on decomposable measures can be found in [Z. Wang and Klir 1992] (pan-integral) and further generalizations in [Murofushi and Sugeno 1991b] and [Grabish, Murofushi and Sugeno 1992]. Both Choquet and Sugano iintegrals are special cases of so called t-conorm integrals [Murofushi and Sugano 1991b] and Choquet-like integrals introduced in [Mesiar 1995].

g-calculus was introduced in [Pap 1993c]. Applications of pseudo-integrals can be found in many fields (see the references at the end), specially in nonlinear differential and difference equations are given in [Pap 1992], [Pap 1993c], [Maslov 1987], [Baccelli, Cohen, Olsder and Quadrat 1992]. The generalized g-derivative was recently introduced by R. Mesiar and A. Marková.

Chapter 9

The Range and Regularity of Null-Additive Set Function

9.1 The Range of Autocontinuous Set Function

9.1.1 Darboux property

Let m be a positive set function defined on a σ-algebra. We will need the following obvious theorem.

Theorem 9.1 *If m is a σ-finite order continuous monotone set function or a finite order continuous set function and if $\{E_\gamma : \gamma \in C\}$ is a disjoint family of subsets of X, then the index set $C_+ = \{\gamma \in C : m(E_\gamma) > 0\}$ is at most countable.*

Theorem 9.2 (Darboux property) *Let m be a finite order continuous, continuous from below and exhaustive set function or a finite fuzzy measure, which is autocontinuous from above. If E and F are two disjoint elements from Σ such that F contains no atoms of m and*

$$m(E) < m(E \cup F),$$

then for every positive real number α between $m(E)$ and $m(E \cup F)$, there exists a set $A \in \Sigma$ such that $A \subset F$ and

$$m(E \cup A) = \alpha.$$

Proof. Let us first suppose that m is an order continuous, continuous from below and exhaustive set function on Σ, which is autocontinuous from above. Let $E \cap F = \emptyset$. We denote by $\mathcal{T}$ a subfamily of Σ such that it contains pairwise disjoint subsets B of F with $m(B) > 0$ and

$$m\Big(E \cup \Big(\bigcup_{B \in \mathcal{T}} B\Big)\Big) \leq \alpha .$$

By Theorem 9.1 each such family $\mathcal{T}$ is at most countable. Let $\mathcal{S}$ denote the class of such families. If $\mathcal{S}_0$ is a chain in $(\mathcal{S}, \subset)$, then there exists an increasing sequence $\{\mathcal{T}_n\}$ from $\mathcal{S}_0$ such that

$$\bigcup_{n=1}^{\infty} \mathcal{T}_n = \bigcup_{\mathcal{T} \in \mathcal{S}_0} \mathcal{T} = \mathcal{T}_0,$$

since $\mathcal{T}_0$ is at most countable. By the continuity from below of m

$$m\Big(E \cup \Big(\bigcup_{B \in \mathcal{T}_0} B\Big)\Big) = \lim_{k \to \infty} m\Big(E \cup \Big(\bigcup_{n=1}^{k} \bigcup_{B \in \mathcal{T}_n} B\Big)\Big) \leq \alpha .$$

Hence $\mathcal{T}_0 \in \mathcal{S}$ and $\mathcal{T} \subset \mathcal{T}_0$ $(\mathcal{T} \in \mathcal{S}_0)$. By the Zorn lemma there exists a maximal element $\mathcal{T}_0$ in $\mathcal{S}$. Let $A = \bigcup_{B \in \mathcal{T}_0} B$. Then we have

$$m(E \cup A) \leq \alpha .$$

We shall prove that $m(E \cup A) = \alpha$. Suppose the contrary, i.e.,

$$m(E \cup A) < \alpha .$$

Now we choose $\epsilon > 0$ such that

$$m(E \cup A) + \epsilon < \alpha . \tag{9.1}$$

By the autocontinuity from above of m there exists $\delta > 0$ such that

$$m(E \cup A \cup C) < m(E \cup A) + \epsilon \tag{9.2}$$

for any $C \in \Sigma$ such that $m(C) < \delta$.

Since m is atomless on $F \setminus A$ and $m(F \setminus A) > 0$ (suppose the contrary, i.e., $m(F \setminus A) = 0$, then by the null-additivity of m would be

$$\alpha < m(E \cup F) = m(E \cup A \cup (F \setminus A)) = m(E \cup A) < \alpha\,,$$

contradiction) there exists by Lemma 6.4 a subset C' of $F \setminus A$ from Σ such that

$$0 < m(C') \leq \delta\,.$$

Hence by (9.1) and (9.1) $m(E \cup A \cup C') < \alpha$. Then $\mathcal{T}_0$ would not be a maximal element of $\mathcal{S}$. Contradiction.

For finite fuzzy measures the proof is the same. □

Remark 9.1 Theorem 9.2 for the fuzzy measure case is true also when E and F are not disjoint. Namely, if $E \cap F \neq \emptyset$ holds, then we reduce this case on the disjoint case using the monotonicity of m and

$$m(E \setminus F) \leq m(E) < \alpha < m(E \cup F)\,.$$

9.1.2 Lyapunov's Type Theorems

We point out first that there exists a nonatomic finite and even finitely additive set function (measure) without the Darboux property, for which Lyapunov theorem does not hold.

Example 9.1 Let μ be a fixed nonatomic measure on Σ and

$$\mathcal{T} = \{A : A \in \Sigma,\ \mu(A) = \mu(X)\}.$$

Let $\mathcal{U}$ be an ultrafilter of subsets of X containing $\mathcal{T}$. Then the measure m_a on Σ defined by

$$m_a(A) = \mu(A) + \lambda_a(A),$$

where for $A \in \Sigma$

$$\lambda_a(A) = \begin{cases} a & \text{if } A \in \mathcal{U}, \\ 0 & \text{if } A \notin \mathcal{U}, \end{cases}$$

for a fixed positive number a, does not satisfy Lyapunov's type theorem even if μ would satisfy it. Namely, we have as a classical result for this case

$$m_a(\Sigma \cap \mathcal{P}(A)) = [0, \mu(A)) \cup (a, a + \mu(A)]$$

for $A \in \mathcal{U}$. We remark that m_a is not order continuous.

Now we shall prove two Lyapunov's type theorems for generally non-additive set functions.

Theorem 9.3 *Let $m\colon \Sigma \to [0,\infty)$ be a finite order continuous, continuous from below, exhaustive and autocontinuous from above set function. If m is atomless and there exists*

$$\max_{A\in\Sigma} m(A) = m(A')$$

for some $A' \in \Sigma$, then the range of m is $[0, m(A')]$.

Proof. Taking $E = \emptyset$ and $F = A'$ in Theorem 9.2 we obtain

$$m(\Sigma) = [0, m(A')] . \qquad \square$$

Theorem 9.4 *Let m be a finite fuzzy measure which is autocontinuous from above, then the range of m is compact. If additionally m is atomless, then $m(\Sigma) = [0, m(X)]$.*

Proof. Since m is monotone the range of m is bounded by $m(X)$.

Now we shall prove that $m(\Sigma)$ is closed. If m is atomless then by Theorem 9.2

$$m(\Sigma) = [0, m(X)]$$

(taking $E = \emptyset$ and $F = X$). For other cases we shall prove first that for each $\epsilon > 0$ there exists $\delta > 0$ such that $m(A) < \delta$ and $m(B) < \delta$ for $A, B \in \Sigma$ imply $m(A \cup B) < \epsilon$. Suppose the contrary. Then there exist $\epsilon > 0$ and two sequences $\{A_n\}$ and $\{B_n\}$ from Σ such that $m(A_n) < \frac{1}{n}$, $m(B_n) < \frac{1}{n}$ and

$$m(A_n \cup B_n) > \epsilon .$$

By autocontinuity from above of m there exist subsequences $\{A_{n_k}\}$ and $\{B_{n_k}\}$ of $\{A_n\}$ and $\{B_n\}$, respectively, such that

$$m\Big(\bigcup_{k=i}^{s} A_{n_k}\Big) < \frac{1}{i} \quad \text{and} \quad m\Big(\bigcup_{k=i}^{s} B_{n_k}\Big) < \frac{1}{i} \tag{9.3}$$

for each $i = 1, 2, \ldots, s$. Let $E_i = \cup_{k=i}^{\infty} A_{n_k}$ and $F_i = \cup_{k=i}^{\infty} B_{n_k}$. Since $\{E_i\}$ and $\{F_i\}$ are two monotone decreasing sequences from Σ we have by (9.3)

$$\lim_{i\to\infty} m(E_i) = 0 \quad \text{and} \quad \lim_{i\to\infty} m(F_i) = 0 .$$

Hence $\lim_{i\to\infty} m(E_i \cup F_i) = 0$. Since by monotonicity of m

$$m(E_i \cup F_i) \geq m(A_{n_i} \cup B_{n_i})$$

we obtain $\lim_{i\to\infty} m(A_{n_i} \cup B_{n_i}) = 0$. Contradiction with the condition $m(A_{n_i} \cup B_{n_i}) > \epsilon$ $(i \in \mathbf{N})$.

The obtained property of m implies by Theorem 5.9, that there exists a submeasure η on Σ such that

$$\lim_{n\to\infty} m(E_n) = 0 \quad \text{iff} \quad \lim_{n\to\infty} \eta(E_n) = 0\,.$$

Now, if m is purely atomic, then X is the union of at most countable many pairwise disjoint atoms. In this case the fuzzy measure space (X, Σ, m) is represented by a simple fuzzy measure space $(\mathbf{N}, \mathcal{P}(\mathbf{N}), v)$ — see also the Section 6.5. Since v is also autocontinuous from above there exists a submeasure η such that

$$\lim_{n\to\infty} \eta(E_n) = 0 \quad \text{iff} \quad \lim_{n\to\infty} v(E_n) = 0\,.$$

Since η is continuous with respect to the countable additive measure g,

$$g(A) = \Sigma_{n\in A} 2^{-n} \qquad (A \in \mathcal{P}(\mathbf{N}))$$

we have that m is continuous also with respect to g. Then $m(\mathcal{P}(\mathbf{N}))$ is compact by the compactness of $(\mathcal{P}(\mathbf{N}), g)$. If m is neither atomless nor purely atomic we divide X into a purely atomic part X_1 and an atomless part X_2.

We shall prove that for an arbitrary convergent sequence $\{s_n\}$ from $m(\Sigma)$ the limit s belongs also to $m(\Sigma)$. We suppose further the non-trivial case: $s_n \neq s$ and $m(X_2) < s$. We have

$$s_n = m(E_n \cup F_n) \quad \text{for} \quad E_n \subset X_1 \quad \text{and} \quad F_n \subset X_2 \qquad (n \in \mathbf{N}).$$

Taking subsequences of $\{s_n\}$ we can reduce all situations to only two cases:

(i) $s_n > s \qquad (n \in \mathbf{N})$,

(ii) $s_n < s \qquad (n \in \mathbf{N})$.

Case (i): If there exists $n \in \mathbf{N}$ such that $m(E_n) < s$, then Theorem 9.2 with Remark 9.1 implies the existence of a subset A of F_n from Σ such that

$m(E_n \cup A) = s$. Hence $s \in m(\Sigma)$. The second subcase $m(E_n) \geq s$ for all $n \in N$ implies

$$s_n = m(E_n \cup F_n) \geq m(E_n) \geq s \qquad (n \in \mathbf{N}).$$

Hence $\lim_{n\to\infty} m(E_n) = s$. Since $m(X_1 \cap \Sigma)$ is closed (see the part of the proof when m was purely atomic), we have $s \in m(X_1 \cap \Sigma)$.

Case (ii): If there exists $n \in \mathbf{N}$ such that $m(E_n \cap X_2) \geq s$, then there exists a subset A of X_2 such that $m(E_n \cup A) = s$ (Theorem 9.2 with Remark 9.1).The second subcase $m(E_n \cup X_2) < s$ $(n \in \mathbf{N})$ implies

$$s > m(E_n \cup X_2) \geq m(E_n \cup F_n) = s_n \qquad (n \in \mathbf{N}).$$

Hence $\lim_{n\to\infty} m(E_n \cup X_2) = s$. Taking X_2 as a 'point' we obtain by the previous part of the proof, where m was purely atomic, that $s \in m(\Sigma)$. □

9.2 Regular Set Functions

Let $\mathbf{T}$ be a locally Hausdorff compact space and $\mathcal{D}$ a class of subsets of $\mathbf{T}$ such that $\emptyset \in \mathcal{D}$. Let $\mathcal{K}$ and $\mathcal{O}$ denote the family of all compact and open sets, respectively.

Definition 9.1 A set function $m: \mathcal{D} \to (-\infty, +\infty)$ is said to be difference regular if for every $A \in \mathcal{D}$ and every $\epsilon > 0$ there exist a compact set $K \subset A$ and an open set $V \supset A$ such that for every set $A' \in \mathcal{D}$, $K \subset A' \subset V$, we have

$$|m(A) - m(A')| < \epsilon\,.$$

Definition 9.2 Let $\mathcal{R}$ be a ring of subsets of $\mathbf{T}$. A set function $m: \mathcal{R} \to (-\infty, +\infty)$ is regular if for every $A \in \mathcal{R}$ and every number $\epsilon > 0$ there exist a compact set $K \subset A$ and an open set $V \supset A$ such that for every set $B \in \mathcal{R}$ with $B \subset V \setminus K$ we have $|m(B)| < \epsilon$.

We compare now both definitions of regularity on a ring $\mathcal{R}$.

Theorem 9.5 *Let $\mathcal{R}$ be a ring of subsets of $\mathbf{T}$. If a set function m, $m: \mathcal{R} \to \mathbf{R}$, is difference regular and superadditive, i.e.,*

$$m(A \cup B) \geq m(A) + m(B) \quad \textit{for every } A, B \in \mathcal{R},\ A \cap B = \emptyset\,,$$

then it is regular.

Proof. Since m is difference regular we can find for $A \in \mathcal{R}$ and $\epsilon > 0$ a compact set $K \subset A$ and an open set $V \supset A$ such that for every $A' \in \mathcal{R}$ with $K \subset A' \subset V$ we have

$$|m(A) - m(A')| < \frac{\epsilon}{2}.$$

For every $B \in \mathcal{R}$ such that $B \subset V \setminus K$ we have

$$B = (A \cup B) \setminus (A \setminus B), \quad A \setminus B \subset A \cup B, \quad K \subset A \cup B \subset V$$

and

$$K \subset A \setminus B \subset V.$$

Since $A \cup B \in \mathcal{R}$ and $A \setminus B \in \mathcal{R}$ we obtain by superadditivity of m

$$m(B) \leq m(A \cup B) - m(A \setminus B) \leq |m(A \cup B) - m(A)| + |m(A) - m(A \setminus B)|.$$

This implies that m is a regular set function. □

Corollary 9.1 *If a set function $m \colon \mathcal{R} \to \mathbf{R}$, $m(\emptyset) = 0$, has difference regular disjoint variation, where the disjoint variation $\overline{m}$ is given by*

$$\overline{m}(E) = \sup_{\pi} \sum_{A \in \pi} |m(A)| \qquad (E \in \mathcal{R})$$

and the supremum is taken over all partitions π of E into a finite number of pairwise disjoint members of $\mathcal{R}$, then m is regular.

Proof. Since $\overline{m}$ is superadditive we can apply Theorem 9.5 for $\overline{m}$. Then the inequality $|m(A)| \leq \overline{m}(A)$ $(A \in \mathcal{R})$ implies our statement. □

To the rest of this section let m be always non-negative monotone set function. Then a set function m is regular if for each set $A \in \mathcal{B}$ and each $\epsilon > 0$ there exist $K \in \mathcal{K}$ and $V \in \mathcal{O}$ such that $K \subset A \subset V$ and

$$m(V \setminus K) < \epsilon.$$

Remark 9.2 Obviously, a non-negative monotone set function m is regular iff

$$\inf\{m(V \setminus K) : K \subset A \subset V,\ K \in \mathcal{K},\ V \in \mathcal{O}\} = 0$$

for each set $A \in \mathcal{B}$.

Propositon 9.1 *A regular positive monotone set function m is $\mathcal{O}$-exhaustive, i.e.,*

$$\lim_{n\to\infty} m(O_n) = 0$$

for each sequence $\{O_n\}$ of open sets from $\mathcal{O}$ which are pairwise disjoint.

Proof. Since

$$\bigcup_{n=1}^{\infty} O_n \in \mathcal{O}$$

and m is regular, we have that for any $\epsilon > 0$ there exists a compact set K such that

$$K \subset \bigcup_{n=1}^{\infty} O_n \qquad \text{and} \qquad m\Big(\bigcup_{n=1}^{\infty} O_n \setminus K\Big) < \epsilon\,.$$

$\{O_n\}$ is an open cover of K and therefore exists $n_0 \in \mathbf{N}$ such that $K \subset \bigcup_{n=1}^{n_0} O_n$. The monotonicity of m implies

$$m(O_k) \le m\Big(O_k \cup \Big(\bigcup_{n=1}^{n_0} O_n \setminus K\Big)\Big) \le m\Big(\bigcup_{n=1}^{\infty} O_n \setminus K\Big)$$

for $k \ge n_0 + 1$. □

The following theorem gives a singleton characterization of atoms of regular null-additive set function.

Theorem 9.6 *Let m be a regular positive monotone null-additive set function. If $A \in \mathcal{B}$ is an atom of m, then there exists a point $a \in A$ such that*

$$m(A) = m(\{a\})\,.$$

Proof. Suppose that $A \in \mathcal{B}$ is an atom of m. Let $\mathcal{K}_1$ be the family of all compact sets $K \subset A$ such that

$$m(A \setminus K) = 0\,.$$

We shall prove that any K from $\mathcal{K}_1$ is an atom of m. Namely, since for any $B \subset K$, $B \in \mathcal{B}$

$$K \setminus B \subset A \setminus B$$

hold, we obtain by the monotonicity of m that either $m(B) = 0$ or $m(K \setminus B) = 0$ and so $m(K) = m((K \setminus B) \cup B) = m(B)$. For both cases by the null-additivity of m we have

$$m(K) = m((A \setminus K) \cup K) = m(A) > 0\,.$$

Let $K_1, K_2 \in \mathcal{K}_1$. Then, since we have

$$m((A \setminus K_1) \cup (A \setminus K_2)) = m(A \setminus (K_1 \cap K_2))$$

and $m(A \setminus K_2) = 0$, we obtain by the null-additivity of m

$$m(A \setminus (K_1 \cap K_2)) = m(A \setminus K_1) = 0\,,$$

i.e., $K_1 \cap K_2 \in \mathcal{K}_1$. We shall prove that

$$K_0 = \bigcap_{K \in \mathcal{K}_1} K$$

is a non-empty compact set. Indeed, if we would suppose the contrary, i.e. $K_0 = \emptyset$, then, since $\mathbf{T}$ is a Hausdorff topological space, some finite subcollection of $\{K\}_{K \in \mathcal{K}_1}$ would exist with the empty intersection. This is impossible, since this finite subcollection would belong to $\mathcal{K}_1$, but it is an atom as an element of $\mathcal{K}_1$, which is non-empty. We shall show that $K_0 \in \mathcal{K}_1$. So for arbitrary but fixed $K \in \mathcal{K}_1$ it has to be that $m(K \setminus K_0) = 0$. If we suppose that this is not true, then for $B \subset K \setminus K_0$ either $m(B) = 0$ or $m((K \setminus K_0) \setminus B) = 0$ and so

$$m(K \setminus K_0) = m(((K \setminus K_0) \setminus B) \cup B) = m(B)\,.$$

For both cases by the supposition $m(K \setminus K_0) > 0$ and so $K \setminus K_0$ would be an atom of m. Since A and K are atoms of m we have $m(A) > 0$ and $m(K_0) = 0$ (namely, $K_0 \subset K$ and $m(K \setminus K_0) > 0$). Null-additivity of m implies

$$m(A) = m((A \setminus K) \cup (K \setminus K_0) \cup K_0) = m(K \setminus K_0)\,.$$

Since we supposed $m(K \setminus K_0) > 0$ and by the preceding equality we obtain by the atomness of A

$$m(A \setminus (K \setminus K_0)) = 0\,.$$

Hence $K \setminus K_0$ have to contain an element of $\mathcal{K}_1$, which is impossible, since K_0 is non-empty. Therefore $m(K \setminus K_0) = 0$. Hence by null-additivity of m

$$m(A) = m((A \setminus K) \cup (K \setminus K_0) \cup K_0) = m(K_0),$$

i.e., $m(A \setminus K_0) = 0$ by the atomness of A. That means $K_0 \in \mathcal{K}_1$.

Now we shall prove that K_0 is a singleton. If we suppose the contrary, i.e., that K_0 contains at least two distinct elements a_1 and a_2, then, since $\mathbf{T}$ is a locally compact Hausdorff topological space, there exists an open neighbourhood V of a_1 such that $\overline{V}$ does not contain a_2. Then we have

$$K_0 = (K_0 \setminus V) \cup (K_0 \cap \overline{V}).$$

Since one of the sets $K_0 \setminus V$ or $K_0 \cap \overline{V}$ has to belong to $\mathcal{K}_1$, but K_0 is the least element of $\mathcal{K}_1$, we obtain a contradiction. Hence there exists $a \in A$ such that

$$m(A) = m(K_0) = m(\{a\}). \qquad \square$$

We have in a special case the following result.

Corollary 9.2 *Let m, $m: \mathcal{B} \to [0,1]$, be a regular S-decomposable measure with respect to an arbitrary but fixed t-conorm S. If $A \in \mathcal{B}$ is an atom of m, then there exists a point $a \in A$ such that*

$$m(A) = m(\{a\}).$$

Corollary 9.3 *Each continuous from above monotone regular Borel null-additive set function m, $m: \mathcal{B}(R) \to [0,\infty]$, which has the property*

$$m((a,b)) = g(b-a)$$

on each finite interval (a,b) for some continuous at 0 function g with $g(0) = 0$, is non-atomic.

Proof. Suppose that the theorem is not true. Then there exists an atom $A \in \mathcal{B}(R)$ of m. Hence by Theorem 9.6 there exists an element a from A such that $m(A) = m(\{a\})$. So we would have

$$m(\{a\}) = m\Big(\bigcap_{n=1}^{\infty}\Big(a-\frac{1}{n}, a+\frac{1}{n}\Big)\Big) = \lim_{n\to\infty} m\Big(\Big(a-\frac{1}{n}, a+\frac{1}{n}\Big)\Big) = \lim_{n\to\infty} g\Big(\frac{2}{n}\Big) = 0,$$

i.e., $m(A) = m(\{a\}) = 0$. Contradiction. $\square$

The following theorem says that inner regularity and autocontinuity from above implies regularity.

Theorem 9.7 *Let m be a positive set function which is autocontinuous from above, $\mathcal{K}$-exhaustive and satisfies the equality*

$$m(A) = \sup\{m(K) : K \in \mathcal{K},\ K \subset A\} \qquad (A \in \mathcal{B}). \tag{9.4}$$

Then m is a regular set function.

Proof. If we suppose that the theorem is not true, then there exists a set $A \in \mathcal{B}$ and a number $\epsilon > 0$ such that

$$m(V \setminus K) > \epsilon \tag{9.5}$$

for each $K \in \mathcal{K}$, $V \in \mathcal{O}$, $K \subset A \subset V$. Let us fix such sets K_0 and V_0. The equality (9.4) implies that there exist

$$C_1, D_1 \in \mathcal{K}, \quad C_1 \subset V_0 \setminus A, \quad D_1 \subset A \setminus K_0$$

such that

$$m(C_1) \geq \frac{1}{2} m(V_0 \setminus A) \quad \text{and} \quad m(D_1) \geq \frac{1}{2} m(A \setminus K_0).$$

Let $V_1 = V_0 \setminus C_1$ and $K_1 = K_0 \cup D_1$. It is obvious that $V_1 \in \mathcal{O}$, $K_1 \in \mathcal{K}$ and $K_1 \subset A \subset V_1$. Then (9.5) implies

$$m(V_1 \setminus A_1) > \epsilon .$$

Repeating this after n-steps we obtain sets

$$C_n, D_n \in \mathcal{K}, \quad C_n \subset V_{n-1} \setminus A, \quad D_n \subset A \setminus K_{n-1},$$

such that

$$m(C_n) \geq \frac{1}{2} m(V_{n-1} \setminus A), \quad m(D_n) \geq \frac{1}{2} m(A \setminus K_{n-1}). \tag{9.6}$$

In this way we have obtained two sequences $\{C_n\}$ and $\{D_n\}$ of pairwise disjoint sets from $\mathcal{K}$. By $\mathcal{K}$-exhaustivity of m we have

$$\lim_{n\to\infty} m(C_n) = 0, \quad \lim_{n\to\infty} m(D_n) = 0 .$$

Hence by (9.6)

$$\lim_{n\to\infty} m(V_n \setminus A) = 0 \quad \text{and} \quad \lim_{n\to\infty} m(A \setminus K_{n-1}) = 0 . \tag{9.7}$$

In this way we have obtained decreasing sequences $\{V_n \setminus A\}$ and $\{A \setminus K_{n-1}\}$ from $\mathcal{B}$ (namely, $V_n \subset V_{n-1}$ and $K_n \supset K_{n-1}$) such that (9.7) holds and

$$(V_0 \setminus A) \cap (A \setminus K_0) = \emptyset .$$

Hence

$$\lim_{n\to\infty} m(V_n \setminus K_n) = \lim_{n\to\infty} m((V_n \setminus A) \cup (A \setminus K_n)) = 0 .$$

So we obtain a contradiction with (9.5). □

9.3 Notes and References

The classical Lyapunov theorem is obtained in [Lyapunov 1940]. The Darboux property is investigated for non-additive set functions in [Dobrakov 1974] and [Drewnowski 1978]. The presented approach is taken from [Pap 1994a]. Regularity of non-additive set functions was investigated in [Dinculeanu 1967], [Dobrakov 1974]. The presented results are taken from [Pap 1993a].

Chapter 10

Radon-Nikodym Type Theorems and Representation Theorems

One of the most important theorems in classical measure theory is the Radon-Nikodym theorem, which ensures the representation of a measure ν which is continuous with respect to a given measure μ by an integral of a function $f = f(\nu)$, i.e., $\nu(A) = \int_A f\, d\mu$ $(A \in \Sigma)$. Unfortunately, such a theorem cannot be directly translated to the general non-additive case. To see that, it is enough to consider the finite case. Suppose that $|X| = n$. Then the dimension of all real valued set functions on $\mathcal{P}(X)$ is $2^n - 1$, while the dimension of the set $\mathbf{F}$ of all functions from X into $\mathbf{R}$ is n.

So it is not realistic to expect such a correspondence between $\mathbf{F}$ and all set functions.

10.1 Radon-Nikodym Theorem for the Choquet Integral

10.1.1 The Finite Case

Suppose that X is finite. We introduce a correspondence between $\mathbf{F}$ and all set functions.

Definition 10.1 For two real-valued set functions m and v, we say that v is totally absolutely continuous with respect to m if for every $A_1, \ldots, A_n \in \mathcal{P}(X)$ the condition

$$m\Big(\bigcup_{i=1}^{n} A_i\Big) = \sum_{\{I: I \subset \{1,2,\ldots,n\}, I \neq \emptyset\}} (-1)^{|I|+1} m\Big(\bigcap_{i \in I} A_i\Big)$$

implies

$$v\Big(\bigcup_{i=1}^{n} A_i\Big) = \sum_{\{I: I\subset\{1,2,\dots,n\}, I\neq\emptyset\}} (-1)^{|I|+1} v\Big(\bigcap_{i\in I} A_i\Big).$$

We will use the σ-algebra notation Σ, since in the finite case we take $\mathcal{A} = \mathcal{P}(X)$.

Propositon 10.1 *If v is totally absolutely continuous with respect to m, then*

$$a_A^m = 0 \quad (A \in \Sigma \setminus \{\emptyset\}) \quad \textit{implies} \quad a_A^v = 0 \quad (A \in \Sigma \setminus \{\emptyset\}),$$

where $\{a_A^m\}_{A\in\Sigma\setminus\{\emptyset\}}$ and $\{a_A^v\}_{A\in\Sigma\setminus\{\emptyset\}}$ are coefficients of the representations in Theorem 4.1, respectively.

Proof. Follows directly by the representation of coefficients in Theorem 4.1. □

The converse is not true.

Example 10.1 Let $X = \{1,2,3\}$ and m is given by $a_A^m = 1$ for $A \neq X$ and $a_X^m = -1$. Then we have for arbitrary but fixed real-valued set function v that $a_A^v = 0$, whenever $a_A^m = 0$, i.e., v is totally absolutely continuous with respect to m.

But on the other hand if we take $A_1 = \{1\}$ and $A_2 = \{2,3\}$, then we have

$$m(\{1,2,3\}) = m(A_1) + m(A_2).$$

It is obvious that there is a set function v for which such equality does not hold.

We have the following Radon-Nikodym type theorem.

Theorem 10.1 *If Σ is finite and m and v are such real-valued set functions defined on Σ that v is totally absolutely continuous with respect to m, then there exists a set function $g\colon \Sigma \setminus \{\emptyset\} \to R$ such that for every $f \in \mathbf{F}$*

$$(C)\int_X f\,dv = \sum_{A\in\Sigma\setminus\{\emptyset\}} a_A^m g(A) \min_{x\in A} f(x)$$

(integral is the asymmetric Choquet integral).

Specially, for every $B \in \Sigma$ we have

$$v(B) = \sum_{A\subset B, A\in\Sigma\setminus\{\emptyset\}} a_A^m g(A).$$

Proof. We define the function $g\colon \Sigma \setminus \{\emptyset\} \to R$ in the following way

$$g(A) = \begin{cases} \frac{a_A^v}{a_A^m} & \text{for } a_A^m \neq 0, \\ 1 & \text{otherwise.} \end{cases}$$

Hence the finite version of Theorem 10.6 implies the desired conclusion. □

10.1.2 The Chain Condition

Let m be a non-negative finite monotone set function with $m(\emptyset) = 0$ on a paving $\mathcal{D}$ and let f be a non-negative measurable simple function on $(X, \mathcal{D})$. Then f can be written in the form

$$f = \sum_{i=1}^{n} a_i \cdot \chi_{A_i}$$

where $A_1 \subset A_2 \subset \cdots \subset A_n$, and $a_i \geq 0$, $i = 1, 2, \ldots, n$. If $f\chi_A$ is $\mathcal{D}$-measurable for all $A \in \mathcal{D}$, then it is easy to see that

$$v(A) = (C)\int_A f\,dm = \sum_{i=1}^{n} a_i \cdot m(A \cap A_i) = \sum_{i=1}^{n} a_i \cdot m_{A_i}(A),$$

where $m_{A_i}(A) = m(A \cap A_i)$. We can rewrite this as

$$v = \sum_{i=1}^{n} a_i \cdot m_{A_i},$$

where $\{A_i\}$ is a chain of subsets in $\mathcal{D}$.

Hence if v is induced by a simple function, then there is a chain $\{A_i\}$ such that v is a linear combination of the measures $\{m_{A_i}\}$. Let $\mathcal{D}$ be a σ-lattice. Let $\mathcal{C}$ be a chain in $\mathcal{D}$, by $\mathcal{U}_{\mathcal{C},m}$ we denote the system of all monotone set functions v induced by m and a $\mathcal{C}$-measurable simple function. We can write

$$\mathcal{U}_{\mathcal{C},m} = \left\{v : \exists \{A_i\}_{i=1}^{n} \subset \mathcal{C},\ a_i \geq 0,\ i = 1, \ldots, n,\ v = \sum_{i=1}^{n} a_i \cdot m_{A_i}\right\}.$$

Let v and τ be from $\mathcal{U}_{\mathcal{C},m}$ let $v = \sum_{i=1}^{n} a_i m_{A_i}$ and let $\tau = \sum_{j=1}^{k} b_j m_{B_j}$. We introduce a partial ordering on $\mathcal{U}_{\mathcal{C},m}$ in the following way: we say that $v \leq \tau$ if and only if $\sum_{i=1}^{n} a_i \chi_{A_i} \leq \sum_{j=1}^{k} b_j \chi_{B_j}$. We introduce

$$\overline{\mathcal{U}}_{\mathcal{C},m} = \{v : \exists \{v_n\} \subset \mathcal{U}_{\mathcal{C},m},\ v_n \nearrow v, \ \text{ is finite on } \mathcal{D}\}.$$

The previous considerations leads us to the following Radon-Nikodym type theorem.

Theorem 10.2 *Let m and v be continuous non-negative set functions on $(X, \mathcal{D})$. Let $\mathcal{D}$ be a σ-lattice. Then there exists a non-negative m-integrable function f on $(X, \mathcal{D})$ such that*

$$v(A) = (C)\int_A f\,dm$$

for every $A \in \mathcal{D}$ if and only if there exists a chain $\mathcal{C} \subset \mathcal{D}$ such that $v \in \mathcal{U}_{\mathcal{C},m}$.

Proof. Let f be a non-negative m-integrable function on $(X, \mathcal{D})$, and let

$$v(A) = (C)\int_A f\,dm$$

for every A in $\mathcal{D}$. Let

$$\mathcal{C} = \{f^{-1}[a, \infty) : a > 0\}.$$

Then $\mathcal{C}$ is a chain in $\mathcal{D}$ with $A \cap \mathcal{C} \subset \mathcal{D}$ for all A in $\mathcal{D}$. Therefore $\{f_n\}$, defined by

$$f_n = \frac{1}{n}\sum_{i=1}^{2^n} \chi_{f^{-1}[\frac{i}{n},\infty)} \qquad (n \in \mathbf{N}),$$

is a sequence of non negative $\mathcal{C}$-measurable simple functions with $f_n \nearrow f$. Clearly $f_n \cdot \chi_A \nearrow f \cdot \chi_A$ for every $A \in \mathcal{D}$. Since

$$(C)\int_A f_n m \leq (C)\int_A f\,dm < \infty$$

by the monotone convergence Theorem 7.13 we have $v_n(A) \nearrow v(A)$ for every $A \in \mathcal{D}$. Since $v_n \in \mathcal{U}_{\mathcal{C},m}$ and $v_n \nearrow v$ we get that $v \in \overline{\mathcal{U}}_{\mathcal{C},m}$.

Let now v be from some $\overline{\mathcal{U}}_{\mathcal{C},m}$. Then there exists a sequence $\{v_n\}$

$$v_n = \sum_{i=1}^{k_n} a_{n,i} m(A_{n,i})$$

in $\mathcal{U}_{\mathcal{C},m}$ with $v_m \nearrow v$ and v is finite on $\mathcal{D}$. Take

$$f_n = \sum_{i=1}^{k_n} a_{n,i}\chi_{A_{n,i}}.$$

Then $\{f_n\}$ is a sequence of non-negative integrable functions. Let $f = \bigvee_{n=1}^{\infty} f_n$. Since $\mathcal{D}$ is a σ-lattice, f is measurable. Then $f_n\chi_A \nearrow f\chi_A$ for every A in $\mathcal{D}$. Since

$$(C)\int_A f_n\,dm = v_n(A) \le v(A) < \infty\,,$$

by the monotone convergence Theorem 7.13 we have that f is an integrable function and

$$v(A) = \lim_n v_n(A) = \lim_{n\to\infty}(C)\int_A f_n\,dm = (C)\int_A f\,dm\,.$$

So we get that

$$v(A) = (C)\int_A f\,dm\,.$$

□

10.2 Representation of the Choquet Integral

10.2.1 Min Minus Min Representation

As a consequence of the representation of a set function from the space CV as a difference of two totally monotone set functions (in Section 4.3) we will obtain a representation of the Choquet integral.

Let $\mathbf{F}$ be the set of all bounded measurable functions and $\mathbf{F}_0$ the subset of all simple functions.

We will give two theorems without proof, whose origin is in game theory.

Definition 10.2 If m is a real valued set function, then the core of m is given by

$$C(m) = \{\mu : \mu \text{ is a finitely additive measure, } \mu(A) \ge m(A)\ (A \in \Sigma),\\ \mu(X) = m(X)\}\,.$$

Theorem 10.3 *For every convex set function m we have $C(m) \neq \emptyset$.*

Theorem 10.4 *A monotone set function m is convex if and only if*

(i) $C(m) \neq \emptyset$;

(ii) for every $f \in \mathbf{F}_0$ ($f \in \mathbf{F}$)

$$(C)\int f\,dm = \min_{\mu \in C(m)} \int f\,d\mu .$$

The very useful representation of the Choquet integral in (ii) cannot be extended on non-convex set functions. But using the Theorem 4.3 in which for finite Σ, a set function $m \in CV$ can be uniquely represented in the form $m = v_1 - v_2$, where v_1 and v_2 are totally monotone set functions. We obtain by Theorem 10.4 the representation

$$(C)\int_X f\,dm = \min_{\mu \in C^+(v_1)} \int_X f\,d\mu - \min_{\mu \in C^-(v_2)} \int_X f\,d\mu$$

for all $f \in \mathbf{F}$.

This representation is not unique, since neither v_1, v_2 nor C^+, C are unique (even C^+ and C^- cannot be cores of totally monotone set functions).

In a quite analogous way, using Theorem 4.6, we can prove the general case.

Theorem 10.5 *If $m \in CV$, then there exist two sets of finitely additive measures, C^+ and C^-, which are convex and closed in weak $*$-topology and such that for all $f \in \mathbf{F}_0$*

$$(C)\int_X f\,dm = \min_{\mu \in C^+} \int_X f\,d\mu - \min_{\mu \in C^-} \int f\,d\mu .$$

10.2.2 Mean of Mins and Min of Means Representations

Suppose first that Σ is finite. Therefore we have for the elementary belief function (see Chapter 4)

$$(C)\int f\,dm_A = \min\{f(x) : x \in A\} .$$

Since by Theorem 4.1 the space of all non-additive set functions, which vanishes on empty set is spanned by elementary belief functions we obtain for an arbitrary real valued set function m on Σ and any real valued function f on X

$$(C)\int f\,dm = \int fd\Big(\sum_{A\in\Sigma\setminus\{\emptyset\}} a_A m_A\Big) = \sum_{A\in\Sigma\setminus\{\emptyset\}} a_A\Big(\int f\,dm_A\Big)$$
$$= \sum_{A\in\Sigma\setminus\{\emptyset\}} a_A\big(\min_{x\in A} f(x)\big)\,.$$

We can interpret this result so that every Choquet integral with respect to any real valued set function is some average over minima. Namely, the Choquet integral of f with respect to additive m ($a_A = 0$ for all A with $|A| \geq 2$) is an average of the values of f. In another special case, when m is a belief function ($a_A \geq 0$ for all $A \in \Sigma \setminus \{\emptyset\}$ and $\sum_{A\in\Sigma\setminus\{\emptyset\}} a_A = 1$), the Choquet integral of f with respect to m is a weighted average over all minima of the function f.

We will give also the corresponding result for the general (infinite) case. For that we will need the following result.

Lemma 10.1 *The subset CCV of CV consisting of set functions m for which μ_m is σ-additive signed measure is a Banach subspace of CV.*

Proof. If $\{m_n\}$ is a Cauchy sequence from CCV, then the pointwise limit m by Theorem 4.5 belongs to CV and $\lim_{n\to\infty} \|m_n - m\|_c = 0$. We have to prove $m \in CCV$, i.e., μ_m is σ-additive.

Taking the corresponding sequence $\{\mu_{m_n}\}$ to $\{m_n\}$ we have

$$\overline{\mu_m - \mu_{m_n}} = \|\mu_m - \mu_{m_n}\|_c = \|\mu_{(m-m_n)}\| = \|m - m_n\|_c\,.$$

Letting $n \to \infty$ we obtain that μ_m as the limit in variation norm of the sequence of σ-additive measures is itself σ-additive. □

Theorem 10.6 *If $m \in CV$ and it is such that the corresponding μ_m is a σ-additive signed measure, i.e, $m \in CCV$ and $f\colon X \to \mathbf{R}$ is a bounded measurable function then*

$$(C)\int_X f\,dm = \int_{A\setminus\{\emptyset\}} \big(\inf_{x\in A} f(x)\big)\,d\mu_n(A)\,.$$

Proof. It is enough to prove the theorem for non-negative functions. Therefore we have for $f \geq 0$

$$\begin{aligned}(C)\int_X f\,dm &= \int_0^\infty m(\{x : f(x) \geq r\})\,dr \\ &= \int_0^\infty \Big(\int_{\mathcal{A}\setminus\{\emptyset\}} m_A(\{x : f(x) \geq r\})\,d\mu_n(A)\Big)\,dr\,. \qquad (10.1)\end{aligned}$$

Since we want to exchange the order of integrals we have to prove that the function

$$(r, A) \mapsto m_A(\{x : f(x) \geq r\})$$

defined on $[0, +\infty] \times \mathcal{A} \setminus \{\emptyset\}$ is $\mathcal{B}([0, +\infty)) \times \sigma(\mathcal{J})$-measurable, where $\mathcal{B}([0, +\infty))$ is the family of all Borel subset of $[0, +\infty)$.

Since for every $r \in [0, +\infty)$ we have

$$\{A : f(x) \geq r \quad \text{for all } x \in A\} = \{x : f(x) \geq r\}$$

and $\{x : f(x) \geq r\} \in \mathcal{A}$, it is easy to prove that the set

$$\{(r, A) : f(x) \geq r \quad \text{for all } x \in A\} = \big\{(r, A) : \inf_{x\in A} f(x) \geq r\big\}$$

is $\mathcal{B}([0, +\infty)) \times \sigma(\mathcal{J})$-measurable. Now, we can apply the classical Fubini's theorem on the last integral in (10.1). Hence

$$\begin{aligned}(C)\int_X f\,dm &= \int_{\mathcal{A}\setminus\{\emptyset\}} \int_0^\infty \mu_A(\{x : f(x) \geq r\})\,dr\,d\mu_m(A) \\ &= \int_{\mathcal{A}\setminus\{\emptyset\}} \big(\inf_{x\in A} f(x)\big)\,d\mu_m(A)\,. \end{aligned}$$

□

Since any totally monotone set function is convex we obtain by Theorem 10.4 as a consequence of the preceding theorem 'Mean of Mins' and 'Min of Means' representations of the Choquet integral with respect to the totally monotone set function.

Corollary 10.1 *If $m \in CCV$ is totally monotone and $f \in \mathbf{F}_0$, then we have*

$$(C)\int_X f\,dm = \int_{\mathcal{A}\setminus\{\emptyset\}} \big(\inf_{x\in A} f(x)\big)\,d\mu_m(A) = \min_{\mu\in C(m)} \int_X f\,d\mu\,.$$

Remark 10.1 A representation of monotone non-negative continuous set function m from Section 4.4 implies another representation of Choquet integral for non-negative measurable function f

$$(C)\int f\,dm = \int I(F)\,d\mu\,,$$

where $I(f)(y) = \sup\{r : y \in H(\{x : f(x) > r\})\}$ $(y \in Y)$ and the integral in the right side is the Lebesgue integral.

10.3 Radon-Nikodym Theorem for Sugeno Integral

We suppose that m is a finite positive monotone continuous set function on a σ-algebra Σ of subsets of a set X. Using the results on the Sugeno integral from Chapter 7 we easily obtain the following proposition.

Propositon 10.2 *For a measurable function $f\colon X \to [0,1]$ with $v_f(x) < \infty$, where*

$$v_f(E) = (S)\int_E f\,dm \qquad (E \in \Sigma)\,,$$

we have

$$m(E \cap \{x : f(x) > v_f(E)\} \le v_f(E)$$
$$\le m(E \cap \{x : f(x) \ge v_f(E)\})$$

for all $E \in \Sigma$.

Propositon 10.3 *For a measurable function $f\colon X \to [0,1]$ we have*

$$v_f(E \cap F_s) \ge s \wedge m(E \cap F_s)$$

for all $E \in \Sigma$ and $s \ge 0$, where $F_s = \{x : f(x) \ge s\}$.

Proof. This follows from the definition of the Sugeno integral for an arbitrary but fixed F_s

$$v_f(E \cap F_s) = (S)\int_{E\cap F_s} f\,dm = \sup_{\alpha\in[0,1]}[\alpha \wedge m(E \cap F_\alpha \cap F_s)]$$
$$= \max[\sup_{\alpha\le s} \alpha \wedge m(E \cap F_s), \sup_{\alpha>s}\{\alpha \wedge m(E \cap F_s)\}$$
$$= \max\{(s \wedge m(E \cap F_s), \sup_{\alpha>s}[\alpha \wedge m(E \cap F_\alpha]\} \ge s \wedge m(E \cap F_s)\,.$$

□

Theorem 10.7 *Let m and v be two monotone and continuous real valued set functions defined on Σ such that $v \leq m$. Then there exists a measurable function $f\colon X \to [0, \infty)$ such that*

$$v(E) = (S) \int_E f\, dm \qquad (E \in \Sigma)$$

if and only if there exists a decreasing family

$$\{B_\alpha\}_{\alpha \in [0,+\infty]}, \quad B_\alpha \in \Sigma \qquad (\alpha \in [0, +\infty]),$$

such that

(i) $v(E \cap B_\alpha) \geq \alpha \wedge m(E \cap B_\alpha)$ *and*

(ii) $v(E) \leq m(E \cap B_{v(E)})$

for every $E \in \Sigma$ and every $\alpha \in [0, +\infty]$.

Proof. Necessity. Taking $B_\alpha = F_\alpha$ $(\alpha \geq 0)$ in Proposition 10.2 and 10.3 we obtain the properties (i) and (ii).

Sufficiency. We define the function f by

$$f(x) = \sup\{\alpha \in [0, +\infty] : x \in B_\alpha\}.$$

Since

$$\{x : f(x) > \alpha\} = \bigcup_{n \in \mathbf{N}} B_{\alpha + \frac{1}{n}}$$

for every $n \in \mathbf{N}$ we obtain that f is measurable. Hence

$$(S) \int_E f\, dm = \sup_{\alpha \in [0,+\infty]} [\alpha \wedge m(E \cap B_\alpha)].$$

therefore by conditions (i) and (ii)

$$v(E) = \int_E f\, dm \qquad (E \in \Sigma).$$

□

10.4 Daniell-Greco-Stone Representation of Functionals

In this section a Daniell-Stone type representation of monotone autocontinuous from above functional by integral with respect to monotone autocontinuous from above set function is given.

10.4.1 Outer Set Function

Let $\mathcal{L}$ be a lattice of subsets of the given set X. For any monotone set function $m : \mathcal{L} \to [0, \infty]$ we have defined the outer set function m^* on $\mathcal{P}(X)$ by

$$m^*(E) = \inf\{m(F) : E \subset F \in \mathcal{L}\}.$$

We shall need the following property of m^* in the proof of the main result.

Propositon 10.4 *If m is a monotone and autocontinuous from above set function on a lattice $\mathcal{L}$ of subsets of a given set X, then the corresponding outer set function m^* is also a monotone and autocontinuous set function on $\mathcal{P}(X)$.*

Proof. Monotonicity of m^* is obvious. Let $A, B \in \mathcal{P}(X)$. If $m^*(A) = \infty$, the statement is true. Let $\epsilon > 0$. By the definition of m^* we choose A_1' from $\mathcal{L}$ such that $A \subset A_1'$ and

$$m(A_1') \leq m^*(A) + \epsilon\,. \tag{10.2}$$

Since m is autocontinuous from above there exists $\delta = \delta(\epsilon, A_1') > 0$ such that

$$m(A_1' \cup B) < m(A_1') + \frac{\epsilon}{2}$$

whenever $m(B) < \delta$, $B \in \mathcal{L}$.

Let $B \in \mathcal{P}(X)$ be such that $m^*(B) < \frac{\delta}{2}$. Then by the definition of m^* we have that for $0 < \delta' < \frac{\delta}{2}$ there exists $B_1' \in \mathcal{L}$ such that

$$m(B_1') \leq m^*(B) + \delta'\,.$$

Hence

$$m(B_1') < \delta\,.$$

Then by the autocontinuity from above of m

$$m(A_1' \cup B_1') < m(A_1') + \frac{\epsilon}{2}. \tag{10.3}$$

Now we have by the monotonicity of m^*, (10.3) and (10.2)

$$m^*(A \cup B) \leq m^*(A_1' \cup B_1') = m(A_1' \cup B_1') < m(A_1') + \epsilon \leq m^*(A) + \epsilon\,. \quad \square$$

10.4.2 Non-Linear Functionals

Let $\mathbf{F}$ be a family of functions $f: X \to [0, \infty]$ with the properties:

$af,\ f \wedge a,\ f - f \wedge a \in \mathbf{F} \quad (f \in \mathbf{F},\ a \in [0, \infty))$ **(Stone condition)**,

$f \wedge g,\ f \vee g \in \mathbf{F} \quad (f, g \in \mathbf{F})$ **(Lattice condition)**.

The family of all upper level sets of the function f is denoted by $\mathcal{U}_f$, i.e.

$$\mathcal{U}_f = \big\{\{x : f(x) > t\} : t \in [0, \infty]\big\} \cup \big\{\{x : f(x) \geq t\} : t \in [0, \infty]\big\}.$$

A class of functions $\mathbf{F}_c \subset \mathbf{F}$ is comonotonic (common monotonic) if

$$\bigcup_{f \in \mathbf{F}_c} \mathcal{U}_f$$

is a chain. A class of functions $\mathbf{F}_c$ is comonotonic iff each pair of functions from $\mathbf{F}_c$ is comonotonic. Equivalent condition for a pair of functions f and g to be comonotonic is that there is no pair $x_1, x_2 \in X$ such that $f(x_1) < f(x_2)$ and $g(x_1) > g(x_2)$.

Let M be a functional $M: \mathbf{F} \to [0, \infty]$. We list the following properties of M for every $f, g \in \mathcal{F}$ and every $a \in [0, \infty)$:

(F1) $f \leq g \Rightarrow M(f) \leq M(g)$ (monotonicity);

(F2) $M(f + g) = M(f) + M(g)$
for comonotonic f and g such that $f + g \in \mathbf{F}$ (comonotonic additivity);

(F3) $\lim_{a \downarrow 0} M(f - f \wedge a) = M(f)$ (lower marginal continuity);

(F4) $\lim_{a \to \infty} M(f \wedge a) = M(f)$ (upper marginal continuity);

(F5) $(\forall \epsilon > 0)(\exists \delta > 0)(\forall g \in \mathbf{F})(\forall f \in \mathbf{F})$
$(M(g) < \delta \Rightarrow M(f \vee g) < M(f) + \epsilon)$ (autocontinuity from above);

(F6) $M(f \vee g) = M(f)$ whenever $M(g) = 0$ (null-additivity);

(F7) $M(f \vee g) + M(f \wedge g) \leq M(f) + M(g)$ (submodularity);

(F8) $M(f \vee g) + M(f \wedge g) \geq M(f) + M(g)$ (supermodularity).

Remark 10.2 Conditions (F1) and (F2) imply the positive homogeneity of M, i.e., $M(af) = aM(f)$ for any $a \in [0, \infty)$.

Remark 10.3 Conditions (F1) and (F5) imply condition (F6).

10.4.3 The Representation

Definition 10.3 For a given class $\mathbf{F}$ of functions $f, f\colon X \to [0,\infty]$, a monotone set function $m\colon \mathcal{P}(X) \to [0,\infty]$ and a functional $M : \mathbf{F} \to [0,\infty]$ we say that m represents M if we have

$$M(f) = (C)\int f\, dm \qquad (f \in \mathbf{F}).$$

Theorem 10.8 *If M is a functional $M : \mathbf{F} \to [0,\infty]$ with the properties (F1)–(F5), then there exists a monotone autocontinuous from above set function m, $m\colon \mathcal{P}(X) \to [0,\infty]$, which represent M.*

Proof. We introduce two monotone set functions for any $E \in \mathcal{P}(X)$

$$\alpha(E) = \sup\{M(f) : f \in \mathbf{F},\ f \leq \chi_E\}$$
$$\beta(E) = \inf\{M(f) : f \in \mathbf{F},\ \chi_E \leq f\}.$$

We shall prove that the set function $m = \beta$ is autocontinuous from above and it represents the functional M.

We define for $f\colon X \to [0,\infty]$

$$G_f = \{(x,t) : (x,t) \in X \times [0,\infty),\ t < f(x)\}.$$

Since $\mathbf{F}$ satisfies the lattice condition we have

$$G_f \cup G_g = G_{f\vee g}, \quad G_f \cap G_g = G_{f\wedge g}.$$

Hence $\mathcal{L} = \{G_f : f \in \mathbf{F}\}$ is a lattice of subsets of the set X. We introduce a set function v on $\mathcal{L}$

$$v(G_f) = M(f) \quad (f \in \mathbf{F}).$$

By

$$G_f \subset G_g \Leftrightarrow f \leq g$$

we obtain that v is a monotone set function. Now we shall use the property (F5) of M. If we suppose $v(G_f) < \delta$, then also $M(f) < \delta$ holds.Then the condition (F5) implies

$$v(G_f \cup G_g) = v(G_{f\vee g}) = M(f \vee g) < M(f) + \epsilon = v(G_f) + \epsilon,$$

i.e., v is autocontinuous from above.

By Proposition 10.4 the outer set function v^* of v is also monotone and autocontinuous from above. Then the restriction of v^* to $\mathcal{P}(X)$ given by $v^*(G_{\chi_E})$ $(E \in \mathcal{P}(X))$ is also monotone and autocontinuous from above, since

$$G_{\chi_E} \leq G_{\chi_F} \Leftrightarrow E \subset F ,$$

we have $G_{\chi_{E \cup F}} = G_{\chi_E} \cup G_{\chi_F}$. The set function v^* is equal to the set function m. Namely, we have for every $E \in \mathcal{P}(X)$

$$\begin{aligned} m(E) &= \inf\{M(f) : \chi_E \leq f \in \mathbf{F}\} = \inf\{v(G_f : G_{\chi_E} \subset G_f \in \mathcal{L}\} \\ &= v^*(G_{\chi_E}) . \end{aligned}$$

Now we shall prove that m represents the functional M. For that purpose we take for a function $f \in \mathbf{F}$ its approximation by functions $f_n(x)$

$$f_n(x) = \frac{1}{2^n} \sum_{i=1}^{n2^n - 1} \chi_{x > \frac{i}{2^n}} .$$

Since we have

$$x \wedge \frac{i+1}{2^n} - x \wedge \frac{i}{2^n} \leq \chi_{x > \frac{i}{2^n}} \leq x \wedge \frac{i}{2^n} - x \wedge \frac{i-1}{2^n} ,$$

we obtain using Remark 10.1

$$f \wedge \frac{i+1}{2^n} - f \wedge \frac{i}{2^n} \leq \alpha\Big(f > \frac{i}{2^n}\Big) \leq m\Big(f > \frac{i}{2^n}\Big) \leq 2^n M\Big(f \wedge \frac{i}{2^n} - f \wedge \frac{i-1}{2^n}\Big).$$

Since the functions $f \wedge \frac{i}{2^n} - f \wedge \frac{i-1}{2^n} \in \mathbf{F}$ are comonotonic we obtain using the property (F2) of M and summing up the last inequalities

$$\begin{aligned} M\Big(\sum_{i=1}^{n2^n - 1} \Big(f \wedge \frac{i+1}{2^n} - f \wedge \frac{i}{2^n}\Big)\Big) &\leq \frac{1}{2^n} \sum_{i=1}^{n2^n - 1} m\Big(f > \frac{i}{2^n}\Big) \\ &\leq M\Big(\sum_{i=1}^{n2^n - 1} \Big(f \wedge \frac{i}{2^n} - f \wedge \frac{i-1}{2^n}\Big)\Big). \end{aligned}$$

Hence

$$M\Big(f \wedge n - f \wedge \frac{1}{2^n}\Big) \leq (C) \int f_n \, dm \leq M\Big(f \wedge \Big(n - \frac{1}{2^n}\Big)\Big) \leq M(f) .$$

Since the functions $(f - f \wedge 1) \wedge (n-1)$ and $(f \wedge 1 - (f \wedge 1) \wedge \frac{1}{2^n})$ are comonotonic and the following representation is true

$$f \wedge n - f \wedge \frac{1}{2^n} = (f \wedge n - f \wedge 1) + \left(f \wedge 1 - f \wedge \frac{1}{2^n}\right)$$
$$= (f - f \wedge 1) \wedge (n-1) + \left(f \wedge 1 - (f \wedge 1) \wedge \frac{1}{2^n}\right),$$

we obtain by the property (F2) of M

$$M\left(f \wedge n - f \wedge \frac{1}{2^n}\right) = M((f - f \wedge 1) \wedge (n-1)) + M\left(\left(f \wedge 1 - (f \wedge 1) \wedge \frac{1}{2^n}\right)\right).$$

Hence by the properties (F3) and (F4) of M

$$\lim_{n \to \infty} M\left(f \wedge n - f \wedge \frac{1}{2^n}\right) = M(f - f \wedge 1) + M(f \wedge 1) = M(f),$$

i.e., $M(f) = (C) \int f_n \, dm$.

Now we shall prove that

$$\lim_{n \to \infty} (C) \int f_n \, dm = (C) \int f \, dm.$$

Namely, by the inequality

$$f \wedge n - f \wedge \frac{1}{2^n} \le f_n \le f$$

we obtain

$$\int_{\frac{1}{2^n}}^{n} m(\{x : f(x) \ge t\}) \, dt = (C) \int \left(f \wedge n - f \wedge \frac{1}{2^n}\right) dm \le (C) \int f_n \, dm$$
$$< (C) \int f \, dm = \int_0^\infty m(\{x : f(x) \ge t\}) \, dt.$$

Taking $n \to \infty$ we obtain the desired equality. □

Corollary 10.2 *Let* **B** *be the family of bounded functions* $f \colon X \to [0, \infty)$ *which satisfy Stone conditions and* $f + 1 \in$ **B** *for* $f \in$ **B**. *Then for each functional* M *on* **B** *which satisfies the conditions* (F1) *and* (F2) *there exists a monotone and autocontinuous from above set function* m *on* $\mathcal{P}(X)$ *which represents* M.

Corollary 10.3 *If* M *is a functional* $M \colon$ **F** $\to [0, \infty]$ *with the properties* (F1)–(F4) *and* (F7) (*or* (F8)), *then there exists a monotone submodular (supermodular) autocontinuous from above set function* $m \colon \mathcal{P}(X) \to [0, \infty]$ *which represent* M.

10.5 Notes and References

The Radon-Nikodym theorem for the Choquet integral with the notion of totally absolute continuity in the finite case was obtained in [Gilboa and Schmeidler 1992a]. The chain characterization of the Radon-Nikodym theorem presented in subsection 10.1.2 was obtained in [Mesiar and Šipoš 1993].

The representation of the Choquet integral was prepared using [Gilboa and Schmeidler 1992a] and [Gilboa and Schmeidler 1992b].

Radon-Nikodym theorem for Sugeno integral was obtained in [Sugeno 1974] and in the presented form in [Candeloro 1987] (for special (MS) integral see [Sugeno and Murofushi 1987]).

The representations in the sense of Daniell-Stone of non-additive functionals with integrals with respect to monotone non-additive set functions were investigated by [Anger 1977], and G. H. Greco investigations can be found in [Denneberg 1994]. The presented representation is our new result.

Open problem: Let m be monotone and null-additive set function. Is m^* null-additive too?

Chapter 11

k-Triangular Set Functions

11.1 Diagonal Theorem for Triangular Set Function

In this section we shall present a version of the very useful Diagonal theorem for triangular set functions. Namely, diagonal theorems as extract and generalization of the classical 'sliding hump' method are very useful in the proofs of generalized fundamental theorems of functional analysis and measure theory.

Let $\mathcal{R}$ be a ring subset of a set. Let $m\colon \mathcal{R} \to [0,\infty]$ be a triangular set function, i.e., k-triangular set function for $k = 1$.

For each set $A \subset X$, let

$$\hat{m}(A) = \sup\{m(B) : B \in \mathcal{R},\ B \subset A\}.$$

Definition 11.1 We say that m is α-bounded ($0 \le \alpha < \infty$) if $\inf_n m(B_n) \le \alpha$ for each disjoint sequence $\{B_n\}$ in $\mathcal{R}$.

If $\{A_n\}$ is a sequence of sets and $I \subset \mathbf{N}$, we write A_I for $\cup_{n\in I} A_n$.

Lemma 11.1 *Let $m\colon \mathcal{R} \to [0,\infty]$ be α-bounded. Then $\hat{m}$ is α-bounded on the subsets of X. Moreover, if $\{B_n\}$ is a disjoint sequence of subset of X and if $\beta > \alpha$, then there exists an infinite set $I \subset \mathbf{N}$ such that $\hat{m}(B_I) < \beta$.*

Proof. The first assertion is obvious. For the second, we represent $\mathbf{N}$ as an infinite disjoint union of infinite sets. □

Theorem 11.1 (Diagonal theorem for triangular set functions) *Let $\{m_n\}$ be a sequence of α-bounded triangular set functions on the ring $\mathcal{R}$ ($0 \le \alpha < \infty$), let $\{B_n\}$ be a disjoint sequence in $\mathcal{R}$, and let $\{\delta_n\}$ be a sequence in $[0,\infty]$. Then either there exists an infinite set $I \subset \mathbf{N}$ such that* *(i)* $\hat{m}_n(B_I \backslash B_n) < \delta_n$ $(n \in I)$,

or there exists a finite set $J \subset \mathbf{N}$ *which satisfies (i) with* $I = J$ *and*

(ii) $\hat{m}_n(B_J) \geq \delta_n - \alpha$ *for infinitely many* $n \in \mathbf{N}$.

Proof. If $\delta_n \geq \alpha$ for infinitely many $n \in \mathbf{N}$, we can take $J = \emptyset$. Suppose $\delta_n > \alpha$ for all $n \in \mathbf{N}$ large enough. Let us call $I \subset \mathbf{N}$ a nice set if it satisfies (i).

Suppose that no finite nice sets $I \subset \mathbf{N}$ satisfy (ii). We shall construct a sequence $\{I_n\}$ of infinite sets and a sequence $\{n_k\}$ in $\mathbf{N}$ as follows. Let $I_0 = \mathbf{N}$, and let n_1 be any natural number with $\delta_{n_1} > \alpha$. Note that $\{n_1\}$ is a nice set. Let $k \in \mathbf{N}$, and suppose that infinite sets $I_0 \supset \cdots \supset I_{k-1}$ and a nice set $\{n_1, \ldots, n_k\}$ have been chosen in such a way that

$$n_j \in I_{j-1} \quad \text{for} \quad 1 \leq j \leq k\,, \tag{11.1}$$

$$\hat{m}_{n_j}\Big(\bigcup_{i=1}^{j-1} B_{n_i}\Big) < \delta_{n_j} - \alpha \quad \text{for} \quad 1 \leq j \leq k, \quad \text{and} \tag{11.2}$$

$$\hat{m}_{n_j}(B_{I_j}) < \delta_{n_j} - \hat{m}_{n_j}\Big(\bigcup_{i=1}^{j-1} B_{n_i}\Big) \quad \text{for} \quad 1 \leq j \leq k-1\,. \tag{11.3}$$

By (11.2), the right-hand side of (11.3) with $j = k$ is larger than α, since each m_n is α-bounded, it follows from Lemma 11.1 that I_{k-1} contains an infinite set I_k which satisfies (11.3) for $j = k$. Moreover, no finite nice sets J satisfy (ii) and $\{n_1, \ldots, n_k\}$ is a nice set. So $I_n \cap (n_k, \infty)$ has an element n_{k+1} which satisfies (11.2) with $j = k+1$. The arguments in the next paragraph will show that $\{n_1, \ldots, n_{k+1}\}$ is a nice set.

Define $I = \{n_k : k \in \mathbf{N}\}$. To show that I satisfies (i), pick any $j \in \mathbf{N}$. Then $k > j$ implies $n_k \in I_{k-1} \subset I_j$ by (11.1). It follows by the subadditivity of $\hat{m}_{n_j}$ that

$$\hat{m}_{n_j}(B_I \setminus B_{n_j}) \leq \hat{m}_{n_j}\Big(\bigcup_{i=1}^{j-1} B_{n_j}\Big) + \hat{m}_{n_j}(B_{I_j}) < \delta_{n_j}\,,$$

which completes the proof. □

Corollary 11.1 (Weber's version of the Nikodym boundedness principle) *Let* $\mathcal{R}$ *be a* σ*-algebra, and let* $\mathcal{T}$ *be a set of* α*-bounded triangular set functions on* $\mathcal{R}$ *for* $\alpha < \infty$, *such that*

$$\sup\{m(B) : m \in \mathcal{T}\} < \infty$$

for each $B \in \mathcal{R}$. *Then* $\mathcal{T}$ *is uniformly bounded.*

Proof. Suppose that the theorem is not true. Then there exists a disjoint sequence $\{B_n\}$ in $\mathcal{R}$ and a sequence $\{m_n\}$ in $\mathcal{T}$ such that $m_n(B_n) > 2n$ for each $n \in \mathbf{N}$. Replacing $\mathcal{R}$ by a smaller ring, we may suppose $\mathcal{R} = \{B_E : E \subset \mathbf{N}$ and $B_E \in \mathcal{R}\}$.

If $J \subset \mathbf{N}$ is a finite set and $\hat{m}_n(B_J) > n - \alpha$ for infinitely many $n \in \mathbf{N}$, then J contains a set F such that $m_n(B_F) > n-\alpha$ for infinitely many $n \in \mathbf{N}$. Hence $\sup_n m_n(B_F) = \infty$, a contradiction. It follows from Theorem 11.1 with $\delta_n = n$ that there exists an infinite set $I \subset \mathbf{N}$ such that $\hat{m}_n(B_{I\setminus\{n\}}) < n$ for each $n \in \mathbf{N}$. Since $\mathcal{R}$ is a σ-algebra, we may suppose $B_I \in \mathcal{R}$. But then $n \in I$ implies

$$m_n(B_I) \geq m_n(B_n) - m_n(B_{I\setminus\{n\}}) > 2n - n = n\,.$$

Contradiction. □

Corollary 11.2 *Let α, β, $\gamma \in [0,\infty]$, and let $\{m_n\}$ be a sequence of α-bounded triangular set functions on the ring $\mathcal{R}$.*

(i) Suppose that for each disjoint sequence $\{B_n\}$ in $\mathcal{R}$ and each subsequence $\{m_{n_k}\}$ of $\{m_n\}$, we have

$$\inf\{m_{n_k}(B_E) : k \in E \subset \mathbf{N} \quad \textit{and} \quad B_E \in \mathcal{R}\} \leq \beta, \quad \textit{and}$$
$$\underline{\lim}_{j\to\infty}\{\underline{\lim}_{k\to\infty} m_{n_k}(B_j)\} \leq \gamma\, \overline{\lim}\, m_n(B_j)\,.$$

Then the $\{m_n\}$ is uniformly $(\alpha+\beta+\gamma)$-bounded.

(ii) If $\mathcal{R}$ is a σ-algebra and if

$$\psi(A) = \overline{\lim_{n\to\infty}}\, m_n(A)$$

is β-bounded, then the sequence $\{m_n\}$ is uniformly $(\alpha+2\beta)$-bounded.

Proof. For (ii), we may suppose $\beta = \gamma$. Suppose to the contrary that m_n is not uniformly $(\alpha+\beta+\gamma)$-bounded. Then there exists a disjoint sequence $\{B_n\}$ in $\mathcal{R}$, $p > \alpha+\beta+\gamma$, and a sequence $\{n_k\}$ in $\mathbf{N}$ such that $m_{n_k}(B_k) > p$ for each $k \in \mathbf{N}$. Since each m_n is α-bounded and $p > \alpha$, we may suppose $n_1 \leq n_2 \leq \ldots$.

Case 1. Suppose $\mathbf{N}$ contains an infinite set I such that $\hat{m}_{n_k}(B_{I\setminus\{k\}}) < (p+\alpha+\gamma-\beta)/2$ for each $k \in I$. Then $k \in E \subset I$ implies

$$\begin{aligned} m_{n_k}(B_E) &\geq m_{n_k}(B_k) - m_{n_k}(B_{E\setminus\{k\}}) \\ &> p - (p+\alpha+\gamma-\beta)/2 \\ &= (p-\alpha-\gamma+\beta)/2 > \beta\,. \end{aligned} \tag{11.4}$$

Hence, $\{m_{n_k}\}$ does not satisfy the first requirement in (i). Also (11.4) shows that $\psi(B_E) \geq (p - \alpha - \gamma + \beta)/2$ whenever $E \subset I$ is an infinite set with $B_E \in \mathcal{R}$. Therefore, if $\mathcal{R}$ is a σ-algebra, then ψ is not β-bounded.

Case 2. Suppose that **N** contains no infinite sets I as in Case 1. Then, by Theorem 11.1 with $\delta_k = \delta = (p + \alpha + \gamma - \beta)/2$, each infinite subset of **N** contains a finite set J and an infinite set I such that $k \in I$ implies

$$\hat{m}_{n_k}(B_J) \geq (p + \alpha + \gamma - \beta)/2 - \alpha = (p - \alpha + \gamma - \beta)/2 > \gamma .$$

An inductive application of this fact yields a disjoint sequence $\{F_j\}$ of finite subsets of **N**, a sequence $\{A_j\}$ in $\mathcal{R}$, and a subsequence $\{m_{m_j}\}$ of $\{m_{n_k}\}$ such that $A_j \subset B_{F_j}$ and

$$m_{m_k}(A_j) \geq (p - \alpha + \gamma - \beta)/2 > \gamma \qquad \forall k \geq j \geq 1 .$$

Hence the second requirement in (i) is not satisfied, and ψ is not γ-bounded. □

In the next section we shall obtain as easy consequences of Theorem 11.1 some diagonal theorems. We shall illustrate their usefulness in functional analysis and measure theory.

11.2 Diagonal Theorems and Some of Their Applications

11.2.1 Diagonal Theorems

Let **S** be a semigroup with triangular functional f (see Section 2.2). Let (x_{ij}) be an infinite matrix (indexed by $\mathbf{N} \times \mathbf{N}$) with entries in **S**. For each sequence $\{x_j\}$ in **S** and each $I \subset \mathbf{N}$, write $f(\sum_{j \in I} x_j)$ for $\lim_{n \to \infty} \sup f(\sum_{j \in I \cap [1,n]} x_j)$.

Corollary 11.3 (Mikusiński, Antosik, Pap, diagonal theorem) *Suppose* $\lim_{j \to \infty} f(x_{ij}) = 0$ $(i \in \mathbf{N})$. *Then there exists an infinite set I and a set $J \subset I$ such that*

(a) $\sum_{j \in I} f(x_{ij}) < \infty$ $(i \in \mathbf{N})$, *and*
(b) $f(\sum_{j \in J} x_{ij}) \geq f(x_{ij})/2$ $(i \in I)$.

Proof. By (i), we may assume that the (x_{ij}), satisfies $\sum_{j\in\mathbf{N}} f(x_{ij}) < \infty$ for each $i \in \mathbf{N}$. Define

$$m_i(E) = f\Big(\sum_{j\in E} x_{ij}\Big) \qquad (i \in \mathbf{N},\ E \subset \mathbf{N}).$$

Then each m_i is a continuous (hence exhaustive triangular set function on $\mathcal{P}(\mathbf{N})$). If $I \subset \mathbf{N}$ is any set such that $\hat{m}_i(I\backslash\{i\}) < m_i(\{i\})/2$ whenever $i \in I$, then $i \in I$ implies

$$m_i(I) \geq m_i(\{i\}) - m_i(I\backslash\{i\}) \geq m_i(\{i\})/2.$$

Therefore the desired result follows from Diagonal Theorem 11.1 for triangular set functions with $\delta_i = m_i(\{i\})/2$ and $\alpha = 0$. □

Let G be a quasi-normed group, i.e., a group endowed with a functional $|\cdot|$ such that it satisfies the conditions:

(i) $|0| = 0$;

(ii) $|x| = |-x|$;

(iii) $|x+y| \leq |x| + |y|$.

Corollary 11.4 (Antosik diagonal theorem) *Let $\beta, \gamma \in [0,\infty)$, and suppose that*

(i) $\lim_{j\to\infty} |x_{ij}| = 0 \quad (i \in \mathbf{N})$,

(ii) for each finite set $F \subset \mathbf{N}$, $\overline{\lim}_{j\to\infty} |\sum_{j\in F} x_{ij}| \leq \beta$, and suppose that each principle infinite submatrix of (x_{ij}) has a further principle submatrix (y_{ij}) such that

(iii) $\underline{\lim}_{i\to\infty} |\sum_{j\in\mathbf{N}} y_{ij}| \leq \gamma$.

Then

$$\overline{\lim_{i\to\infty}}\, |x_{ij}| \leq \beta + \gamma.$$

Proof. Taking an infinite submatrix of (x_{ij}), we may suppose that $\sum_{j\in\mathbf{N}} |x_{ij}| < \infty$ for each $i \in \mathbf{N}$ and $\lim_{i\to\infty} |x_{ii}|$ exists in $[0,\infty]$. Let m_i be the continuous triangular set function on $\mathcal{P}(\mathbf{N})$ defined as in the proof of the last theorem.

Fix any $\beta' > \beta$. Then (ii) implies that there exists no finite set $J \subset \mathbf{N}$ such that $\hat{m}_i(J) \geq \beta'$ for infinitely many $i \in \mathbf{N}$. So, by Diagonal Theorem 11.1 with $\alpha = 0$ and $\delta_i = \beta'$, there exists an infinite set $I \subset \mathbf{N}$ such that $\hat{m}_i(I\backslash\{i\}) < \beta'$ for each $i \in I$. Then, $i \in I' \subset I$ implies

$$|x_{ii}| = m_i(\{i\}) \leq m_i(I') + m_i(I'\backslash\{i\}) < m_i(I') + \beta' .$$

Then by (iii) we obtain that $\lim |x_{ii}| \leq \gamma + \beta'$. □

As an immediate consequence we obtain a diagonal theorem (which we will denote by DT) given in the next theorem.

Theorem 11.2 *Let $(G, |\cdot|)$ be a quasi-normed group, and (x_{ij}) an infinite matrix in G such that for every increasing sequence $\{m_i\}$ in* N *there exists a subsequence $\{n_i\}$ of $\{m_i\}$ such that*

$$\lim_{i\to\infty} x_{n_i n_j} = 0 \quad \textit{for all} \quad j \in \mathbf{N}, \quad \textit{and} \quad \lim_{i\to\infty} \sum_{j=1}^{\infty} x_{n_i n_j} = 0 .$$

Then $\lim_{i\to\infty} x_{ii} = 0$.

As a consequence we obtain the basic matrix theorem.

Theorem 11.3 *Let $(G, |\cdot|)$ be a quasi-normed group, and (x_{ij}) an infinite matrix in G such that $\lim_{i\to\infty} x_{ij} = x_j$ exists for all $j \in$* N *and for every increasing sequence $\{m_j\}$ in* N *there exists a subsequence $\{n_j\}$ of $\{m_j\}$ such that $\{\sum_{j=1}^{\infty} x_{in_j}\}$ is a Cauchy sequence.*

Then $\lim_{i\to\infty} x_{ij} = x_j$. uniformly with respect to $j \in$ N.

Proof. If we suppose that the theorem is not true, then there exists $\epsilon > 0$ and a subsequence $\{s_i\}$ of natural numbers such that $\sup_j |x_{s_i j} - x_j| > \epsilon$. By (I) we can choose sequences $\{i_k\}$ and $\{j_s\}$ such that the matrix (y_{ks}) for $y_{ks} = x_{i_k j_s} - x_{i_{k+1} j_s}$ satisfies the conditions of DT, but $|y_{kk}| > \epsilon$ $(k \in \mathbf{N})$, (taking eventually a submatrix). Contradiction. □

11.2.2 Some Applications

We now give some applications of diagonal theorems. We do not intend to formulate theorems in full generality, but to present the method of proofs. We will use some standard notions from functional analysis in this section.

First two theorems from measure theory.

Nikodym convergence. *Let $\{\mu_n\}_{n\in\mathbf{N}}$ be a sequence of countable additive functions from a σ-algebra Σ to a topological group G and let $\{E_n\}$ be a sequence of pairwise disjoint sets E_n from Σ.*

If $\mu_n(E) \to 0$ for $E \in \mathcal{R}$, then $\mu_n(E_n) \to 0$. Applying the DT to the matrix $\{\mu_i(E_j)\}$, we get the assertion.

Nikodym boundedness. *If X is a topological vector space and $\{\mu_n(E)\}$ is a bounded sequence for each $E \in \Sigma$, then $\{\mu_n(E_n)\}$ is a bounded sequence, for each pairwise disjoint sequence $\{E_n\}$ from Σ.*

Applying the DT to the matrix $\{\alpha_i\mu_i(E_j)\}$ where $\{\alpha_j\}$ is a scalar sequence converging to zero, we get the assertion.

Continuous convergence. *Assume that f_n for $n \in \mathbf{N}$ are additive and continuous mappings from a complete metric group G to a topological group Y, $x_n \in G$ for $n \in \mathbf{N}$ and $x_n \to 0$.*

If $f_n(x) \to 0$ for each $x \in G$, then $f_n(x_n) \to 0$. We remark that due to the completeness of G, for each subsequence $\{u_n\}$ of $\{x_n\}$ there are a subsequence $\{v_n\}$ of $\{u_n\}$ and $v \in G$ such that $v_1 + v_2 + \cdots = v$. Applying the DT to the matrix $\{f_i(x_j)\}$, we get $f_i(x_i) \to 0$.

Equicontinuity. *If Y is a topological vector space and $\{f_n(x)\}$ is a bounded sequence for each $x \in G$, then $f_n(x_n) \to 0$, $\{x_n\}$ is the same as in the previous case.*

Let $\{k_n\}$ be a sequence of positive integers such that $k_n \to \infty$ and $k_n x_n \to 0$. Applying the DT to the matrix $\{k_i^{-1} f_i(k_j x_j)\}$ we get $f_n(x_n) \to 0$.

Joint continuity. *If G is a metric group and at the same time X is a linear space such that the multiplication αx is separately continuous, then the multiplication is jointly continuous.*

Assume that $\alpha_n \to 0$ and $x_n \to 0$. Applying the DT to the matrix $\{\alpha_j x_j\}$ we get $\alpha_i x_i \to 0$.

Weak boundedness. *Let X be a normed space, X' be the dual of X, $x_n \in X$ and $x'_n \in X'$ for $n \in \mathbf{N}$. We note that X' is a Banach space. If $\{\langle x_n, x'\rangle\}$ (where $\langle x_n, x'_n\rangle = x'_n(x_n)$) is a bounded sequence for each $x' \in X$ and $\|x'_n\| \le 1$ for $n \in \mathbf{N}$, then $\{\langle x_n, x'_n\rangle\}$ is a bounded sequence.*

Applying the DT to the matrix $\{\langle\sqrt{\alpha_i}x_i, \sqrt{\alpha_i}x'_j\rangle\}$, where $\{\alpha_i\}$ is a numerical sequence converging to zero, we get $\alpha_i \langle x_i, x'_i\rangle \to 0$.

Weak $*$-boundedness. *If X is a Banach space, $\{\langle x, x'_n\rangle\}$ us a bounded sequence for each $x \in X$ and $\|x_n\| \le 1$ for $n \in \mathbf{N}$, then $\{\langle x_n, x'_n\rangle\}$ is a bounded sequence.*

Applying the DT to the matrix $\{\sqrt{\alpha_j}x_j, \sqrt{\alpha_i}x_i\rangle\}$ where $\{\alpha_i\}$ is a numerical sequence converging to zero, we get $\alpha_i \langle x_i, x_i'\rangle \to 0$.

Orlicz-Pettis. *If $\|x_n'\| \le 1$ for $n \in \mathbf{N}$ and for each increasing sequence of positive integers k_j there exists an $x \in X$ such that*

$$\sum_{j=1}^{\infty} \langle x_{k_j}, x'\rangle = \langle x, x'\rangle$$

for each $x' \in X$, then $\langle x_n, x_n'\rangle \to 0$.

We may assume that X is separable and $\langle x, x_n'\rangle \to \langle x, x'\rangle$ for each $x \in X$ and an x' in X'. Otherwise we would take a suitable subsequence of $\{x_n'\}$. Applying the DT to the matrix $\{\langle x_j, x_i' - x'\rangle\}$ we get $\langle x_i, x_i' - x'\rangle \to 0$, or equivalently, $\langle x_i, x_i'\rangle \to 0$ for $\langle x_i, x'\rangle \to 0$.

The method of proving theorems by the DT leads to various generalizations of classical results and to new results. In particular this can be achieved by considered sequences with a summability property, the so called K-sequences.

Definition 11.2 A sequence $\{x_n\}$ of elements x_n in a topological group G is said to be a K-sequence in G if for each subsequence $\{u_n\}$ of $\{x_n\}$ there are a subsequence $\{v_n\}$ of $\{u_n\}$ and $v \in G$ such that

$$v_1 + v_2 + \cdots = v\,.$$

K-sequences in G converge to zero. Sequences of which almost all elements equal zero are K-sequences. Subsequences of K-sequence are K-sequences in G. If Q is the group of rational numbers, then $\{10^{-10^n}\}$ is not a K-sequence in Q. Sequences converging to zero in complete metric groups are K-sequences. The converse is not true, i.e., it may happen that sequences converging to zero in a normed space are K-sequences but the space is not complete . An interesting property of K-sequences is that they may converge to zero with respect to a stronger topology than the initial topology, e.g., K-sequences with respect to the pointwise convergence in $C[0,1]$ (continuous functions on the interval $[0,1]$), converge to zero uniformly.

Now using DT we can generalize the classical theorems in functional analysis.

Continuous convergence. *Assume that the functions f_n for $n \in \mathbf{N}$ are additive and sequentially continuous mappings from a topological group G to a topological group Y and $x_n \in G$ for $n \in \mathbf{N}$. If $f_n(x) \to 0$ as $n \to \infty$ for $x \in G$ and $\{x_n\}$ as a K-sequence in G, then $f_n(x_n) \to 0$.*

Applying the DT to the matrix $\{f_i(x_i)\}$ we get the assertion.

Uniform boundedness. *If Y is a topological vector space, $\{f_n(x)\}$ is a bounded sequence for each $x \in G$ and $\{x_n\}$ is a K-sequence in G, then $\{f_n(x_n)\}$ is a bounded sequence.*

Applying the DT to the matrix $\{\alpha_i f_i(x_j)\}$ where $\{\alpha_i\}$ is a scalar sequence converging to zero we get the assertion.

K-weak sequences. *Assume that $\{x_n\}$ is a K-sequence with respect to the weak topology in a normed space X and $\{x'_n\}$ is a sequence of functionals x'_n on X such that $\|x'_n\| \leq 1$ for $n \in \mathbf{N}$. Then $\langle x_n, x'_n\rangle \to 0$.*

We may assume that X is separable and $\langle x, x'_n\rangle \to \langle x, x'\rangle$ as $n \to \infty$ for each $x \in X$ and an $x' \in X'$. Applying the DT to the matrix $\{\langle x_j, x_i - x'\rangle\}$ we get $\langle x_i, x'_i - x'\rangle \to 0$, or equivalently, $\langle x_i, x'_i\rangle \to 0$ for $\langle x_i, x'\rangle \to 0$.

Continuous convergence. *Assume that a group $C(X, Y)$ of sequentially continuous mappings from a topological group X to a topological group Y is endowed with the topology of pointwise convergence, $\{f_n\}$ is a K-sequence in $C(X, Y)$ and $x_n \to x$ in X. Then $f_n(x_n) \to 0$.*

Applying the DT to the matrix $\{f_j(x_i) - f_j(x)\}$ we get $f_i(x_i) - f_i(x) \to 0$, or equivalently, $f_i(x_i) \to 0$ for $f_i(x) \to 0$.

Joint continuity. *Assume that X, Y and Z are topological groups, $B\colon X \times Y \to Z$ is a bi-additive and separately continuous mapping, $x_n \to 0$ in X and $y_n \to 0$ in Y. If $\{x_n\}$ (resp. $\{y_n\}$) is a K-sequence in X (resp. in Y), then $B(x_n, y_n) \to 0$.*

Applying the DT to the matrix $\{B(x_j, y_i)\}$ (resp. $\{B(x_i, y_j)\}$) we get the assertion.

We will use in this and next subsetctions some notions from functional analysis (see [Horvath 1966]).

11.2.3 The Adjoint Theorem

If (X, τ) is a topological vector space, a sequence $\{x_k\}$ in X is said to be a τ-K-sequence if every subsequence of $\{x_k\}$ has a further subsequence $\{x_{n_k}\}$ such that the subseries $\sum_k x_{n_k}$ is τ-convergent to an element of X. A topological vector space (X, τ) is said to K-space if every sequence which converges to 0 is a τ-K-sequence. A subset A of X is said to be τ-K-bounded if for every sequence $\{x_n\} \subseteq A$ and every scalar sequence $\{t_n\}$ which converges to 0, the sequence $\{t_n x_n\}$ is a τ-K-sequence.

Throughout this subsection let E and F be Hausdorff locally convex topological vector spaces and $T\colon E \to F$ a linear mapping. The domain of the adjoint operator, T', is defined to be

$$D(T') = \{y' \in F' : y'T \in E'\}$$

and

$$T'\colon D(T') \to E' \quad \text{is defined by} \quad T'y' = y'T\,.$$

Let $\sigma(E, E')$ and $\beta(E, E')$ denote weak and strong topology on E, respectively.

Now we have the adjoint theorem.

Theorem 11.4 *T' is sequentially continuous with respect to the relative $\sigma(F', F)$ (weak) topology on $D(T')$ and the topology on E' of uniform convergence on $\sigma(E, E')$-K-convergent sequences. In particular, T' is bounded with respect to these topologies.*

Proof. Let $\{y_j'\} \subseteq D(T')$ be $\sigma(F', F)$ convergent to 0 and let $\{x_i\} \subseteq E$ be a $\sigma(E, E')$-K-sequence. Consider the matrix

$$M = [\langle T'y_i', x_j\rangle] = [\langle y_i', Tx_j\rangle]\,.$$

It is easily checked that M is a K-matrix, so by the basic matrix theorem

$$\lim_{i\to\infty} \langle T'y_i', x_j\rangle = 0$$

uniformly for $j \in \mathbf{N}$. □

In particular if E is a normed vector space we obtain an interesting consequence. First, we require

Lemma 11.2 *Let E be normed vector space. A subset B of X' is norm bounded iff B is uniformly bounded on sequences which are norm convergent to* 0.

Proof. It is obvious that a normed bounded subset B of X' is uniformly bounded on sequences which are norm convergent to 0. Suppose that the converse is not true, i.e. there exists a sequence $\{x'_n\}$ from B with $\{x'_n\} > n^2$. Pick $x_n \in X$ such that $\{x_n\} = 1$ and

$$|\langle x'_n, x_n \rangle| > \{x'_n\} - \frac{1}{n}.$$

Then we have $\{\frac{x_n}{n}\} \to 0$ as $n \to \infty$. But

$$\left| \left\langle x'_n, \frac{x_n}{n} \right\rangle \right| > n - \frac{1}{n^2},$$

contradicting the hypothesis. □

Corollary 11.5 *Let E be a normed K-space. Then T' carries weak* bounded subsets of $D(T')$ to norm bounded subsets of E'. In particular, T' is norm-bounded.*

Proof. In a normed K-space E any sequence which converges to 0 is norm K convergent and so $\sigma(E, E')$-K-convergent. Hence by Theorem 11.4 and Lemma 11.4 we obtain the desired conclusion. □

$\beta^*(E', E)$ is the topology on E' of uniform convergence on $\beta(E, E')$ bounded subsets of E. $K(E', E)$ is the locally convex topology on E' of uniform convergence on the $\sigma(E, E')$-K-bounded subsets of E. The topology $\beta^*(E', E)$ is stronger than $K(E', E)$ (see the following results).

Lemma 11.3 *If $A \subseteq E$ is absolutely convex, bounded and sequentially complete, then A is K-bounded.*

Proof. Let $\{x_j\}$ be a sequence from A and $\{t_j\}$ be a sequence of scalars such that $t_j \to 0$. Given any subsequence of $\{t_j\}$ pick a further subsequence $\{t_{n_j}\}$ such that

$$\sum_{j=1}^{\infty} |t_{n_j}| \le 1.$$

Let p be a continuous semi-norm on E. The partial sums $S_k = \sum_j^k t_{n_j} x_{n_j}$ belong to A and form a Cauchy sequence in A since if $k > \ell$

$$p(S_k - S_\ell) \leq \sum_{j=\ell+1}^{k} |t_{n_j}| p(x_{n_j}) \leq \sup_{x \in A} p(x) \sum_{j=\ell+1}^{k} |t_{n_j}| .$$

Hence the series $\sum_{j=1}^{\infty} t_{n_j} x_{n_j}$ converges in A since A is sequentially complete. □

Lemma 11.4 *If $B \subset E'$ is $\sigma(E', E)$-K bounded, then B is $\beta(E', E)$ bounded.*

Proof. Let $A \subseteq E$ be $\sigma(E, E')$ bounded. It suffices to show $\{\langle x_i', x_i \rangle\}$ is bounded whenever $\{x_i'\} \subseteq B$ and $\{x_i\} \subseteq A$. Let $\{t_i\}$ be a sequence of positive real numbers which tends to zero. Consider the matrix

$$M = \left[\langle \sqrt{t_j} x_j', \sqrt{t_i} x_i \rangle\right] .$$

Since $\{\sqrt{t_i} x_i\}$ is $\sigma(E, E')$ convergent to 0 and $\{\sqrt{t_j} x_j'\}$ is $\sigma(E', E)$-K convergent, M is a K-matrix. Then by the Basic Matrix Theorem 11.3, $t_i < x_i', x_i > \rightarrow 0$ as $i \rightarrow \infty$. Hence $\{\langle x_i', x_i \rangle\}$ is bounded. □

We shall need the following known result.

Theorem 11.5 *A subset B of E is $\sigma(E, E')$ bounded iff B is $\beta^*(E, E')$ bounded.*

Now we can prove

Theorem 11.6 *T' carries weak* bounded subsets of $D(T')$ into $\beta^*(E', E)$ bounded sets.*

Proof. Let $B \subseteq D(T')$ be weak* bounded. We claim $T'B$ is $\sigma(E', E)$ bounded. Let $x \in E$. Since B is weak* bounded we have

$$\sup_{y' \in B} |\langle T'y', x \rangle| = \sup_{y' \in B} |\langle y', Tx \rangle| < \infty .$$

By Theorem 11.5 the set $T'B$ is $\beta^*(E', E)$ bounded. □

If E is an $\mathcal{A}$-space, i.e. every bounded subset of E is K-bounded, then we have the following two results.

Theorem 11.7 *If $D(T') = F'$ (equivalently, if T is $\sigma(E, E')$-$\sigma(F, F')$-continuous), then T' is $\beta^*(E, E')$-$\beta(F, F')$-continuous.*

Proof. Let $B \subseteq F'$ be $\sigma(F', F)$-bounded. Then TB is $\beta^*(E', E)$ bounded by Theorem 11.6. Since $(T'B)^0$ is the basic $\beta^*(E, E')$ neighbourhood and B_0 is the basic $\beta(F, F')$ neighbourhood, we have that the equality $(T'B)_0 = T^{-1}(B_0)$ gives the result. □

Corollary 11.6 *If E is an $\mathcal{A}$-space and $D(T') = F'$, then T is $\beta^*(E, E')$-$\beta(F, F')$-continuous.*

11.2.4 The 'Fatness' Condition

In this subsection we present an application of diagonal theorems on the relation of (Saeki's) fatness condition with the geometry of the Banach space.

Definition 11.3 A Banach space X satisfies the fatness condition if for each constant $C > 0$ there exists a positive integer k such that given $x_1, \ldots, x_k \in X$, $\|x_i\| \geq 1$, for each $i = 1, \ldots, k$, there exists $F \subseteq \{1, \ldots, k\}$ such that $\|\sum_{i \in F} x_i\| \geq C$.

Let m and v be two triangular set functions defined on Σ. v is m-absolutely continuous if and only if v given $\epsilon > 0$, there exists $\delta > 0$ such that $A \in \Sigma$ and $\hat{m}(A) < \delta$ implies $v(A) < \epsilon$.

If $\mathcal{T}$ is a set of triangular set functions and $\psi_{\mathcal{T}}(A) = \sup\{v(A) : v \in \mathcal{T}\}$ for each $A \in \Sigma$, then $\psi_{\mathcal{T}}$ is a triangular set function. $\mathcal{T}$ is equi-exhaustive, equi-continuous, or equi-absolutely continuous iff $\psi_{\mathcal{T}}$ is exhaustive, continuous, or m absolutely continuous, respectively. For vector measure w and a collection of such concepts in the definitions above with v replaced by $A \mapsto \|w(A)\|$, for example $\hat{w}(A) = \{\|w(B)\| : B \subset A,\ B \in \Sigma\}$.

Theorem 11.8 *Let m be a countable subadditive triangular set function, let X be a Banach space with fatness condition, and let $\{m_n\}$ be a sequence of finitely additive X-valued measures such that there exists a positive finite constant γ*

$$\overline{\lim_{n \to \infty}}\, \|m_n(A)\| \leq \gamma \underline{\lim_{n \to \infty}}\, \|m_n(A)\| < \infty \tag{11.5}$$

for each $A \in \Sigma$. Then we have:

(a) If each m_n is m absolutely continuous, then $\{m_n\}$ is equi-m-absolutely continuous.

(b) If each m_n is exhaustive, then $\{m_n\}$ is equi-exhaustive and bounded.

(c) If each m_n is continuous (i.e. countable additive), then $\{m_n\}$ is equi-continuous and bounded.

Proof. Suppose that the left-hand side of 11.5, denoted by ψ, is not exhaustive. Then there exists a disjoint sequence $\{B_k\}$ in Σ and $\beta > 0$ such that $\psi(B_k) > \beta$ for every k. Given $C < \infty$, let k be the natural number furnished by the fatness condition. By our assumption on $\{B_k\}$ and (11.5), there exists $n \geq 1$ such that $\beta < \gamma\|m_n(B_j)\|$ for $j = 1, \ldots, k$. By our choice of k, we can find $F \subset \{1, 2, \ldots, k\}$ such that

$$\Big\|m_n\Big(\bigcup_{j\in F} B_j\Big)\Big\| = \Big\|\sum_{j\in F} m_n(B_j)\Big\| \geq C\beta/\gamma .$$

Since $C < \infty$ is arbitrary, it follows that the $\{m_n\}$ is not uniformly bounded. This fact, combined with (11.5) and Corollary 11.1, shows that if each m_n is exhaustive then ψ is exhaustive. Hence (b) and (c) with $\mathcal{T} = \{m_n\}$ follows by Corollary 11.2.

(a) Suppose that each m_n is m-absolutely continuous but the $\{m_n\}$ is not equi-m-absolutely continuous. Then the above $\{B_s\}$ can be chosen to satisfy $\sum_1^\infty \hat{m}(B_s) < \infty$. But X is supposed to satisfy the fatness condition, so (11.5) ensures that there exists a disjoint sequence $\{F_s\}$ of finite subsets of $\mathbf{N}$ and a subsequence $\{m_{n_s}\}$ of $\{m_n\}$ such that $\|m_{n_s}(\cup_{j\in F_s} B_j)\| > k$ for each $s > 1$. Since each m_{n_s} is m-absolutely continuous and $\hat{m}(\cup_{j\in F_s} B_j) \to 0$ as $k \to \infty$, it follows from diagonal Theorem 11.1 for set functions that there exists $I \subset \mathbf{N}$ such that $\|m_{n_s}(\cup_{j\in I} B_j)\| > s/2$ for infinitely many s. This contradicts (11.5), which establishes (a). □

Definition 11.4 A Banach space X has cotype q (≥ 2) if there is a constant $K_q > 0$ such that for each $n \geq 1$, $x_1, \ldots, x_n \in X$, we have

$$\Big(\sum_{k=1}^n \|x_k\|^q\Big)^{1/q} \leq K_q \int_0^1 \Big\|\sum_{k=1}^n r_k(t)x_k\Big\|\, dt$$

where $\{r_n\}$ denotes the Rademacher sequence on $[0, 1]$.

Definition 11.5 A Banach space X contains the ℓ_∞^n's uniformly if there is a constant $\lambda > 1$ such that for each $n \geq 1$ there is an n-dimensional subspace E_n of X and isomorphism $v_n : \ell_\infty^n \to E_n$ such that $\|v_n\|\|v_n^{-1}\| < \lambda$.

Using Theorem 11.8 the following theorem can be proved.

Theorem 11.9 *Let X be any Banach space. The following are equivalent:*

(a) X has cotype.

(b) X satisfies the fatness condition.

(c) If $\{m_n\}$ is a sequence of exhaustive (respectively, countable additive) X-valued vector measures on a σ-algebra Σ such that there exists $\gamma \geq 1$ such that for each $A \in \Sigma$ we have $\limsup_{n\to\infty} \|m_n(A)\| \geq \gamma \liminf_{n\to\infty} \|m_n(A)\| < \infty$, then $\{m_n\}$ is uniformly exhaustive (respectively, uniformly countable additive).

(d) X does not contain the ℓ_∞^n's uniformly.

11.3 The Brooks-Jewett Type Theorem

The classical Nikodym convergence theorem says that the limit of the sequence of scalar measures on a σ-algebra is again a measure. Further generalization the Brooks-Jewett theorem says that the limit of the sequence of finitely additive scalar and exhaustive set functions on a σ-algebra is again additive and exhaustive. In this section we will prove a type of Brooks-Jewett theorem for k-triangular set functions, defined on algebras with some weak σ-conditions.

It is known that for additive set functions the Brooks-Jewett theorem is equivalent with the Nikodym convergence theorem, and even more with the Vitali-Hahn-Saks theorem which says that the limit of a sequence of signed measures which are continuous with respect to a measure is also continuous with respect to this measure.

11.3.1 Algebras with Some Additional Properties

Let $\mathcal{A}$ be an algebra of subsets of a nonempty set X.

Let m be a set function from $\mathcal{A}$ into $[0, +\infty]$, or into a commutative semigroup **S** with a neutral element 0 and endowed with a triangular functional f.

Definition 11.6 An algebra $\mathcal{A}$ has the (BJ) (Brooks-Jewett) property with respect to the class $\mathcal{K}$ of set functions if every sequence $\{m_n\}$ from $\mathcal{K}$, which is such that

$$\lim_{n\to\infty} m_n(A) = m_0(A)$$

exists for each $A \in \mathcal{A}$, is uniformly exhaustive.

The classical result is that a σ-algebra Σ has the (BJ) property with respect to the class of all measures on Σ.

Further generalization is that a σ-algebra Σ has the (BJ) property with respect to the class of additive exhaustive set functions with values in real numbers, or in a topological group (on a special lattice), or in a uniform semigroup (σ-additive).

Definition 11.7 An algebra $\mathcal{A}$ has the sequential completeness property — SCP if each disjoint sequence $\{E_n\}$ from $\mathcal{A}$ has a subsequence $\{E_{n_j}\}$, whose union is in $\mathcal{A}$.

Definition 11.8 An algebra $\mathcal{A}$ has the subsequentional interpolation property — SIP if each subsequence $\{A_{j_n}\}$ of each disjoint sequence $\{A_j\}$ from $\mathcal{A}$ has a subsequence $\{A_{j_{n_k}}\}$ and there exists a set $B \in \mathcal{A}$ such that

$$A_{j_{n_k}} \subset B \qquad (k \in \mathbf{N}).$$

and $A_j \cap B = \emptyset$ for $j \in \mathbf{N} \setminus \{j_{n_k} : k \in \mathbf{N}\}$.

11.3.2 Convergence Theorem

Let $k \in [1, +\infty)$.

Definition 11.9 A set function m from an algebra $\mathcal{A}$ into $[0, +\infty]$ is k-subadditive if $m(\emptyset) = 0$ and for each disjoint set A and B from $\mathcal{A}$ holds

$$m(A \cup B) \leq m(A) + km(B).$$

As we have seen in Subsection 2.2.8 a set function m from an algebra $\mathcal{A}$ into $[0, +\infty)$ is k-triangular if it is k-subadditive and for each disjoint set A and B from $\mathcal{A}$ holds

$$m(A) - km(B) \leq m(A \cup B).$$

We shall need the following two lemmas in the proof of Theorem 11.10.

Lemma 11.5 *Let $\mathcal{A}$ satisfy SCP and let $\{m_n\}, m_n\colon \mathcal{A} \to [0, +\infty)$, be a sequence of k-subadditive exhaustive set functions. Then each disjoint sequence $\{E_n\}$ of sets from $\mathcal{A}$ contains a subsequence $\{E_{k_n}\}$ such that each m_n is k-countable subadditive on the algebra*

$$\mathcal{A}(E) = \{A \cap E : A \in \mathcal{A}\},$$

where $E = \bigcup_{n=1}^{\infty} E_{k_n} \in \mathcal{A}$, *i.e., for each disjoint sequence* $\{A_i\}$ *from* $\mathcal{A}(E)$ *such that* $\bigcup_{n=1}^{\infty} A_i \in \mathcal{A}(E)$ *holds*

$$m_n\Big(\bigcup_{i=1}^{\infty} A_i\Big) \leq m_n(A_1) + k\sum_{i=2}^{\infty} m_n(A_i) \qquad (n \in \mathbf{N}).$$

Proof. The semivariation $\widehat{m}_n$ of m_n is also exhaustive.
This implies the exhaustivity of the monotone subadditive set function

$$m = \sum_{n=1}^{\infty} \frac{\min(1, \widehat{m}_n)}{2^n}.$$

For submeasure m we can easily obtain (Drewnowski's Lemma) a subsequence $\{E_{n_k}\}$ of $\{E_n\}$ such that m is order continuous on the σ-algebra Σ generated by $\{E_{n_k}\}$. Namely, this construction goes by repeating the procedure of dividing the set $\mathbf{N}$ on two disjoint infinite sets so that the series of values of m on elements of the sequence $\{E_n\}$ indexed by one of these infinite subset of $\mathbf{N}$ is less than a power of $1/2$. Taking in each step a lower bound of the chosen set of indices we obtain the desired sequence $\{n_k\}$ for the subsequence of $\{E_n\}$. We can assume that $E = \bigcup_{n=1}^{\infty} E_{k_n} \in \mathcal{A}$. Then by the definition of m we obtain that each $\widehat{m}_n$ is also order continuous on Σ. Now, let $\{A_i\}$ be a disjoint sequence from $\mathcal{A}(E)$ such that $\bigcup_{n=1}^{\infty} A_n \in \mathcal{A}(E)$. We have

$$m_n\Big(\bigcup_{i=1}^{\infty} A_i\Big) \leq m_n\Big(\bigcup_{i=1}^{s} A_i\Big) + k m_n\Big(\bigcup_{i=s+1}^{\infty} A_i\Big)$$
$$\leq m_n(A_1) + k\sum_{i=2}^{s} m_n(A_i) + \widehat{m}_n\Big(\bigcup_{i=s+1}^{\infty} A_i\Big).$$

Letting $s \to \infty$, we obtain the countable k-subaddivity of each m_n. □

We can easily prove the following generalization of this lemma given as an extract from the elementary proof of the Antosik-Mikusiński-Pap Diagonal Theorem (see Subsection 11.2.1).

Lemma 11.6 *If* $x_{ij} \in [0,+\infty)$ $(i,j \in \mathbf{N})$ *is such that*

$$\lim_{j\to\infty} x_{ij} = 0 \qquad (i \in \mathbf{N})$$

$$\lim_{i\to\infty} x_{ij} = 0 \qquad (j \in \mathbf{N})$$

and $x_{ij} > 0$ $(i \in \mathbf{N})$ *and* $k \geq 1$, *then there exists a sequence* $\{i_n\}$ *of natural numbers and a sequence* $\{\epsilon_n\}$ *of positive real numbers such that*

$$k\sum_{k=1}^{n-1} x_{i_n i_k} = \left(\frac{1}{2} - \epsilon_n\right) x_{i_n i_n}\,, \tag{11.6}$$

$$k^2 x_{i_n i_{n+q}} < 2^{-q}\epsilon_n x_{i_n i_n} \qquad (n,q \in \mathbf{N}), \tag{11.7}$$

where $\sum_{k=1}^{0} = 0$.

Theorem 11.10 *Let* $\mathcal{A}$ *satisfy* SCP *and* $\{m_n\}$, *let* $m_n\colon \mathcal{A} \to [0,+\infty)$ $(n \in \mathbf{N})$ *be a sequence of k-triangular exhaustive set functions. If*

$$\lim_{n\to\infty} m_n(E) = m_0(E)$$

exists for each $E \in \mathcal{A}$ *and* m_0 *is exhaustive, then* $\{m_n\}_{n=0}^{\infty}$ *is uniformly exhaustive and* m_0 *is k-triangular.*

Proof. Suppose that $\{m_n\}_{n=0}^{\infty}$ is not uniformly exhaustive. We can assume by passing to a subsequence if necessary, that there exists $\epsilon > 0$ and a disjoint sequence $\{A_n\}$ from $\mathcal{A}$ such that

$$m_n(A_n) > 2\epsilon \qquad (n \in \mathbf{N})\,. \tag{11.8}$$

Since m_0 is exhaustive there exists a subsequence $\{E_n\}$ of $\{A_n\}$ (by SCP of $\mathcal{A}$ we can assume that $\bigcup_{n=1}^{\infty} E_n \in \mathcal{A}$) such that

$$\widehat{m}_0\Big(\bigcup_{n=1}^{\infty} E_n\Big) < \frac{\epsilon}{2}\,. \tag{11.9}$$

By (11.8) we have

$$m_n(E_n) > 2\epsilon \qquad (n \in \mathbf{N}). \tag{11.10}$$

Let $x_{ij} = m_i(E_j)/i$, $(i,j \in \mathbf{N})$. We have by exhaustivity of each m_i, $\lim_{j\to\infty} x_{ij} = 0$ $(i \in \mathbf{N})$. We have also $\lim_{i\to\infty} x_{ij} = 0$ $(j \in \mathbf{N})$. So we can apply Lemma 11.6 on (x_{ij}) $(i, j \in \mathbf{N})$. Then we obtain a sequence $\{i_n\}$ of natural numbers and a sequence $\{\epsilon_n\}$ of positive real numbers such that (11.6) and (11.7) hold.

Since $\mathcal{A}$ has (SCP) by Lemma 11.5 there exists a subsequence $\{E_{i_{n_s}}\}$ of $\{E_n\}$ such that $A = \bigcup_{s=1}^{\infty} E_{i_{n_s}} \in \mathcal{A}$ and $m_{i_{n_s}}$ $(s \in \mathbf{N})$ are k-countable subadditive on $\mathcal{A}(A)$.

Taking $y_{sr} = x_{i_{n_s} i_{n_r}}$ we have by (11.6) and (11.7)

$$k \sum_{r=1}^{s-1} y_{sr} \leq \left(\frac{1}{2} - \epsilon'_s\right) y_{ss} \tag{11.11}$$

$$k^2 y_{s+q} < 2^{-q} \epsilon'_s y_{ss} \qquad (s, q \in \mathbf{N}), \tag{11.12}$$

where $\epsilon'_s = \epsilon_{n_s}$.

Using the k-countable subadditivity of $m_{i_{n_s}}$ $(s \in \mathbf{N})$ on $\mathcal{A}(A)$ we obtain

$$m_{i_{n_s}}(A) \geq m_{i_{n_s}}(E_{i_{n_s}}) - k \sum_{r=1}^{s-1} m_{i_{n_s}}(E_{i_{n_s}}) - k^2 \sum_{r=s+1}^{\infty} m_{i_{n_s}}(E_{i_{n_r}}) \quad (s \in \mathbf{N}).$$

Hence by (11.11) and (11.12) we obtain

$$i_{n_s}^{-1} m_{i_{n_s}}(A) \geq y_{ss} - k \sum_{r=1}^{s-1} y_{sr} - k^2 \sum_{r=s+1}^{\infty} y_{sr} \geq \frac{y_{ss}}{2} \qquad (s \in \mathbf{N}).$$

This implies

$$m_{i_{n_s}}(A) \geq m_{i_{n_s}}(E_{i_{n_s}})/2 \qquad (s \in \mathbf{N}).$$

Hence by (11.10)

$$m_{i_{n_s}}(A) > \epsilon \qquad (s \in \mathbf{N}).$$

This is a contradiction with

$$\lim_{s \to \infty} m_{i_{n_s}}(A) = m_0(A) \leq \widehat{m_0}(A) \leq \hat{m}_0\left(\bigcup_{n=1}^{\infty} E_n\right) < \frac{\epsilon}{2},$$

where we have used (11.9). □

Using Theorem 11.10 we can obtain a more general statement.

Theorem 11.11 *All assumptions and statements as in Theorem 11.10, but now the algebra $\mathcal{A}$ satisfies (SIP) instead of (SCP).*

By Theorem 11.10, $\mathcal{A}$ with (SIP) has (BJ) property with respect to the family of k-triangular set functions for which m_0 in Definition 11.6 is exhaustive.

Let $\mathbf{S}$ be a commutative semigroup with a neutral element 0, which is endowed with a triangular functional f.

Definition 11.10 A set function $m \colon \mathcal{A} \to \mathbf{S}$ is k-triangular for $k \in [1, \infty)$ if

$$f(m(A)) - kf(m(B)) \le f(m(A \cup B)) \le f(m(A)) + kf(m(B))$$

holds for each $A, B \in \mathcal{A}$ such that $A \cap B = \emptyset$.

Corollary 11.7 *Let $\mathcal{A}$ satisfy (SIP) and let $\{m_n\}$, $m_n \colon \mathcal{A} \to \mathbf{S}$ $(n \in \mathbf{N})$ be a sequence of k-triangular exhaustive set functions, i.e.,* $\lim_{j\to\infty} f(m_n(E_j)) = 0$ *for each disjoint sequence $\{E_j\}$ from $\mathcal{A}$. If*

$$\lim_{n\to\infty} f(m_n(E)) = \nu(E)$$

exists for each $E \in \mathcal{A}$ and ν is exhaustive, then $\{m_n\}$ is uniformly exhaustive.

Proof. Let $\nu_n(E) = f(m_n(E))$ $(E \in \mathcal{A})$. Then applying Theorem 11.11 on $\{\nu_n\}$ we obtain the desired statement. □

11.4 Brooks-Jewett Theorem on Orthomodular Lattice

If A is a set, we denote by $\mathcal{P}(A)$, $\mathcal{F}(A)$, and $\mathcal{I}(A)$ respectively the set of all subsets of A, the set of finite subsets of A and the set of infinite subsets of A. Let $\mathbf{N}' = \mathbf{N} \cup \{0\}$.

11.4.1 Orthomodular Lattice

Definition 11.11 $\mathbf{L} = (\mathbf{L}, \le, ', \mathbf{0}, \mathbf{1})$ is an orthomodular poset if it is a non-empty partially ordered set (poset, for short), with an unary operation on $\mathbf{L}$ such that

(i) $\mathbf{0}$ and $\mathbf{1}$ are respectively the smallest and the largest element of $\mathbf{L}$,

(ii) $'$ is an orthocomplementation on $\mathbf{L}$, i.e., $'$ is idempotent, decreasing and $a \wedge a' = \mathbf{0}$, $a \vee a' = \mathbf{1}$ holds for every $a \in \mathbf{L}$.

(iii) If $a, b \in \mathbf{L}$ and $a \leq b'$, then $a \vee b$ exist,

(iv) if $a, b \in \mathbf{L}$ and $a \leq b$, then $b = a \vee (b \wedge a')$.

$\mathbf{L}$ is an orthomodular lattice if $\mathbf{L}$ is also a lattice, i.e., for $a, b \in \mathbf{L}$, $a \vee b$ and $a \wedge b$ exists.

A pair (a, b) of elements of $\mathbf{L}$ is said to be orthogonal, denoted by $a \perp a$, if $a \leq b'$.

A subset $\mathbf{L}'$ of $\mathbf{L}$ is said to be orthogonal if $a, b \in \mathbf{L}'$ and $a \neq b$ implies $a \perp b$.

Let $a, b \in \mathbf{L}$. We say that a commutes with b (the pair (a, b) is compatible), in symbols aCb, if $a \wedge b$ and $a \wedge b'$ exist and $a = (a \wedge b) \vee (a \wedge b')$. It is easy to deduce that also b commutes with a.

Definition 11.12 (i) An orthomodular poset is said to be σ-orthocomplete if the supremum of every countable orthogonal subset of $\mathbf{L}$ exists.

(ii) An orthomodular lattice $\mathbf{L}$ has the subsequential completeness property (SCP) if for every orthogonal sequence $\{a_i\}_{i \in \mathbf{N}'}$ in $\mathbf{L}$ there exists $M \in \mathcal{I}(\mathbf{N}')$ such that the supremum of $\{a_i : i \in M\}$ exists.

(iii) An orthomodular lattice $\mathbf{L}$ has the subsequential interpolations property (SIP) if for every countable orthogonal subset K of $\mathbf{L}$ and every $K_0 \in \mathcal{I}(K)$, there exist $b \in \mathbf{L}$ and $H \in \mathcal{I}(K_0)$ such that

$$a \leq b \quad \text{for every} \quad a \in H, \qquad a \leq b' \quad \text{for every} \quad a \in K \setminus H\,.$$

Definition 11.13 A subset G of a lattice $\mathbf{L}$ is said to be a sublattice of $\mathbf{L}$ if $a \wedge b \in G$ and $a \vee b \in G$, for every $a, b \in G$.

A sublattice G of $\mathbf{L}$ is said to be orthomodular if

(i) the supremum of G exists and belongs to G;

(ii) $a' \wedge (\vee_{b \in G} b) \in G$ for every $a \in G$.

If we have $\mathbf{1} = \vee_{a \in G} a$ we obtain the definition of a subalgebra of an orthomodular lattice.

Propositon 11.1 *Let G be an orthomodular sublattice of the orthomodular lattice* **L**. *If we define* $*: G \to G$ *by*

$$a \mapsto a^* = a' \wedge (\vee_{b\in G} b),$$

then $(G, \leq, *, \mathbf{0}, \mathbf{1}^* = \vee_{b\in G} b)$ *is an orthomodular lattice.*

Proof. We note first that (i) and (ii) of the Definition 11.11 are trivially verified.

We shall prove that $*$ is an orthocomplementation on G. We have

$$(a^*)'=(a^*)' \wedge \mathbf{1}^*=(\mathbf{1}^* \wedge a')' \wedge \mathbf{1}^*=(a \vee (\mathbf{1}^*)') \wedge \mathbf{1}^*=a \vee (\mathbf{1}^* \wedge (\mathbf{1}^*)')=a;$$
$$a \vee a^* = a \vee (\mathbf{1}^* \wedge a') = \mathbf{1} \wedge (a \vee a') = \mathbf{1}^*.$$

Moreover, we have

$$a \leq b \Rightarrow a' \wedge \mathbf{1}^* \geq b' \wedge \mathbf{1}^*$$

and $a \wedge a^* = \mathbf{0}$.

It remains to show that G satisfies (iv). Let $a, b \in G$ such that $a \leq b$. Then

$$a \vee (a^* \wedge b) = a \vee (a' \wedge \mathbf{1}^* \wedge b) = a \vee (a' \wedge b) = b. \qquad \square$$

We remark that if G is an orthomodular sublattice of the orthomodular lattice **L**, a pair of elements of G is orthogonal in $G = (G, \leq, *, \mathbf{0}, \mathbf{1}^*)$ if and only if it is orthogonal in **L**.

Let **L** be an orthomodular lattice and G a sublattice of **L**. We say that G is a σ-orthocomplete sublattice of **L** if the supremum of every countable orthogonal subset of G exists and belongs to G. Sequences can be taken with indices in $\mathbf{N}'$ or $\mathbf{N}$.

Definition 11.14 We say that G is a SCP-sublattice (resp. a SIP-sublattice) of **L** if, for every orthogonal sequence $\{a_n\}$ in G there exists a subsequence $\{a_{n_i}\}$ of $\{a_n\}$ and an orthomodular sublattice G^* of **L** such that

(i) $a_{n_i} \in G^* \subseteq G$ for every $i \in \mathbf{N}'$,

(ii) G^* is an orthomodular lattice with the (SCP) (resp. with the (SIP)).

Obviously if G is a σ-orthocomplete sublattice of the orthomodular lattice **L**, for every orthogonal sequence $\{a_n\}$ in G, the set $\{\vee_{n\in \mathbf{N}^0} a_n : \mathbf{N}^0_a \in \mathcal{P}(\mathbf{N}))\}$ is a σ-complete Boolean algebra contained in G, so G is a (SCP)-sublattice and therefore a (SIP)-sublattice of **L**. An orthomodular lattice with the subsequential completeness property (resp. with the subsequential interpolation property) is a SCP-sublattice (resp. a SIP-sublattice) of itself. Note that if $\mathbf{L}$ is an algebra of subsets of a set S and G is a SCP- (resp. SIP-) sublattice of $\mathbf{L}$, then G is a (SCP-) (resp. SIP-) lattice.

We will denote by $\mathbf{S}$ a commutative semigroup with a neutral element 0 and endowed with a triangular functional f, i.e., which satisfies the conditions $f(0) = 0$ and $f(x) - f(y) \leq f(x+y) \leq f(x) + f(y)$ for all $x, y \in \mathbf{S}$. If $\{x_n\}$ is a sequence in $\mathbf{S}$, we write $\lim_n x_n = 0$ if and only if $\lim_n f(x_n) = 0$.

11.4.2 k-Triangular Function

Let $k \in [1, +\infty)$. Let m be a function $\mathbf{L}$ to $[0, +\infty)$.

Definition 11.15 We say that m is a k-triangular function if $m(\mathbf{0}) = 0$ and

$$m(a) - km(b) \leq m(a \vee b) \leq m(a) + km(b) \quad \text{when} \quad a, b \in \mathbf{L} \quad \text{and} \quad a \perp b .$$

Propositon 11.2 *Let m be a function from $\mathbf{L}$ to $[0, +\infty)$ such that $m(\mathbf{0}) = 0$. Then m is a k-triangular function if and only if*

$$|m(a) - m(b)| \leq km(a \wedge b') + km(b \wedge a') \qquad (a, b \in \mathbf{L} \ \ \textit{and} \ \ aCb) .$$

Proof. For the sufficiency, we recall that if $a \perp b$ then aCb and also

$$(a \vee b)Ca, \quad (a \vee b) \wedge a' = b, \quad a \wedge (a \vee b)' = \mathbf{0} .$$

Therefore

$$|m(a \vee b) - m(a)| \leq km(b) + km(\mathbf{0}) .$$

Concerning the necessity, note that if aCb, then $a = (a \wedge b) \vee (a \vee b')$, $b = (a \wedge b) \vee (b \wedge a')$ and $a \wedge b$, $a \wedge b'$, $b \wedge a'$ are pairwise orthogonal elements in $\mathbf{L}$, hence

$$m(a \wedge b) - km(a \wedge b') \leq m(a) \leq m(a \wedge b) + km(a \wedge b')$$
$$m(a \wedge b) - km(b \wedge a') \leq m(b) \leq m(a \wedge b) + km(b \wedge a')$$

and therefore

$$|m(a) - m(b)| \leq km(a \wedge b') + km(b \wedge a') . \qquad \square$$

A function from $\mathbf{L}$ to $\mathbf{S}$ is said to be k-triangular ($k \in [1, +\infty)$) if $m(\mathbf{0}) = 0$ and

$$f(m(a)) - kf(m(b)) \leq f(m(a \vee b)) \leq f(m(a)) + kf(m(b)) ,$$

when $a, b \in \mathbf{L}$ and $a \perp b$.

Let m be a function from $\mathbf{L}$ to $\mathbf{S}$ (resp. from $\mathbf{L}$ to $[0, +\infty)$).

Definition 11.16 We say that m is exhaustive (or s-bounded) if $\lim_i m(a_i) = 0$ for every orthogonal sequence $\{a_i\}$ in $\mathbf{L}$. A family $\{m_k\}_{k\in K}$ of exhaustive functions from $\mathbf{L}$ to X (resp. to $[0, +\infty)$) is said to be uniformly exhaustive if for every orthogonal sequence $\{a_i\}$ in $\mathbf{L}$ we have $\lim_{i\to\infty} m_k(a_i) = 0$, uniformly in $k \in K$.

Propositon 11.3 *(i) Let $\mathbf{L}$ be an orthomodular lattice, m a function from $\mathbf{L}$ to $\mathbf{S}$ (resp. to $[0, +\infty)$). Then m is exhaustive if and only if for every orthogonal sequence $\{a_i\}_{i\in\mathbf{N}'}$ in $\mathbf{L}$, we have $\lim_i m(x \wedge a_i) = 0$, uniformly on $x \in \mathbf{L}$.*

(ii) Let $\mathbf{L}$ be an orthomodular lattice with the (SIP), m an exhaustive function from $\mathbf{L}$ to $[0, +\infty)$. If $\{a_i\}_{i\in\mathbf{N}'}$ is an orthogonal sequence in $\mathbf{L}$, for every $\epsilon > 0$ and every $M \in \mathcal{I}(\mathbf{N})$, there exist $\mathbf{N}^0 \in \mathcal{I}(M)$ and $b \in \mathbf{L}$ such that $a_i \leq b$ for every $i \in \mathbf{N}^0$, $a_i \leq b'$ for every $i \in \mathbf{N}' \setminus \mathbf{N}^0$, $m(x \wedge b) < \epsilon$ for every $x \in \mathbf{L}$.

(iii) Let $\mathbf{L}$ be an orthomodular lattice with the (SIP) and $\{m_n\}$ a sequence of exhaustive functions from $\mathbf{L}$ to $[0, +\infty)$. If $\{a_i\}_{i\in\mathbf{N}'}$ is an orthogonal sequence in $\mathbf{L}$ there exists a decreasing sequence $\{d_i\}$ in $\mathbf{L}$ and a decreasing sequence $\{\mathbf{N}_i\}_{i\in\mathbf{N}'}$ in $\mathcal{I}(\mathbf{N}')$ such that, for every $i \in \mathbf{N}$

$$a_j \leq d_i \quad \text{for every } j \in \mathbf{N}_i,$$
$$a_j \leq d_i' \text{ for every } j \in \mathbf{N}' \setminus \mathbf{N}_i,\ \mathbf{N}_i \subseteq \mathbf{N}_{i-1} \setminus \{\min \mathbf{N}_{i-1}\},\ (\mathbf{N}_1 = \mathbf{N}')$$
$$m_p(x \wedge d_i) < 1/i + 1 \quad \text{for every } p \leq i \quad \text{and for every } x \in \mathbf{L}\,.$$

Propositon 11.4 *Let $\mathbf{L}$ be an orthomodular lattice, m a function from $\mathbf{L}$ to $[0, +\infty)$ (resp. to $\mathbf{S}$). Then m is exhaustive if and only if, for every orthogonal sequence $\{a_i\}_{i\in\mathbf{N}'}$ in $\mathbf{L}$ and for every $\epsilon > 0$, there exists $I \in \mathcal{I}(\mathbf{N}')$ such that, for every $J \in \mathcal{I}(\mathbf{N}')$ such that $J \cap I = \emptyset$, $m(\vee_{i\in J} a_i) < \epsilon$ (resp. $f(m(\vee_{i\in J} a_i) < \epsilon)$).*

Proof. For the proof of the non-trivial implication, suppose the contrary. Then there exists an orthogonal sequence $\{a_i\}_{i\in\mathbf{N}'}$ and an $\epsilon > 0$ for which the statement fails. By induction, we can construct a disjoint sequence $\{I_n\}_{n\in\mathbf{N}'}$ in $\mathcal{I}(\mathbf{N}')$ with $I_{-1} = \{0\}$, such that $m(\vee_{i\in I_n} a_i) \geq \epsilon$ (resp. $f(m(\vee_{i\in I_n} a_i) \geq \epsilon)$) for every $n \in \mathbf{N}'$. Since the sequence $\{\vee_{i\in I_n} a_i\}_{n\in\mathbf{N}'}$ is orthogonal, this contradicts the exhaustivity of m. □

Propositon 11.5 *Let* **L** *be an orthomodular lattice and let* m *be a* k*-triangular and exhaustive function from* **L** *to* $[0,+\infty)$. *Then, for every orthogonal sequence* $\{a_i\}_{n\in\mathbf{N}'}$ *in* **L** *and for every* $\Delta \in \mathcal{P}(\mathbf{N}')$, $\{m(\vee_{i\in F} a_i))\}_{F\in\mathcal{F}(\Delta)}$ *is a Cauchy net.*

Proof. Let $\Delta \in \mathcal{P}(\mathbf{N}')$. For every $I, J \in \mathcal{F}(\Delta)$, $a_i C a_j$ holds for every $(i,j) \in I \times J$, and therefore $\vee_{i\in I} a_i C \vee_{j\in J} a_j$. Moreover, for every $I, J \in \mathcal{F}(\Delta)$ such that $I \subseteq J$, we obtain, focusing on $\wedge_{i\in I} a_i'$,

$$\begin{aligned}(\vee_{j\in J} a_j) \wedge (\vee_{i\in I} a_i)' &= (\vee_{j\in J} a_j) \wedge (\wedge_{i\in I} a_i') = \vee_{j\in J}(a_j \wedge (\vee_{i\in I} a_i'))\\ &= \vee_{j\in J\setminus I}(a_j \wedge (\wedge_{i\in I} a_i')) \vee \vee_{j\in I}(a_j \wedge (\wedge_{i\in I} a_i')) = \vee_{j\in J\setminus I} a_j\,.\end{aligned}$$

For every $\epsilon > 0$, by Proposition 11.3, it follows that there exists $I_0 \in \mathcal{F}(\Delta)$ such that $J \cap I_0 = \emptyset$, $m(\vee_{i\in J} a_i) < \epsilon/2k$; then, for every $J', J'' \in \mathcal{F}(\Delta)$ including I_0, we have

$$\begin{aligned}&|m(\vee_{i\in J'} a_i) - m(\vee_{i\in J''} a_i)|\\ &\quad\leq |m(\vee_{i\in J'} a_i) - m(\vee_{i\in I_0} a_i)| + |m(\vee_{i\in I_0} a_i) - m(\vee_{i\in J''} a_i)|\\ &\quad\leq km((\vee_{i\in J'} a_i) \wedge (\vee_{i\in I_0} a_i)') + km((\vee_{i\in J''} a_i) \wedge (\vee_{i\in I_0} a_i)')\\ &\quad= km(\vee_{i\in J'\setminus I_0} a_i) + km(\vee_{i\in J''\setminus I_0} a_i) < \epsilon\end{aligned}$$

and the proof is complete. □

Theorem 11.12 *Let* **L** *be an orthomodular lattice and* m *a* k*-triangular and exhaustive function from* **L** *to* $[0,+\infty)$. *For every orthogonal sequence* $\{a_i\}_{i\in\mathbf{N}'}$ *in* **L** *the function* $\lambda\colon \mathcal{P}(\mathbf{N}') \to [0,+\infty)$ *defined by*

$$\Delta \mapsto \lim_{F\subset\mathcal{F}(\Delta)} m(\vee_{i\in F} a_i)$$

is a k*-triangular function.*

Proof. By Proposition 11.5, λ is a well defined function from $P(\mathbf{N}')$ to $[0,+\infty)$; moreover obviously $\lambda(\mathbf{0}) = 0$. We shall prove that, for every $\Delta_1, \Delta_2 \in \mathcal{P}(\mathbf{N}')$ such that $\Delta_1 \cap \Delta_2 = \emptyset$,

$$|\lambda(\Delta_1 \cup \Delta_2) - \lambda(\Delta_1)| \leq k\lambda(\Delta_2)\,.$$

For this, we observe that, for every $F \in \mathcal{F}(\Delta_1 \cup \Delta_2)$, obviously, $(\vee_{i\in F} a_i)$ $C(\vee_{i\in F\cap\Delta_1} a_i)$ and moreover

$$(\vee_{i\in F} a_i) \wedge (\vee_{i\in F\cap\Delta_1} a_i)' = \vee_{i\in F\cap\Delta_2} a_i\,.$$

Then we can write, for every $F \in \mathcal{F}(\Delta_1 \cup \Delta_2)$

$$
\begin{aligned}
&|\lambda(\Delta_1 \cup \Delta_2) - \lambda(\Delta_1)| \\
&\qquad \le |\lambda(\Delta_1 \cup \Delta_2) - m(\vee_{i\in F} a_i)| + |m(\vee_{i\in F} a_i) - m(\vee_{i\in F\cap\Delta_1} a_i)| \\
&\qquad\quad + |m(\vee_{i\in F\cap\Delta_1} a_i) - \lambda(\Delta_1)| \\
&\qquad \le |\lambda(\Delta_1 \cup \Delta_2) - m(\vee_{i\in F} a_i)| + km(\vee_{i\in F\cap\Delta_2} a_i) \\
&\qquad\quad + |m(\vee_{i\in F\cap\Delta_1} a_i) - \lambda(\Delta_1)|\,.
\end{aligned}
$$

We can choose, for every $\epsilon > 0$, a finite subset F of $\Delta_1 \cup \Delta_2$, such that

$$
\begin{aligned}
|\lambda(\Delta_1 \cup \Delta_2) - m(\vee_{i\in F} a_i)| &\le \epsilon/4\,, \\
|\lambda(\Delta_1) - m(\vee_{i\in F\cap\Delta_1} a_i)| &< \epsilon/4\,, \\
m(\vee_{i\in F\cap\Delta_2} a_i) &< \lambda(\Delta_2) + \frac{\epsilon}{2k}.
\end{aligned}
$$

Hence for every $\epsilon > 0$,

$$|\lambda(\Delta_1 \cup \Delta_2) - \lambda(\Delta_1)| < \epsilon + k\lambda(\Delta_2)\,.$$ □

Theorem 11.13 *Let* $\mathbf{L}$ *be an orthomodular lattice with the subsequential interpolation property,* $\{m_n\}_{n\in\mathbf{N}'}$ *a sequence of k-triangular and exhaustive functions from* $\mathbf{L}$ *to* $[0,+\infty)$, $\{a_i\}_{i\in\mathbf{N}'}$ *an orthogonal sequence in* $\mathbf{L}$. *Then there exists a subsequence* $\{a_{r_i}\}$ *of* $\{a_i\}_{n\in\mathbf{N}'}$ *an infinite subset* $\mathbf{N}^0$ *of* $\mathbf{N}'$ *and a subring* $\mathcal{G}$ *of* $\mathcal{P}(\mathbf{N}')$ *with the* (SCP) *such that*

(i) $\{i\} \in \mathcal{G}$ *for every* $i \in \mathbf{N}^0$,

(ii) for every $n \in \mathbf{N}'$, *the function* $\lambda_n\colon \mathcal{G} \to [0,+\infty)$ *defined by*

$$\Delta \mapsto \lim_{F\in\mathcal{F}(\Delta)} m_n(\vee_{i\in F} a_{r_i})$$

is a k-triangular and exhaustive function,

(iii) for every $\Delta \in \mathcal{G}$, *there exists* $c_\Delta \in \mathbf{L}$ *such that*

$$\lambda_n(\Delta) = m_n(c_\Delta) \qquad (n \in \mathbf{N}')\,.$$

Proof. Let $\{d_i\}_{i\in\mathbf{N}'}$ be a decreasing sequence in $\mathbf{L}$ and $\{\mathbf{N}_i\}_{i\in\mathbf{N}'}$ a decreasing sequence in $\mathcal{I}(\mathbf{N}')$ which fulfill the conditions of (iii) in Proposition 11.3 and let $r_i = \min \mathbf{N}_i$.

For the subsequence $\{a_{r_i}\}_{i\in\mathbf{N}'}$ there exists $\mathbf{N}^0 \in \mathcal{I}(\mathbf{N}')$ and $a \in \mathbf{L}$ such that

$$\begin{aligned} a_{r_i} &\le a \quad \text{for every } i \in \mathbf{N}^0, \\ a_{r_i} &\le a \quad \text{for every } i \in \mathbf{N}' \setminus \mathbf{N}^0, \\ d_i \wedge d'_{i+1} \wedge a'_{r_i} &\le a_i \quad \text{for every } i \in \mathbf{N}'. \end{aligned}$$

The subset $\mathcal{G}$ of $\mathcal{P}(\mathbf{N}')$ is defined by the properties:

(a) $\Delta \in \mathcal{P}(\mathbf{N}')$ and

(b) there exists $b \in \mathbf{L}$ such that

$$a_{r_i} \le b \quad \text{for every } i \in \Delta, \quad a_{r_i} \le b' \quad \text{for every } i \in \mathbf{N}' \setminus \Delta$$

is a Boolean ring with the (SCP) such that $\{i\} \in \mathcal{G}$ for every $i \in \mathbf{N}^0$. Then, if we consider for every $n \in \mathbf{N}'$, the function $\lambda_n : \mathcal{G} \to [0, +\infty)$ defined by

$$\Delta \mapsto \lim_{F\in\mathcal{I}(\Delta)} m_n(\vee_{i\in F} a_{r_i}).$$

By Theorem 11.12 λ_n is a k-triangular function. Moreover, we can prove, using Proposition 11.3 that, for every $n \in \mathbf{N}$, the function λ_n is also exhaustive.

We will prove (iii). For this, for every $\Delta \in \mathcal{G}$, let $b_\Delta \in \mathbf{L}$ such that

$$a_{r_i} \le b_\Delta \quad \text{for every } i \in \Delta, \quad a_{r_i} \le b' \quad \text{for every } i \in \mathbf{N}' \setminus \Delta$$

and write $c_\Delta = d_0 \wedge a \wedge b_\Delta$. For every $q \in \mathbf{N}^0$, we have

$$c_\Delta = d_0 \wedge a \wedge b_\Delta = \Big(\vee_{\substack{i<q\\ i\in\Delta}} a_{r_i}\Big) \vee (a \wedge b_\Delta \wedge b_q).$$

Therefore we have, by the k-triangular of m_n,

$$\Big| m_n(a \wedge b_\Delta \wedge d_0) - m_n\Big(\vee_{\substack{i<q\\ i\in\Delta}} a_{r_i}\Big)\Big| \le k m_n(a \wedge b_n \wedge d_q).$$

For every $\epsilon > 0$ let $\bar{q} \in \mathbf{N}$, such that $\bar{q} > n$, $\bar{q} > 2k/\epsilon$ and, for every $q \ge \bar{q}$

$$\Big| m_n\Big(\vee_{\substack{i<q\\ i\in\Delta}} a_{r_i}\Big) - \lambda_n(\Delta)\Big| < \frac{\epsilon}{2}, \quad m_n(x \wedge d_q) < \frac{1}{q} + 1 \qquad (x \in \mathbf{L}).$$

Then we obtain

$$\begin{aligned}
&|m_n(a \wedge b_\Delta \wedge d_0) - \lambda_n(\Delta)| \\
&\quad \leq \left| m_n(a \wedge b_\Delta \wedge d_0) - m_n\left(\vee_{\substack{i<q \\ i\in\Delta}} a_{r_i}\right)\right| + \left| m_n\left(\vee_{\substack{i<q \\ i\in\Delta}} a_{r_i}\right) - \lambda_n(\Delta)\right| \\
&\quad \leq k m_n(a \wedge b_\Delta \wedge d_q) + \frac{\epsilon}{2} + \frac{\epsilon}{2} - \epsilon\,,
\end{aligned}$$

so that $\lambda_n(\Delta) = m_n(a \wedge b_\Delta \wedge d_0)$. □

Theorem 11.14 *Let* $\mathbf{L}$ *be an orthomodular lattice with the* (SIP) $\{m_n\}_{n\in\mathbf{N}}$ *a sequence of* k*-triangular and exhaustive functions from* $\mathbf{L}$ *to* $[0,+\infty)$ *such that* $\lim_{n\to+\infty} m_n(a) = m_0(a)$ *for every* $a \in \mathbf{L}$ *and* m_0 *is an exhaustive function then* m_0 *is* k*-triangular and* $\{m_n\}_{n\in\mathbf{N}'}$ *is uniformly exhaustive.*

Proof. Suppose that $\{m_n\}_{n\in\mathbf{N}'}$ is not uniformly exhaustive. Then we may assume (by passing to a subsequence if necessary) that there exists an $\epsilon > 0$, and an orthogonal sequence $\{a_i\}_{i\in\mathbf{N}'}$ in $\mathbf{L}$ such that $m_i(a_i) > \epsilon$ for every $i \in \mathbf{N}'$.

By Theorem 11.13 there exists a subsequence $\{a_{r_i}\}_{i\in\mathbf{N}'}$, an infinite subset $\mathbf{N}^0$ of $\mathbf{N}'$ and a subring $\mathcal{G}$ of $\mathcal{P}(\mathbf{N}')$ with the subsequential completeness properties such that $\{i\} \in \mathcal{G}$ for every $i \in \mathbf{N}^0$ and the functions $\lambda_n\colon \mathcal{G} \to [0,+\infty)$ defined by

$$\Delta \mapsto \lim_{F\in\mathcal{F}(\Delta)} m_n(\vee_{i\in F} a_{r_i}) \qquad (n \in \mathbf{N}')$$

are k-triangular and exhaustive functions such that $\lambda_0(\Delta) = \lim_n \lambda_0(\Delta) = \lim_n \lambda_n(\Delta)$ for every $\Delta \in \mathcal{G}$.

Let $\{i_j\}_{j\in\mathbf{N}'}$ be a strictly increasing sequence in $\mathbf{N}^0$ such that $A = \bigcup_{j\in\mathbf{N}}\{i_j\} \in \mathcal{G}$ and let $\mathcal{G}' = \{\Delta \in \mathcal{G} : \Delta \subset A\}$. Then, by Theorem 11.10 $\{\lambda_n : \mathcal{G}', n \in \mathbf{N}'\}$ is uniformly exhaustive; in particular, we have, for every $n \in \mathbf{N}'$ and for some $j \in \mathbf{N}'$,

$$m_n(a_{r_j}) = \lambda_n(\{i_j\}) < \epsilon\,,$$

a contradiction. □

Corollary 11.8 *Let* $\mathbf{L}$ *be an orthomodular lattice with the subsequential interpolation property and let* $\{m_n\}_{n\in\mathbf{N}}$ *be a sequence of* k*-triangular exhaustive* *functions form* **L** *to* **S**. *If* $\lim_{n\to+\infty} f(m_n(a)) = \gamma_0(a)$ *exists for each* $a \in \mathbf{L}$ *and* γ_0 *is an exhaustive function, then* $\{m_n\}_{n\in\mathbf{N}'}$ *is uniformly exhaustive.*

Proof. It suffices to apply Theorem 11.14 on the sequence $\{f(m_n)\}_{n\in\mathbf{N}}$.□

11.5 Dieudonné Type Theorems

The Nikodym boundedness theorem says that the family of measures on a σ-algebra is uniformly bounded if and only if it is pointwise bounded.

As is well-known, the Nikodym boundedness theorem for measures in general fails for algebras of sets. But there are uniform boundedness theorems in which the initial boundedness conditions are on some subfamilies of a given σ-algebra; those subfamilies may not be σ-algebras. A famous theorem of Dieudonné states that for compact metric spaces the pointwise boundedness of a family of Borel regular measures on open sets implies its uniform boundedness on all Borel sets. We shall generalize Dieudonné's theorem on a wider class of set functions. The class of finitely additive regular Borel set functions gives nothing new, because each finitely additive regular Borel set function (also in the case of vector measures) is necessarily countable additive.

We prove a generalized Dieudonné type theorem for the class of k-triangular set functions. Using some modifications we obtain also a generalization of Dieudonné's theorem for semigroup valued set functions.

11.5.1 Real k-Triangular Set Functions

Let $\mathcal{B}$ be the collection of all Borel sets of a Hausdorff locally compact topological space **T**.

It is obvious that a k-triangular set function m with regular variation is itself regular.

We formulate the main theorem.

Theorem 11.15 *Let* $\mathcal{M}$ *be a family of k-triangular set functions defined on* $\mathcal{B}$ *with difference regular disjoint variations. If the set* $\{m(O) : m \in \mathcal{M}\}$ *is bounded for every open set* O, *then*

$$\{m(B) : m \in \mathcal{M}, B \in \mathcal{B}\}$$

is a bounded set.

Remark 11.1 We shall assume in the following proofs that $\mathbf{T}$ is a compact Hausdorff space. Namely, we can replace $\mathbf{T}$ with an Alexandrov one point ω compactification $\mathbf{T} \cup \{\omega\}$, taking $m(\omega) = 0$ $(m \in \mathcal{M})$.

Corollary 11.9 *Let $\mathcal{M}$ be a family of regular scalar measures defined on $\mathcal{B}$. If the set $\{|m(O)| : m \in \mathcal{M}\}$ is bounded for every open set O, then*

$$\{|m(B)| : m \in \mathcal{M},\ B \in \mathcal{B}\}$$

is a bounded set.

Proof. Let $\nu(B) = |m(B)|$ $(B \in \mathcal{B},\ m \in \mathcal{M})$. It is obvious that the family of all such set functions ν satisfies the conditions of Theorem 11.15 since $\overline{\nu} = \overline{m}$ is also regular (for real valued set functions this follows from $|m| = m^+ + m^-$ by Jordan decomposition and for complex valued set functions from the inequality $|m| \leq |m_1| + |m_2|$, where $m = m_1 + im_2$). So we apply Theorem 11.15. □

In the proof of Theorem 11.15 we will need the following lemma.

Lemma 11.7 *Let m be a k-triangular set function defined on $\mathcal{B}$ with regular variation. Then m is k-σ-subadditive on each sequence of disjoint open sets $\{O_n\}$, i.e.,*

$$m\Big(\bigcup_{j=1}^{\infty} O_j\Big) \leq k \sum_{j=2}^{\infty} m(O_j) + m(O_1).$$

Proof. First we will prove that m is order continuous on open sets, i.e., for each sequence $\{U_n\}$ of open sets such that $U_j \supset U_{j+1}$ $(j \in \mathbf{N})$ and $\bigcap_{j=1}^{\infty} U_j = \emptyset$, we have

$$\lim_{j \to \infty} m(U_j) = 0.$$

For each $\epsilon > 0$ there exists a sequence of compact sets $\{K_n\}$ such that $K_j \subset U_j$ and

$$\overline{m}(U_j \setminus K_j) < \epsilon/2^j k \qquad (j \in \mathbf{N}). \tag{11.13}$$

Then there exists $n_0 \in \mathbf{N}$ such that $\bigcap_{j=1}^{n} K_j = \emptyset$ for all $n \geq n_0$. Let $n \geq n_0$. Then we have

$$m(U_n) = m\Big(U_n \setminus \bigcap_{j=1}^{n} K_j\Big) = m\Big(\bigcup_{j=1}^{n} (U_n \setminus K_j)\Big) \leq \overline{m}\Big(\bigcup_{j=1}^{n} (U_n \setminus K_j)\Big).$$

Hence, since $\overline{m}$ is k-subadditive (i.e. $\overline{m}(A \cup B) \leq \overline{m}(A) + k\overline{m}(B)$ for every pair A, B of not necessarily disjoint sets from $\mathcal{B}$), and nondecreasing, we obtain by (11.13) for $k \geq 1$

$$m(U_n) \leq k \sum_{j=1}^{n} \overline{m}(U_j \setminus K_j) < \epsilon$$

for all $n \geq n_0$. Now, let $\{O_n\}$ be a sequence of disjoint open sets. Then we have

$$m\Big(\bigcup_{j=1}^{\infty} O_j\Big) \leq k \sum_{j=1}^{n} m(O_j) + m\Big(\bigcup_{j=n+1}^{\infty} O_j\Big) + m(O_1)\,.$$

Taking $n \to \infty$ we obtain

$$m\Big(\bigcup_{j=1}^{\infty} O_j\Big) \leq k \sum_{j=2}^{\infty} m(O_j) + m(O_1)\,.$$ □

Proof of Theorem 11.15. For simplicity we will suppose that $k = 1$. The general case is proved in an analogous way. It suffices to prove that every point in $\mathbf{T}$ belongs to an open set O on which holds

$$\sup\{m(A) : A \subset O \ (A \in \mathcal{B}),\ m \in \mathcal{M}\} < \infty\,. \tag{11.14}$$

Suppose that this is not true. Then there exists a point $x \in \mathbf{T}$ such that (11.14) does not hold for every open set O such that $x \in O$. We shall prove that there exists a sequence of pairwise disjoint open sets $\{E_n\}$ and a sequence $\{m_n\}$ from $\mathcal{M}$ such that $m_i(E_i) > i$ $(i \in \mathbf{N})$. For any open set O such that $x \in O$ there exists a Borel set $B \subset O$ and $m_1 \in \mathcal{M}$ such that

$$m_1(B) > 4 + 2 \sup_{m \in \mathcal{M}} m(\{x\})\,. \tag{11.15}$$

It is easy to prove that the preceding supremum is finite. Since m_1 has regular variation, by Corollary 9.1 there exists a compact set $K \subset B$ and an open set $O' \subset O$, $B \subset O'$ such that $\mu_1(B') < 1$ for each $B' \subset O' \setminus K$. We have by the subadditivity of m_1

$$m_1(K) + m_1(B \setminus K) \geq m_1(B)\,.$$

Using the preceding inequality, the inequality $m_1(B \setminus K) < 1$ and (11.15), we obtain

$$m_1(K) > 3 + 2 \sup_{m \in \mathcal{M}} m(\{x\}).$$

Let $K_1 = K \cup \{x\}$. Then the last inequality implies (directly for $x \in K$) by the triangularity of m_1 (for $x \notin K$) that

$$m(K_1) > 3 + \sup_{m \in \mathcal{M}} m(\{x\}).$$

By the regularity of m_1 there exists an open set U such that $O \supset U \supset K_1$ and $m_1(B'') < 1$ for every $B'' \subset U \setminus K_1$. The preceding inequality together with the inequality

$$m_1(U) \geq m_1(K_1) - m_1(U \setminus K_1)$$

implies

$$m_1(U) > 2 + \sup_{m \in \mathcal{M}} m(\{x\}). \tag{11.16}$$

Again by the regularity of m_1 there exists an open set W such that $\{x\} \subset W \subset U$ and

$$m_1(B''') < 1 \tag{11.17}$$

for every $B''' \subset W \setminus \{x\}$.

Let H be an open set such that $x \in H \subset \overline{H} \subset W$ (where $\overline{H}$ is the closure of the set H). Then we have

$$m_1(\overline{H}) \leq \sup_{A \subset H \setminus \{x\}} m_1(A) + m_1(\{x\}) \leq \sup_{B \subset W \setminus \{x\}} m_1(B) + m_1(\{x\}).$$

Hence by (11.17) we obtain

$$m_1(\overline{H}) < 1 + \sup_{m \in \mathcal{M}} m(\{x\}). \tag{11.18}$$

Let $E_1 = U \setminus \overline{H}$. Then we have $E_1 \subset O$ and $E_1 \cap \overline{H} = \emptyset$. By the inequality

$$m_1(E_1) + m_1(\overline{H}) \geq m_1(U),$$

(11.16) and (11.18) we obtain $m_1(E_1) > 1$. Using the preceding procedure, taking in the inequality (11.15) the constant 5 instead of 4 and taking into account the facts that $x \in H$ and the family $\mathcal{M}$ is not bounded on H, we obtain open sets E_2, H_1 ($H_1 \subset H$) and $m_2 \in \mathcal{M}$ such that $E_2 \cap H_1 = \emptyset$, $x \in H_1$ and $m_2(E_2) > 2$. We have $E_1 \cap E_2 = \emptyset$. Continuing this procedure we obtain a sequence $\{m_i\}$ from $\mathcal{M}$ and a sequence $\{E_i\}$ of pairwise disjoint open sets such that

$$m_i(E_i) > i \qquad (i \in \mathbf{N}). \tag{11.19}$$

We shall prove that m_i ($i \in \mathbf{N}$) are exhaustive on the sequence $\{E_n\}$ of disjoint open sets, i.e.,

$$\lim_{j\to\infty} m_i(E_j) = 0 \qquad (i \in \mathbf{N}). \tag{11.20}$$

Since $\bigcup_{j=1}^{\infty} E_j$ is an open set and $\overline{m}_i$ are regular, for $\epsilon > 0$ by Corollary 9.1 there exists a compact set $K' \subset \bigcup_{j=1}^{\infty} E_j$ such that $m_i(C) < \epsilon$ for each $i \in \mathbf{N}$ and each $C \subset \bigcup_{j=1}^{\infty} E_j \setminus K'$. Since $\{E_j\}$ is an open cover of K', there exists $n_0 \in \mathbf{N}$ such that $K' \subset \bigcup_{j=1}^{n_0} E_j$. Then we have for $m \geq n_0 + 1$

$$m_i(E_m) \leq \sup_{C'} m_i(C') \leq \sup_{C} m_i(C) < \epsilon \qquad (i \in \mathbf{N})$$

where $C' \subset E_m \cup (\bigcup_{j=1}^{n_0} E_j \setminus K')$ and $C \subset \bigcup_{j=1}^{\infty} E_j \setminus K'$. So we obtain (11.20).

Let $x_{ij} = m_i(E_j)/i$. We have by (11.20) $\lim_{j\to\infty} x_{ij} = 0$ ($i \in \mathbf{N}$). We obtain by the boundedness assumption of the theorem that $\lim_{i\to\infty} x_{ij} = 0$ ($j \in \mathbf{N}$). Applying Lemma 11.6 for the infinite matrix $[x_{ij}]$ ($i, j \in \mathbf{N}$) we obtain a sequence $\{i_n\}$ from $\mathbf{N}$ and a sequence $\{\epsilon_n\}$ of positive real numbers such that

$$\sum_{k=1}^{n-1} x_{i_n i_k} = (1/2 - \epsilon_n) x_{i_n i_n}, \tag{11.21}$$

$$x_{i_n i_{n+q}} < 2^{-q} \epsilon_n x_{i_n i_n} \qquad (n \in \mathbf{N}). \tag{11.22}$$

Using the triangularity of m_{i_n} ($n \in \mathbf{N}$) and Lemma 11.7 we obtain

$$m_{i_n}\Big(\bigcup_{k=1}^{\infty} E_{i_k}\Big) \geq m_{i_n}(E_{i_n}) - \sum_{k=1}^{n-1} m_{i_n}(E_{i_k}) - \sum_{k=n+1}^{\infty} m_{i_n}(E_{i_k}) \qquad (n \in \mathbf{N}).$$

Hence by (11.21) and (11.22)

$$i_n^{-1} m_{i_n}\Big(\bigcup_{k=1}^{\infty} E_{i_k}\Big) \geq x_{i_n i_n} - \sum_{k=1}^{n-1} x_{i_n i_k} - \sum_{k=n+1}^{\infty} x_{i_n i_n} \geq x_{i_n i_n}/2 \qquad (n \in \mathbf{N}),$$

i.e.,

$$m_{i_n}\Big(\bigcup_{k=1}^{\infty} E_{i_k}\Big) \geq m_{i_n}(E_{i_n})/2 \qquad (n \in \mathbf{N}).$$

Then by (11.19) we obtain

$$m_{i_n}\Big(\bigcup_{k=1}^{\infty} E_{i_k}\Big) \geq i_n/2 \quad \text{for each} \quad n \in \mathbf{N}.$$

Since $\bigcup_{k=1}^{\infty} E_{i_k}$ is an open set we obtain a contradiction with the boundedness of $\{m_{i_n}\}$ on open sets. □

11.5.2 Semigroup Valued k-Triangular Set Functions

Let $\mathbf{S}$ be a commutative semigroup with a neutral element 0. Let $d\colon \mathbf{S} \to [0, +\infty)$ be a pseudometric such that it satisfies the following condition

$$d(x + x_1, y + y_1) \leq d(x, y) + d(x_1, y_1)$$

for all $x, x_1, y, y_1 \in \mathbf{S}$.

Let $\mathbf{S}$ be endowed with a pseudometric d which satisfies the preceding inequality. Now we can extend the definition of the difference regularity of a set function $\nu\colon \mathcal{B} \to \mathbf{S}$ taking only ν and $d(\nu(A), \nu(A')) < \epsilon$ instead of m and $|m(A) - m(A')| < \epsilon$, respectively.

The pseudometric d induces a triangular functional in the following way:

$$f(x) = d(x, 0) \qquad (x \in \mathbf{S}).$$

The functional f satisfies $(F_1) f(x + y) \leq f(x) + f(y)$, and $(F_2) f(x + y) \geq |f(x) - f(y)|$ for all $x, y \in \mathbf{S}$.

Now we define the variation $\overline{\nu}$ of a set function $\nu\colon \mathcal{R} \to \mathbf{S}$ with $\nu(\emptyset) = 0$ in the following way:

$$\overline{\nu}(E) = \sup_{\pi} \sum_{A \in \pi} f(\nu(A)) \qquad (E \in \mathcal{R}),$$

where the supremum is taken over all partitions π of E into a finite number of pairwise disjoint members of $\mathcal{R}$. It is easy to see that $\overline{\nu}$ is superadditive.

Theorem 11.16 *Let $\mathcal{M}$ be a family of semigroup valued k-triangular set functions with difference regular variations defined on $\mathcal{B}$. If the set $\{f(\nu(O)) : \nu \in \mathcal{M}\}$ is bounded for every open set O, then*

$$\{f(\nu(B)) : \nu \in \mathcal{M},\ B \in \mathcal{B}\}$$

is a bounded set.

Proof. We take $m(B) = f(\nu(B))$ $(B \in \mathcal{B},\ \nu \in \mathcal{M})$ and we apply Theorem 11.15. □

11.5.3 Convergence Theorem

Theorem 11.17 *Let $\{m_n\}$ be a sequence of k-triangular set functions defined on $\mathcal{B}$ with regular disjoint variations. If*

$$\lim_{n\to\infty} m_n(O) = m_0(O)$$

exists for all open sets O and m_0 is exhaustive with respect to the family $\mathcal{O}$ of all open sets, then $\{m_n\}_{n=0}^{\infty}$ is uniformly exhaustive on the whole $\mathcal{O}$.

Proof. We have to prove that $\{m_n\}_{n=0}^{\infty}$ is uniformly exhaustive on the family $\mathcal{O}$ of all open sets in $\mathbf{T}$. We note that m_i $(i \in \mathbf{N})$ are exhaustive on the family $\mathcal{O}$, i.e., if $\{E_n\}$ is a sequence of disjoint open sets then

$$\lim_{j\to\infty} m_i(E_j) = 0 \qquad (i \in \mathbf{N}).$$

This follows by the regularity of $\overline{m}_i$ (see the proof of Theorem 11.15).

Suppose that theorem is not true. Then there exists $\epsilon > 0$ and a disjoint sequence $\{E_n\}$ of open sets such that

$$m_n(E_n) > 2\epsilon, \qquad (n \in \mathbf{N}).$$

Then repeating the proof of Theorem 11.10, only taking instead of Lemma 11.5 Lemma 11.7 and the fact that an arbitrary union of open set is again an open set, we obtain that $\{m_n\}$ is uniformly exhaustive on the family $\mathcal{O}$. □

The assumption on the regularity of variation is not so restrictive, because it turns out that in the case of regular additive set functions (where they must be also regular measures) variations are also regular. So we have as a consequence the classical Dieudonné's theorem.

Corollary 11.10 *Let $\{m_n\}$ be a sequence of regular scalar measures defined on $\mathcal{B}$. If*

$$\lim_{n\to\infty} m_n(O) = m_0(O)$$

exists for all open sets O, then $\{m_n\}$ is uniformly exhaustive on the whole $\mathcal{B}$ and m_0 is exhaustive.

Proof. Let $\nu_n(B) = |m_n(B)|$ $(B \in \mathcal{B},\ n \in \mathbf{N})$. The sequence $\{\nu_n\}$ satisfies the conditions of Theorem 11.17, since $|\nu| = |m|$ is also regular. Applying Theorem 11.17 we obtain that $\{m_n\}$ is uniformly exhaustive on $\mathcal{O}$ and then it can be proved that also on $\mathcal{B}$. □

11.6 Notes and References

The first diagonal theorem was proved by J. Mikusiński [Mikusiński 1970], and the first diagonal theorem for triangular set functions [Pap 1976], then in [H. Weber 1986] and [Saeki 1992]. We presented here Theorem 11.1, the version of the diagonal theorem obtained in [Antosik and Saeki 1992].

Weber has proved Theorem 11.1 in [H. Weber 1986] only for the case

$$\lim_{m\to\infty} \lim_{n\to\infty} \hat{m}_n(B_m) = 0\,.$$

Thus the second possibility in Theorem 11.1 cannot occur in his setting unless $\lim_{i\to\infty} \sup(d_n - \alpha) \le 0$. On the other hand, in [Saeki 1992] this result was obtained only for $\alpha = 0$ and $d_n = \hat{m}_n(B_n)/2$. Corollary 11.2 with $\alpha = \gamma = 0$ is a generalization of the nontrivial part of Theorem 2.4 in [H. Weber 1986]. Actually, a careful examination shows that his proof works if $\gamma = 0$.

The diagonal theorem – Corollary 11.3 was proved for group case in [Antosik 1971] and for semigroup case in [Pap 1974]. The diagonal theorem 11.2 denoted by DT is proved by P. Antosik and the basic matrix Theorem 11.3 was obtained in [Antosik and Swartz 1985].

Many useful applications of diagonal theorems are collected in the books [Pap 1982], [Antosik and Swartz 1985], [C. Swartz 1992]. The adjoint theorem was obtained for pre-Hilbert K-space in [Pap 1985a] and then elaborated with applications to the closed graph theorem in [Pap 1985b], [C. Swartz 1987], [Pap and C. Swartz 1990b], [Pap and C. Swartz 1990c]. [Pap and C. Swartz 1990c].

In Section 11.2.4 we presented the result of [Abraham 1992] on the relation of Saeki's fatness condition [Saeki 1992] with the geometry of the Banach space.

Section 11.3 was taken from by [Pap 1987] and Section 11.4 by [Guariglia 1991]. Section 11.5 is based on [Pap 1986].

An overview of diagonal theorems and their applications can be found in [de Lucia and Pap 1994a]. A similar approach is by Rosenthal's lemma, see [Diestel and Uhl 1977] and [Pap 1982].

Chapter 12

Convergence and Boundedness Theorems on Difference Poset

Nikodym boundedness and convergence theorems were generalized not only with respect to the domain of the measures but also with respect to the range and the properties of the considered measures. So there are vector, group and even semigroup valued measure versions of these theorems. On the other side, the interest for non-additive set functions (submeasures, k-triangular set functions, decomposable measures, null-additive set functions) is growing mainly from the need of many different applications. There are recent investigations of set functions with values in sets endowed only with some topological structures without any algebraic operations.

12.1 Difference-Posets

Very recently, motivated by the investigations in the theory of fuzzy sets Chovanec and Kôpka have introduced a new very useful notion of difference poset (D-poset). This structure is a generalization of quantum logic, MV-algebras, orthoalgebras, the set of all effects, the collection of fuzzy sets.

12.1.1 Definitions and Examples

In this section we shall explain some notions and their properties which we shall need in the following sections.

We now give the definition of the basic structure.

Definition 12.1 A D-poset (difference poset) is a partially ordered set $\mathbf{L}$ with a partial ordering $\leq$, maximal element $\mathbf{1}$, and with a partial binary operation $\ominus: \mathbf{L} \times \mathbf{L} \to \mathbf{L}$, called difference, such that, for $a, b \in \mathbf{L}$, $b \ominus a$ is defined if and only if $a \leq b$, such that the following axioms hold for $a, b, c \in \mathbf{L}$:

(DP_1) $b \ominus a \leq b$;

(DP_2) $b \ominus (b \ominus a) = a$;

(DP_3) $a \leq b \leq c \Rightarrow c \ominus b \leq c \ominus a$ and $(c \ominus a) \ominus (c \ominus b) = b \ominus a$.

The previous axioms imply that there exists also a minimal element $\mathbf{0}$ $(= 1 \ominus 1)$.

Example 12.1 If S is a t-conorm on $[0,1]$, then we can introduce a difference in the following way

$$y -_S x = \inf\{z \in [0,1] : S(x,z) \geq y\}$$

for $x \leq y$ and $x, y \in [0,1]$. If $-_S$ satisfies the condition

$$y -_S (y -_S x) = x \quad \text{if and only if} \quad 0 \leq x \leq y \leq 1,$$

then $([0,1], -_S)$ is a D-poset.

The following basic properties of the operation $\ominus$ can be easily proved by Definition 12.1.

Propositon 12.1 *We have*

(a) $a \ominus \mathbf{0} = a$.

(b) $a \ominus a = \mathbf{0}$.

(c) $a \leq b \Rightarrow b \ominus a = \mathbf{0} \Leftrightarrow b = a$.

(d) $a \leq b \Rightarrow b \ominus a = b \Leftrightarrow a = \mathbf{0}$.

(e) $a \leq b \leq c \Rightarrow b \ominus a \leq c \ominus a$ *and* $(c \ominus a) \ominus (b \ominus a) = c \ominus b$.

(f) $b \leq c,\ a \leq c \ominus b \Rightarrow b \leq c \ominus a$ *and* $(c \ominus b) \ominus a = (c \ominus a) \ominus b$.

(g) $a \leq b \leq c \Rightarrow a \leq (c \ominus (b \ominus a)) \ominus a = c \ominus b$.

For an arbitrary but fixed element $a \in \mathbf{L}$ we define

$$a^{\perp} = \mathbf{1} \ominus a\,.$$

We have:

(i) $a^{\perp\perp} = a$;

(ii) $a \leq b \Rightarrow b^{\perp} \leq a^{\perp}$.

The elements a and b from $\mathbf{L}$ are orthogonal iff $a \leq b^{\perp}$ (or $b \leq a^{\perp}$).

We define a partial binary operation $\oplus$: $\mathbf{L} \times \mathbf{L} \to \mathbf{L}$ for orthogonal elements a and b such that

$$b \leq a \oplus b \quad \text{and} \quad a = (a \oplus b) \ominus b$$

(do not mix here $\oplus$ with the pseudo-addition!). This operation $\oplus$ is commutative and associative.

The notion of D-poset covers many important examples.

Example 12.2 We have seen in the previous chapter (in a special case) that an orthomodular poset is a partially ordered set $\mathbf{O}$ with an ordering $\leq$, the minimal and maximal elements being $\mathbf{0}$ and $\mathbf{1}$, respectively, and with an orthocomplementation $'$: $\mathbf{O} \to \mathbf{O}$ such that:

(OM_1) $(a')' = a \quad (a \in \mathbf{O})$,

(OM_2) $a \vee a' = \mathbf{1} \quad (a \in \mathbf{O})$;

(OM_3) if $a \leq b$, then $b' \leq a'$;

(OM_4) if $a \leq b'$, then $a \vee b \in \mathbf{O}$;

(OM_5) if $a \leq b$, then $b = a \vee (a \vee b')'$.

Taking for $a \leq b$

$$b \ominus a = (a \vee b')',$$

we obtain that the orthomodular poset $\mathbf{O}$ is a D-poset.

Example 12.3 An MV-algebra is a set $\mathbf{M}$ endowed with two binary operations $\oplus$ (do not mix with the pseudo-addition!) and $\odot$, a unary operation $\star$ and two elements 0 and 1 such that, for all $a, b, c \in \mathbf{M}$,

(MV_1) $a \oplus b = b \oplus a$;

(MV_2) $(a \oplus b) \oplus c = a \oplus (b \oplus c)$;

(MV_3) $a \oplus 0 = a$;

(MV_4) $a \oplus 1 = 1$;

(MV_5) $(a^{\star})^{\star} = a$;

(MV_6) $0^{\star} = 1$;

(MV_7) $a \oplus a^{\star} = 1$;

(MV_8) $(a^\star \oplus b)^\star \oplus b = (a \oplus b^\star)^\star \oplus a$;

(MV_9) $a \odot b = (a^\star \oplus b^\star)^\star$.

Taking

$$a \leq b \Leftrightarrow (a \odot b^\star) \oplus b = b$$

and for $a \leq b$

$$b \ominus a = (a \oplus b^\star)^\star,$$

we obtain that the MV-algebra $\mathbf{M}$ is a D-poset.

Example 12.4 An orthoalgebra is a set $\mathbf{A}$ with two particular elements $\mathbf{0}$, $\mathbf{1}$, and with a partial binary operation $\oplus$: $\mathbf{A} \times \mathbf{A} \to \mathbf{A}$ (do not mix with the pseudo-addition!) such that for all $a, b, c \in \mathbf{A}$,

(OA_1) if $a \oplus b \in \mathbf{A}$, then $b \oplus a \in \mathbf{A}$ and $a \oplus b = b \oplus a$;

(OA_2) if $b \oplus c \in \mathbf{A}$ and $a \oplus (b \oplus c) \in \mathbf{A}$, then $a \oplus b \in \mathbf{A}$ and $(a \oplus b) \oplus c \in \mathbf{A}$, and $a \oplus (b \oplus c) = (a \oplus b) \oplus c$;

(OA_3) for any $a \in \mathbf{A}$ there is a unique $b \in \mathbf{A}$ such that $a \oplus b$ is defined, and $a \oplus b = \mathbf{1}$;

(OA_4) if $a \oplus a$ is defined, then $a = \mathbf{0}$.

We have $a \leq b$ iff there exists an element $c \in \mathbf{A}$ such that $a \oplus c$ is defined in $\mathbf{A}$ and $a \oplus c = b$. An element b is the orthocomplement of a (denoted by $a^\perp$) iff b is a (unique) element of $\mathbf{A}$ such that $b \oplus a$ is defined in $\mathbf{A}$ and $a \oplus b = \mathbf{1}$.

Taking for $a \leq b$

$$b \ominus a = (a \oplus b^\perp)^\perp,$$

we obtain that the orthoalgebra $\mathbf{A}$ is a D-poset. We remark that each orthomodular poset (Example 12.2) is an orthoalgebra, but the opposite is not true.

Example 12.5 Let $\mathcal{E}(H)$ be the set of all Hermitian operators T on a Hilbert space H with $O \leq T \leq I$, where O and I are the zero and identity operators, respectively, on H. The set $\mathcal{E}(H)$ is a D-poset, which is not an orthoalgebra.

Example 12.6 Let X be a non-empty set and $\mathcal{F}'$ the family of all fuzzy sets on X, i.e., $\mathcal{F}' = [0,1]^X$. We have for $f, g \in \mathcal{F}'$

$$f \leq g \Leftrightarrow f(x) \leq g(x) \qquad (x \in X).$$

Let $\Phi\colon [0,1] \to [0,\infty)$ be an injective increasing continuous function such that $\Phi(0)=0$. Taking for $f \le g$

$$(g \ominus f)(x) = \Phi^{-1}(\Phi(g(x)) - \Phi(f(x)) \qquad (x \in X)$$

we obtain that $\mathcal{F}'$ is a D-poset.

12.1.2 $\oplus$-orthogonality

L will always denote a D-poset. Let $\{a_1,\dots,a_n\} \subset \mathbf{L}$. We define

$$\begin{aligned} a_1 \oplus \cdots \oplus a_n &= \mathbf{0} \quad \text{for} \quad n=0,\\ a_1 \oplus \cdots \oplus a_n &= a_1 \quad \text{for} \quad n=1,\\ a_1 \oplus \cdots \oplus a_n &= (a_1 \oplus \cdots \oplus a_{n-1}) \oplus a_n \quad \text{for} \quad n \ge 3, \end{aligned}$$

supposing that $a_1 \oplus \cdots \oplus a_{n-1}$ and $a_1 \oplus \cdots \oplus a_n$ exist in **L**. We have

Definition 12.2 A finite subset $\{a_1,\dots,a_n\}$ of **L** is $\oplus$-orthogonal if $a_1 \oplus \dots \oplus a_n$ exists in **L**.

We say that an $\oplus$-orthogonal subset $\{a_1,\dots,a_n\}$ of **L** has an $\oplus$-sum, $\bigoplus_{i=1}^n a_i$, defined by

$$\bigoplus_{i=1}^n a_i = a_1 \oplus \cdots \oplus a_n\,.$$

We remark that the preceding $\oplus$-sum is independent of any permutation of elements.

Definition 12.3 A subset G of **L** is $\oplus$-orthogonal if every finite subset F of G is $\oplus$-orthogonal.

We say that an $\oplus$-orthogonal subset $G=\{a_i : i \in I\}$ of **L** has an $\oplus$-sum in **L**, $\bigoplus_{i\in I} a_i$, if in **L** there exists the join

$$\bigoplus_{i\in I} a_i = \sup\Big\{\bigoplus_{i\in F} a_i : F \;\text{ finite subset of } I\Big\}\,.$$

Any subset of an $\oplus$-orthogonal set is again $\oplus$-orthogonal.

Definition 12.4 An D-poset **L** is complete D-poset ($\sigma(\oplus)$-D-poset) if, for every $\oplus$-orthogonal subset (every countable $\oplus$-orthogonal subset) G of **L**, there exists the $\oplus$-sum in **L**.

Definition 12.5 A D-poset $\mathbf{L}$ is quasi-σ-complete if for every $\oplus$-orthogonal sequence $\{a_i\}$ in $\mathbf{L}$ there exists a subsequence $\{a_i\}_{i\in M}$ such that $\bigoplus_{i\in I} a_i \in \mathbf{L}$ for each $I \subset M$.

We shall give now an example of a $\sigma(\oplus)$-D-poset.

Example 12.7 Let H be any set of real numbers between 0 and 1 where H satisfies the following conditions

(i) $0 \in H$ and $1 \in H$;
(ii) if $x, y \in H$ then $\min(1, x + y) \in H$;
(iii) if $x, y \in H$ then $\max(0, x + y - 1) \in H$;
(iv) if $x \in H$ then $1 - x \in H$.

The operations $\oplus, \odot$ and * are defined as follows

$$x \oplus y = \min(1, x + y),$$
$$x \odot y = \max(0, x + y - 1),$$
$$x^* = 1 - x.$$

The system $(H, \oplus, \odot, ^*, 0, 1)$ is an MV-algebra. If we take $H = [0, 1]$ we obtain a σ-MV-algebra with respect to the operation $\oplus$ and in this way also a $\sigma(\oplus)$-D-poset, since for $x \leq y$ we have that the operation $\ominus$ defined by

$$x \ominus y = (x \oplus y^*)^*$$

gives a σ-D-poset with respect to the operation $\oplus_D$ defined by

$$x \oplus_D y = (y^* \ominus x)^*$$

which coincides with the operation $\oplus$, i.e.,

$$x \oplus_D y = (y^* \ominus x)^* = ((x \oplus (y^*)^*)^*)^* = x \oplus y.$$

We remark that for $H = \{0, 1\}$ we trivially obtain also a $\sigma(\oplus)$-D-poset. But if H is the set of all rational numbers between 0 and 1, then this is an MV-algebra, and so also a D-poset, which is not a $\sigma(\oplus)$-MV-algebra and so also not a $\sigma(\oplus)$-D-poset.

12.2 Nikodym Boundedness Theorem

12.2.1 Boundedness in Uniform Space

Let Y be a uniform space with the uniformity $\mathcal{U}$.

Definition 12.6 A subset B of Y is bounded ($\mathcal{U}$-bounded) if for every $U \in \mathcal{U}$ there exists a finite set $K \subset B$ and a natural number n such that

$$B \subset U^n[K],$$

where $U^1 = U$, $U^n = U \circ U^{n-1}$ ($\circ$-composition of the relations) and $U[K]$ is the set of all $x \in Y$ such that $(x,y) \in U$ for some $y \in K$.

A subset B of a metrisable uniform space $(Y,\mathcal{U})$ is $\mathcal{U}$-bounded if and only if it is d-bounded for every metric d generating the same uniformity $\mathcal{U}$. The following known characterization of $\mathcal{U}$-boundedness will be often used.

Theorem 12.1 *A set $B \subset Y$ is $\mathcal{U}$-bounded if and only if it is d-bounded for every uniformly continuous pseudo-metric d defined on Y.*

We denote with $\mathcal{P}m$ the family of all uniformly continuous pseudometrics defined on $(Y,\mathcal{U})$.

12.2.2 Boundedness Theorem

Let $\mathbf{L}$ be a quasi-σ-D-poset.

Definition 12.7 For $d \in \mathcal{P}m$ the d-semivariation of a function $m\colon \mathbf{L} \to Y$, with respect to a point $x_o \in Y$ is

$$\widetilde{m}_d^{x_o}(b) = \sup\{d(m(c),x_o) : c \le b,\ c \in \mathbf{L}\} \qquad (b \in \mathbf{L}).$$

We define for $d \in \mathcal{P}m$ and $x_o \in Y$, a function $m\colon \mathbf{L} \to Y$

$$\alpha_d^{x_o}(a,m) = \limsup_{n\to\infty}\left\{d(m(a \oplus b),x_o) : \widetilde{m}_d^{x_o}(b) < \frac{1}{n}, b \in \mathbf{L}\right\} \qquad (a \in \mathbf{L}).$$

Definition 12.8 A function $m\colon \mathbf{L} \to Y$ is said to be x_o-exhaustive, for $x_o \in Y$, if for each $d \in \mathcal{P}m$,

$$\lim_{n\to\infty} d(m(a_n),x_o) = 0$$

for each $\oplus$-orthogonal sequence $\{a_n\}$ of elements from $\mathbf{L}$.

Lemma 12.1 *Let Y be a uniform space and $\mathbf{L}$ a quasi-σ-complete D-poset. If $m\colon \mathbf{L} \to Y$ is a x_o-exhaustive function and $\{a_n\}$ a sequence of $\oplus$-orthogonal elements from $\mathbf{L}$, then for each $d \in \mathcal{P}m$ and each $\epsilon > 0$ there exists a $\oplus$-orthogonal subsequence $\{a_{n_i}\}$ of $\{a_n\}$ such that*

$$\widetilde{m}_d^{x_o}\left(\bigoplus_{i\in I} a_{n_i}\right) < \epsilon$$

for any $I \subset \mathbf{N}$.

This is proved by the contradiction with the exhaustivity of the function m.

Theorem 12.2 *Let $\mathcal{M}$ be a family of x_0-exhaustive functions $m\colon \mathbf{L} \to Y$, where $\mathbf{L}$ is a quasi-σ-D-poset and Y is a uniform space. Then the set*

$$\{m(a) : m \in \mathcal{M},\ a \in \mathbf{L}\}$$

is $\mathcal{U}$-bounded if and only if the following conditions hold

(i) For each $d \in \mathcal{P}m$ and each $r \in \mathbf{N}$ there exists $s(r) \in \mathbf{N}$ such that

$$d(m(a), m(b)) > s(r)$$

implies either for $b \le a$

$$d(m(a \ominus b), x_0) > r,$$

or for $a \le b$

$$d(m(b \ominus a), x_0) > r;$$

(ii) For each $d \in \mathcal{P}m$ the set

$$\{\alpha_d^{x_0}(a, m) : m \in \mathcal{M}\}$$

is bounded for each $a \in \mathbf{L}$;

(iii) For each $d \in \mathcal{P}m$

$$\{d(m(a_n), x_0) : m \in \mathcal{M},\ n \in \mathbf{N}\}$$

is bounded for every orthogonal sequence $\{a_n\}$ from $\mathbf{L}$.

Proof. The necessary part of the proof is obvious. Suppose that (i), (ii), and (iii) hold but the set

$$\{m(a) : m \in \mathcal{M},\ a \in \mathbf{L}\}$$

is not $\mathcal{U}$-bounded. Then by Theorem 12.1 and (iii) there exists d from $\mathcal{P}m$ and an orthogonal sequence $\{a_n\}$ from $\mathbf{L}$ such that

$$d(m_n(a_n), x_0) > n \qquad (n \in \mathbf{N}).$$

We take by (ii)

$$r_1 = \sup_n \alpha_d^{x_0}(a_1, m_n) + 1 \in \mathbf{N}. \tag{12.1}$$

Then, there exists $s(r_1) + 1 \in \mathbf{N}$, by (i). We take $n_1 > s(r_1) + 1$. By Lemma 12.1 and (12.1) there exists a subsequence $\{a_n^1\}$ of the sequence $a_{n_1+1}, a_{n_2+2}, \ldots$ such that

$$d\Big(m_{n_1}\Big(a_1 \bigoplus \bigoplus_{i\in I} a_i^1, x_0\Big)\Big) < r_1$$

for arbitrary $I \subset \mathbf{N}$. Now we take

$$r_2 = \sup_n \alpha_d^{x_0}(a_1 \oplus a_{n_1}, m_n) + 1 \in \mathbf{N}$$

by (ii) and we repeat this whole procedure.

Continuing this procedure we obtain two sequences of natural numbers $\{r_k\}$ and $\{n_k\}$, $n_0 = 1$, such that for

$$r_k = \sup_n \alpha_d^{x_0}(\oplus_{i=0}^{k-1} a_{n_i}, m_n) + 1$$

we have

$$d(m_{n_k}(a_{n_k}), x_0) > n_k > s(r_k) + k \qquad (k \in \mathbf{N}) \tag{12.2}$$

and

$$d(m_{n_k}(\oplus_{i=0}^{\infty} a_{n_i} \ominus a_{n_k}), x_0) < r_k \qquad (k \in \mathbf{N}) \tag{12.3}$$

(by (i)).

Taking $k > \sup_j d(m_{n_j}(\oplus_{i=0}^{\infty} a_{n_i}), x_0)$, we obtain by (12.2)

$$\begin{aligned} d(m_{n_k}(\oplus_{i=0}^{\infty} a_{n_i}), m_{n_k}(a_{n_k})) &\geq d(m_{n_k}(a_{n_k}), x_0) - d(m_{n_k}(\oplus_{i=0}^{\infty} a_{n_i}), x_0) \\ &> s(r_k). \end{aligned}$$

Hence by condition (i)

$$d(m_{n_k}(\oplus_{i=0}^{\infty} a_{n_i} \ominus a_{n_k}), x_0) > r_k \qquad (k \in \mathbf{N}).$$

Contradiction with (12.3). □

Remark 12.1 Each k-subadditive set function for some $k \geq 1$ satisfies the condition (i) in Theorem 12.2.

For a specially important case, the orthomodular lattice (Example 12.2), we can relax the conditions in the previous theorem.

Theorem 12.3 *Let* $\mathbf{L}$ *be a* σ*-orthomodular lattice. Let* $\mathcal{M}$ *be a family of functions* $m\colon \mathbf{L} \to Y$. *Then the set*

$$\{m(a) : m \in \mathcal{M},\ a \in \mathbf{L}\}$$

is $\mathcal{U}$*-bounded if and only if the following conditions hold:*

(i) For each $m \in \mathcal{M}$ *and each* $r \in \mathbf{N}$ *there exists* $s(r) \in \mathbf{N}$ *such that for each* $a, b \in \mathbf{L}$

$$d(m(a), m(b)) > s(r) \qquad (d \in \mathcal{P}m,\ m \in \mathcal{M})$$

implies either

$$\widetilde{m}_d^{x_0}((b \vee a')') > r$$

or

$$\widetilde{m}_d^{x_0}((a \vee b')') > r\,;$$

(ii) For each $d \in \mathcal{P}m$ *the set*

$$\{m(a_n) : m \in \mathcal{M},\ n \in \mathbf{N}\}$$

is d*-bounded for every orthogonal sequence* $\{a_n\}$ *from* $\mathbf{L}$.

Proof. The necessity of (i) and (ii) is obvious. Now, suppose that (i) and (ii) hold but the set

$$\{m(a) : a \in \mathbf{L},\ m \in \mathcal{M}\}$$

is not $\mathcal{U}$-bounded. Then for any $x_0 \in Y$ there exists $d \in \mathcal{P}m$, a sequence $\{m_n\}$ from $\mathcal{M}$ and a sequence $\{a_n\}$ of elements from $\mathbf{L}$ such that

$$d(m_n(a_n), x_0) > n \qquad (n \in \mathbf{N})\,. \tag{12.4}$$

On the other side condition (ii) implies that there exists $M > 0$ such that

$$d(m_n(a_1), x_0) < M\,.$$

Hence by (12.4)

$$d(m_n(a_1), m_n(a_n)) \to \infty$$

as $n \to \infty$. Then, for $n \to \infty$, by (i) either

a) $$\sup\{d(m_n(c), x_0) : c \in \mathbf{L},\ c \leq (a_1 \vee a_n')'\} \to \infty$$

or

b) $$\sup\{d(m_n(c), x_0) : c \in \mathbf{L},\ c \leq (a_n \vee a_1')'\} \to \infty\,.$$

Case a). We shall choose a subsequence $\{m_{n_k}\}$ of the sequence $\{m_n\}$ and a sequence $\{e_n\}$ of orthogonal elements from $\mathbf{L}$ such that

$$d(m_{n_k}(e_k), x_0) > k \qquad (k \in \mathbf{N}),$$

a contradiction with (ii).

Let $m_{n_1} = m_1$ and $e_1 = a_1$. We choose n_2 such that $m_{n_2} = m_{i_2}$ where $i_2, i_3, \ldots$ is a sequence of natural numbers and elements $a_{i_2}, a_{i_3}, \ldots$. from the sequence $\{a_n\}$ and elements a_j^1 from $\mathbf{L}$ such that $a_j^1 \leq a_{i_j}$ and

$$d(m_{i_j}((a_1 \vee (a_j^1)')'), x_0) > j, \qquad j = 2, 3, \ldots.$$

We take $e_2 = (a_1 \vee (a_2^1)')'$. Now we repeat this procedure taking the sequences $\{m_{i_j}\}$ and $\{(a_1 \vee (a_j^1)')'\}_{j=2}^{\infty}$. Continuing this procedure we obtain the desired sequences.

Case b). Let $m_{n_1} = m_1$. We choose n_2 such that $m_{n_2} = m_{i_2}$, where we take a sequence $i_2, i_3, \ldots$. of natural numbers, sequence of elements $a_{i_2}, a_{i_3}, \ldots$. from the sequence $\{a_n\}$ and elements a_j^1 $(j = 2, 3, \ldots)$ from $\mathbf{L}$ such that $a_{i_j} \leq a_j^1$ and

$$d(m_{i_j}((a_j^1 \vee a_1')'), x_0) > j, \qquad j = 2, 3, \ldots.$$

Let $b_1 = a_1$ and $b_2 = (a_2^1 \vee a_1')'$. Now we shall repeat the procedure for the sequences $m_{i_2}, m_{i_3}, \ldots$ and $(a_2^1 \vee a_1')', (a_3^1 \vee a_1')', \ldots$. The sequence $\{b_n\}$ is decreasing, i.e., $b_n \geq b_{n+1}$ $(n \in \mathbf{N})$. The condition (i) implies the existence of the subsequences $\{m_{n_{k_i}}\}$ and $\{b_{k_i}\}$ such that

$$(\widetilde{m}_{n_{k_i}})_d^{x_0}((b_{k_i} \vee b_{k_{i-1}}')') > i \qquad (i \in \mathbf{N}).$$

Since we have

$$(b_{k_i} \vee b'_{k_{i-1}})' = b'_k \wedge b_{k_{i-1}},$$

the sequence $\{(b_{k_i} \wedge b'_{k_{i-1}})'\}$ is orthogonal. Hence the preceding inequalities imply the existence of an orthogonal sequence $\{c_i\}$ such that $c_i \leq (b_{k_i} \vee b'_{k_{i-1}})'$ and

$$d(m_{n_{k_i}}(c_i), x_0) > i \qquad (i \in \mathbf{N}).$$

Contradiction with (ii). □

12.3 Convergence Theorem

Theorem 12.4 *Let Y be a uniform space and $\mathbf{L}$ a quasi-σ-complete D-poset. Let $\{m_n\}$ be a sequence of x_0-exhaustive functions m_n, $m_n\colon \mathbf{L} \to Y$, such that they satisfy the following conditions for an element x_o from Y*

(i) for each $d \in \mathcal{P}m$ and for each $\epsilon > 0$, there exists $\delta > 0$ such that

$$d(m_n(a), x_o) < \delta \quad \textit{and} \quad d(m_n(b), x_o) < \delta$$

for $a \leq b$, $a, b \in \mathbf{L}$ $(n \in \mathbf{N})$ implies $d(m_n(b \ominus a), x_o) < \epsilon$;

(ii) for each $d \in \mathcal{P}m$ and for each $\delta > 0$, there exists $\gamma > 0$ such that

$$d(m_n(a), x_o) < \gamma, \; a \in \mathbf{L} \; (n \in \mathbf{N}) \quad \textit{implies} \quad \alpha_d^{x_o}(a, m_n) < \delta \quad (n \in \mathbf{N});$$

(iii) for each $d \in \mathcal{P}m$ we have $lim_{n\to\infty} d(m_n(a), m(a)) = 0$ for each $a \in \mathbf{L}$.

Then m is x_o-exhaustive if and only if m_n $(n \in \mathbf{N})$ are uniformly x_o-exhaustive.

Proof. Let us suppose that m is x_o–exhaustive, but $\{m_n\}$ is not uniformly x_o-exhaustive. Hence there exists $\epsilon > 0$, d from $\mathcal{P}m$ and a $\oplus$-orthogonal sequence $\{a_k\}$ of elements from $\mathbf{L}$ and a subsequence $\{m_{n_k}\}$ such that

$$d(m_{n_k}(a_k), x_o) > \epsilon \tag{12.5}$$

for each $k \in \mathbf{N}$. By (i) we choose $\delta > 0$ corresponding to $\epsilon > 0$. By (ii) we choose $\gamma > 0$ corresponding to $\delta > 0$. Since m is x_o-exhaustive by

Lemma 12.1 there exists a $\oplus$-orthogonal subsequence $\{a_{k_i}\}$ of $\{a_k\}$ such that

$$\widetilde{m}_d^{x_o}\left(\bigoplus_{i\in I} a_{k_i}\right) < \frac{\gamma}{2} \tag{12.6}$$

for each $I \subset \mathbf{N}$. Now, let us denote $m_i = m_{n_{k_i}}$ and $b_i = a_{k_i}$ ($i \in \mathbf{N}$) and $i_1 = 1$.

By (iii) there exists an index i_2 such that

$$d(m_{i_2}(b_{i_1}), m(b_{i_1})) < \frac{\gamma}{2}. \tag{12.7}$$

The inequality

$$d(m_{i_2}(b_{i_1}), m(b_{i_1})) \geq d(m_{i_2}(b_{i_1}), x_o) - d(m(b_{i_1}), x_o)$$

by (12.6) and (12.7) implies

$$d(m_{i_2}(b_{i_1}), x_o) < \gamma. \tag{12.8}$$

Since m_{i_2} is x_o-exhaustive we have by Lemma 12.1 that there exists a $\oplus$-orthogonal subsequence $\{b_i^2\}$ of $\{b_i\}_{i=i_1+1}^{\infty}$ such that

$$(\widetilde{m})_d^{x_o}\left(\bigoplus_{i\in I} b_i^2\right) < \frac{\gamma}{2}$$

for each $I \subset \mathbf{N}$. This implies by (12.8) and (ii)

$$\alpha_d^{x_o}\left(b_{i_1} \oplus \bigoplus_{i\in I} b_i^2, m_{i_2}\right) < \delta$$

for each $I \subset \mathbf{N}$.

Using (12.6) we obtain

$$d(m(b_{i_1} \oplus b_k^2), x_o) < \frac{\gamma}{2}$$

for each $k \in \mathbf{N}$ and by (iii) there exists an index i_3 such that

$$d(m_{i_3}(b_{i_1} \oplus b_k^2), m(b_{i_1} \oplus b_k^2)) < \frac{\gamma}{2}.$$

Hence by the inequality

$$d(m_{i_3}(b_{i_1} \oplus b_k^2), m(b_{i_1} \oplus b_k^2)) \geq d(m_{i_3}(b_{i_1} \oplus b_k^2), x_o) - d(m(b_{i_1} \oplus b_k^2), x_o)$$

we obtain

$$d(m_{i_3}(b_{i_1} \oplus b_{i_2}), x_o) < \gamma, \tag{12.9}$$

where b_{i_2} is chosen from the sequence $\{b_k^2\}$.

Since m_{i_3} is x_o-exhaustive by Lemma 12.1 there exists a $\oplus$-orthogonal subsequence $\{b_i^3\}$ of $\{b_i^2\}_{i=i_2+1}^{\infty}$ such that

$$(\widetilde{m}_{i_3})_d^{x_o}\left(\bigoplus_{i\in I} b_i^3\right) < \frac{\gamma}{2}$$

for each $I \subset \mathbf{N}$. This implies by (12.9) and (ii)

$$d\left(m_{i_3}\left(b_{i_1} \oplus b_{i_2} \oplus \bigoplus_{i\in I} b_i^3\right), x_o\right) < \delta$$

for each $I \subset \mathbf{N}$.

Continuing this procedure we obtain two sequences $\{m_{i_k}\}$ and $\{b_{i_k}\}$. Taking $b = \oplus_{k=1}^{\infty} b_{i_k}$, we choose by (iii) an index k_o such that

$$d(m_{i_{k_o}}(b), x_o) < \eta < \delta. \tag{12.10}$$

This follows from (12.6) and the inequality

$$d(m_{i_{k_o}}(b), m(b)) \geq d(m_{i_{k_o}}(b), x_o) - d(m(b), x_o).$$

Since by (DP_1)

$$b \ominus a \leq b$$

we obtain by the preceding procedure that

$$d(m_{i_{k_o}}(b \ominus a), x_o) < \delta.$$

This, together with (i) and (12.10) implies by (DP_2)

$$\epsilon > d(m_{i_{k_o}}(b \ominus (b \ominus b_{i_{k_o}})), x_o) = d(m_{i_{k_o}}(b_{i_{k_o}}), x_o).$$

Contradiction with (12.5).

The opposite statement follows from (iii). □

12.4 Notes and References

The difference poset (D-poset) was introduced in the papers [Kôpka 1992], [Kôpka and Chovanec 1994], [Dvurečenskij and Riečan 1994]. Traces of the notion of D-poset go back to BCK-algebras [Imai and Iseki 1966], [Iseki and Tanaka 1978] and to [Giuntini and Greuling 1989]. This structure is a generalization of quantum logic [Kalmbach 1983], [Pták and Pulmannová 1991], MV-algebras [Chang 1958] and [Mesiar 1994a], orthoalgebras [Foulis, Greeche and Rüttimann 1992], the set of all effects [Dvurečenskij 1995], and the collection of fuzzy sets [Kôpka 1992].

Set functions with values in sets endowed only with some topological structures without any algebraic operations: metric space, special topological spaces [Klimkin 1989], uniform spaces [Pap 1991a], [Pap 1991c].

The notion of quasi-σ-ring is introduced in [Constantinescu 1984].

The notion of boundedness in uniform space is by [Hejcman 1959]. For a set function m defined on a quasi-σ-ring Σ the previous definition of $\alpha_d^{x_o}$ coincides with that given in the paper of [Pap 1991a]. The general convergence theorem is by [de Lucia and Pap 1995] and the boundedness theorem is a new result of P. de Lucia and E. Pap.

Bibliography

P. Abraham [1992] Saeki's improvement of the Vitali-Hahn-Saks-Nikodym theorem holds precisely for Banach spaces having cotype, *Proc. Amer. Math. Soc.* **116**, 171–179.

L. V. Agafanova and V. M. Klimkin [1974] Nikodym theorem for triangular set functions (in Russian), Siberian Math. J. **15**, 669–674.

V. N. Aleksjuk [1968] Two theorems on the existence of quasibase for a family of quasimeasures (in Russian), *Izv. vish. ucseb. zav.* **6 (73)**, 11–18.

V. N. Aleksjuk [1970] On the weak compactness of a family of quasimeasures (In Russian), *Sibirian Mat. J.* **11**, 723–738.

B. Anger [1977] Representation of capacities, *Math. Ann.* **229**, 245–258.

B. Anger and J. Lembcke [1986] Infinitely subadditive capacities as upper envelopes of measures, *Z. Wahrsch. Theor. Verw. Gebiete* **68**, 403–414.

F. J. Anscombe and R. J. Aumann [1963] A Definition of Subjective Probability, *Ann. Math. Statist.* **34**, 199–205.

P. Antosik [1971] On the Mikusiński diagonal theorem, *Bull. Acad. Polon. Sci. Ser. Math. Astronom. Phys.* **19**, 305–310.

P. Antosik [1978] Sur les suites d'applications, *C. R. Acad. Sci. Paris,* **287 A**, 75–77.

P. Antosik and S. Saeki [1992] Diagonal Theorem for measuroids and Its Applications, preprint.

P. Antosik and C. Swartz [1985] *Matrix Methods in Analysis,* Springer Verlag Lecture Notes in Mathematics 1113, Heidelberg.

R. J. Aumann and L. S. Shapley [1974] *Values of Non-Atomic Games,* Princeton Univ. Press.

F. Baccelli, G. Cohen, G. J. Olsder and J. P. Quadrat [1992] *Synchronization and Linearity: an Algebra for Discrete Event Systems,* John Wiley and Sons, New York.

A. Basile and H. Weber [1986] Topological Boolean rings of first and second category. Separating points for countable families of measures, *Rad. Mat.* **2**, 113–125.

N. Batle and E. Trillas [1979] Entropy and fuzzy integral, *J. Math. Anal. Appl.* **69**, 469–474.

C. Baumont [1989] Theory of possibility as a basis for analyzing business process, *Fuzzy Sets and Systems* **31**, 1–12.

S. K. Berberian [1965] *Measure and Integration,* New York-London.

C. Berg and G. Forst [1975] *Potential theory on locally compact Abelian groups*, (Ergeb. Math. Grenzgeb. 87), Springer-Verlag, Berlin Heidelberg.

M. Berres [1988] λ-additive measures on measure spaces, *Fuzzy Sets and Systems* **27**, 159–169.

A. Billot [1992] From fuzzy set theory of non-additive probabilities: how have economists reacted?, *Fuzzy Sets and Systems* **49**, 75–89.

G. Birkhoff [1967] *Lattice Theory*, 3rd ed., Amer. Math. Soc. Colloq. Publ. vol. 25.

N. Bourbaki [1955] *Topologie générale*, Hermann, Paris.

C. Brook [1984] Decompositions of submeasures, *Canad. J. Math.* **36**, 577–590.

J. K. Brooks and R. S. Jewett [1970] On finitely additive vector measures, *Proc. Nat. Acad. Sci. U.S.A.* **67**, 1294–1298.

A. Buja [1984] Simultaneously Least Favorable Experiments, Part I: Upper Standard Functionals and Sufficiency, *Zeit. für Wahrsch. und verwandte Geb.* **65**, 367–384.

J. Burzyk, Cz. Kliś, Z. Lipecki [1984] On K-property of metric spaces, *Colloq. Math.* **49**, 33–39.

D. Butnariu and E. P. Klement [1992] *Triangular Norm Based measures and games with Fuzzy coalitions*, Kluwer Academic Publishers, Dordrecht.

D. Butnariu and E. Roventa [1992] A measure theoretical approach to the problem of computing production costs, *Fuzzy Sets and Systems* **48**, 305–321.

D. Candeloro [1987] Radon-Nikodym derivatives and conditioning in fuzzy measure theory, *Stochastica* **11**, 107–120.

L. Carleson [1967] *Selected Problems on Exceptional Sets*, van Nostrand, Toronto-London-Melbourne.

C. C. Chang [1958] Algebraic analysis of many valued logics, *Trans. Amer. Math. Soc.*, 467–490.

A. Chateauneuf [1987] Uncertainty aversion and risk aversion in models with non-additive probabilities, *in: Risk, Decision and Rationality, B. R. Munier (ed.)*, D. Reidel Publ. Comp., 615–627.

A. Chateauneuf [1991] On the use of capacities in modeling uncertainty aversion and risk aversion, *J. Math. Econom.* **20**, 343–369.

G. Choquet [1953-1954] Theory of capacities, *Ann. Inst. Fourier* (Grenoble) **5**, 131–292.

J. P. R. Christensen and W. Herer [1975] On the existence of pathological submeasures and the construction of exotic topological groups, *Math. Ann.* **213**, 203–210.

W. Congxin and H. Minghu [1993], On the null-additivity and the uniform autocontinuity of a fuzzy measure, *Fuzzy Sets and Systems* **58**, 243–245.

C. Constantinescu [1984] *Spaces of Measures*, Walter de Gruyter, Berlin-New York.

C. Constantinescu and K.Weber [1985] *Integration Theory, Vol. 1: Measure and Integration*, Wiley-Interscience, New York.

A. B. D'Andrea and P. de Lucia [1992] The Lebesgue decomposition theorem on an orthomodular lattice, *Supl. di Rend. Circ. Mat. Palermo* **Ser II – num. 28**, 379–386.

L. M. de Campos and M. Jorge [1992] Characterization and comparision of Sugeno and Choquet integrals, Fuzzy Sets and Systems **52**, 61–67.

L. M. de Campos, M.T.Lamata and M. J. Bolanos Carmona [1989] Representation of fuzzy measures through probabilities, *Fuzzy Sets and Systems* **31**, 23–36.

L. M. de Campos, M. T. Lamata and S. Moral [1990a] Distances between fuzzy measures through associated probabilities: some applications, *Fuzzy Sets and Systems* **35**, 57–68.

L. M. de Campos, M. T. Lamata and S. Moral [1990b] The concept of conditional fuzzy measure, *Internat. J. Intell. Systems* **5**, 237–246.

M. Delgado and S. Moral [1989] Upper and lower fuzzy measures, *Fuzzy Sets and Systems* **33**, 191–200.

C. Dellacherie [1972] *Capacité et processus stochastiques* (Ergeb. Math. Grenzgeb. 67), Springer-Verlag, Berlin-Heidelberg.

P. de Lucia and P. Morales [1992] A noncommutative version of a Theorem of Marczewski for submeasures, *Studia Math.* **101(2)**, 123–138.

P. de Lucia and P. Morales [1994] Noncommutative decomposition theorems in Riesz spaces, *Proc. Amer. Math. Soc.* **120**, 193–202.

P. de Lucia and E. Pap [1994a] Diagonal Theorems and Their Applications, preprint.

P. de Lucia and E. Pap [1994b] Lebesgue decomposition of null-σ-additive set functions by ideals, *Rend. Circ. Mat. Palermo* (to appear).

P. de Lucia and E.Pap [1995] Nikodym convergence theorem for uniform space valued functions defined on D-poset, *Math. Slovaca* **45**(to appear).

A. P. Dempster [1967] Upper and Lower Probabilities Induced by a Multivalued Mapping, *Ann. Math. Statistics* **38**, 325–339.

A. P. Dempster [1968] A Generalization of Bayesian Inference, *J. Roy. Statist. Soc. Ser. B* **30**, 205–247.

D. Denneberg [1994] *Non-Additive Measure and Integral*, Kluwer Academic Publishers, Dordrecht.

J. J. Diestel and J. J. Uhl [1977] *Vector measures*, Math. Surveys 15, Amer. Math. Soc. (Providence R. I.).

J. Dieudonné [1951] Sur la convergence des suites de measures de Radon, *An. Acad. Brasil. Ciênc.* **23**, 21–38, 277–281.

N. Dinculeanu [1967] *Vector Measures*, Pergamon Press, New York.

I. Dobrakov [1972] On subadditive operators on $C_0(T)$, *Bull. Acad. Polon. Sci. Ser. Math. Astronom. Phys.* **20**, 561–562.

I. Dobrakov [1974] On submeasures I, *Dissertationes Math.* **112**, Warszawa.

I. Dobrakov and J. Farkova [1980] On submeasures II, *Math. Slovaca* **30**, 65–81.

L. Drewnowski [1972a] Topological rings of sets, continuous set functions, integration I, II, III, *Bull. Acad. Polon. Sci. Ser. Math. Astronom. Phys.* **20**, 269–276, 277–286, 439–445.

L. Drewnowski [1972b] Equivalence of Brooks-Jewett, Vitali-Hahn-Saks and Nikodym Theorems, *Bull. Acad. Polon. Ser. Sci. Math. Astronom. Phys.* **20**, 725–731.

L. Drewnowski [1978] On the continuity of certain non-additive set functions, *Colloquium Math.* **38**, 243–253.

D. Dubois and H. Prade [1980] *Fuzzy Sets and Systems: Theory and Applications*, Academic Press, New York.

D. Dubois and H. Prade [1982] A class of fuzzy measures based on triangular norms, *Internat. J. Gen. System* **8**, 43–61.

D. Dubois and H. Prade [1985] Evidence Measures Based on Fuzzy Information, *Automatica - J. IFAC* **21**, 547–562.

D. Dubois and H. Prade [1986] A set-theoretic view of belief function, *Internat. J. Gen. Systems* **12**, 193–226.

D. Dubois and H. Prade [1988] *Possibility Theory*, Plenum Press, New York.

D. Dubois and H. Prade [1992] When upper probabilities are possibility measures, *Fuzzy Sets and Systems* **49**, 65–74.

N. Dunford and J. T. Schwartz [1958] *Linear operators, Part 1*, Interscience, New York.

A. Dvurečenskij [1991] Regular measures and completeness of inner product spaces, "Contributions to General Algebras", vol. 7, Hölder-Pichler-Tempski Verlag, 137–147.

A. Dvurečenskij [1993] *Gleason's Theorem and Applications*, Kluwer Academic Publishers, Dordrecht, Boston, London, Ister Science Press, Bratislava.

A. Dvurečenskij [1995] Tensor product of difference posets, *Trans. Amer. Math. Soc.* (to appear).

A. Dvurečenskij and S. Pulmannová [1994a] Difference posets, effects, and quantum measurements, *Internat. J. Theoret. Phys.* **33**, 819–850.

A. Dvurečenskij and S. Pulmannová [1994b] *D*-test spaces and difference posets, *Rep. Math. Phys.* **34**, 151–170.

A. Dvurečenskij and B. Riečan [1991] Fuzzy quantum models, *Internat. J. Gen. Systems* **20** , 39–54.

A. Dvurečenskij and B. Riečan [1994] Decompositions of measures on orthoalgebras and difference posets, *Internat. J. Theoret. Phys.* **33**, 1387–1402.

D. Ellsberg [1961] Risk, Ambiguity and the Savage Axioms, *Journal of Economics* **75**, 643–669.

A. M. Fadem [1977] *Economics of Space and Time: The Measure-Theoretic Foundations of Social Science*, Iowa State Univ. Press, Ames, Iowa.

K. Falconer [1990] *Fractal Geometry*, John Wiley and Sons, New York.

T. L. Fine [1973] *Theories of Probability: An Examination of Foundations*, Academic Press, New York.

P. C. Fishburn [1970] *Utility Theory for Decision Making*, John Wiley and Sons, New York.

D. J. Foulis, R. J. Greechie and G. T. Rüttimann [1992] Filters and supports in orthoalgebras, *Internat. J. Theoret. Phys.* **31**, 787–807.

F. J. Freniche [1984] The Vitali-Hahn-Saks theorem for Boolean algebras with the subsequential interpolation property, *Proc. Amer. Math. Soc.* **92**, 262–266.

F. S. Garcia and P. G. Alvarez [1986] Two families of fuzzy integrals, *Fuzzy Sets and Systems* **18**, 67–81.

I. Gilboa [1987] Expected Utility Theory with Purely Subjective Non-Additive Probabilities, *J. Math. Econom.* **16**, 635–88.

I. Gilboa [1989a] Expectation and Variation in Multi-Period Decisions, *Econometrica* **57**, 1153–1169.

I. Gilboa [1989b] Additivizations of Non-Additive Measures, *Math. Oper. Res.* **14**, 1–17.

I. Gilboa [1989c] Duality in Non-Additive Expected Utility Theory, *Ann. Oper. Res.* **19**, 405–414.

I. Gilboa and E. Lehrer [1991] Global Games, *Internat J. Game Theory* **20**, 129–147.

I. Gilboa and D. Schmeidler [1989] Maxmin Expected Utility with Non-Unique Prior, *J. Math. Econom.* **18**, 141–153.

I. Gilboa and D. Schmeidler [1991] Updating Ambigous Beliefs, *J. Econom. Theory* (to appear)

I. Gilboa and D. Schmeidler [1992a] Additive Representations of Non-Additive Measures and the Choquet Integral, Discussion Paper No. 985.

I. Gilboa and D. Schmeidler [1992b] Canonical Representation of Set Functions, Discussion Paper No. 986.

R. Giuntini and H. Greuling [1989] Toward a Formal Language for Unsharp Properties, *Found. Physics* **19**, 931–945.

J. S. Golan [1992] *The Theory of Semirings with Applications in Mathematics and Theoretical Computer Sciences*, Pitman Monographs and Surveys in Pure and Appl. Maths. 54, Longman, New York.

M. Grabish, T. Murofushi and M. Sugeno [1992] Fuzzy measure of fuzzy events defined by fuzzy integrals, *Fuzzy Sets and Systems* **50**, 293–313.

J. W. Guan and D. A. Bell [1991] *Evidence and Its Applications*, Vol. 1, North-Holland, New York.

E. Guariglia [1990a] On Dieudonne's Boundedness Theorem, *J. Math. Anal. Appl.* **145**, 447–454.

E. Guariglia [1990b] Uniform boundedness theorems for k-triangular set functions, *Acta Sci. Math. (Szeged)* **54**, 391–407.

E. Guariglia [1991] k-triangular functions on an orthomodular lattice and the Brooks-Jewett Theorem, *Rad. Mat.* **Vol. 7**, 241–251.

M. M. Gupta and E. Sanchez [1982] *Fuzzy Information and Decision Processes*, North-Holland, New York.

N. S. Guselnikov [1975] On the extension of quasi-Lipshitz set functions (in Russian), *Mat. zametki* **17**, 21–31.

M. A. S. Guth [1988] Uncertainty analysis of rule-based expert systems with Dempster-Shafer mass assignment, *Internat. J. Intell. Systems* **3**, 123–139.

E. D. Habil [1993] Orthoalgebras and Noncommutative Measure Theory, Ph. D. Thesis, Manhattan, Kansas.

I. Hacking [1975] *The Emergence of Probability*, Cambridge University Press, Cambridge.

L. O. Hall and A. Kandel [1986] *Designing Fuzzy Expert Systems*, Verlag TÜV Rheiland, Köln.

P. R. Halmos [1950] *Measure Theory*, Springer-Verlag, New York.

R. Haydon [1981] A non reflexive Grothendick space that does not contain l_∞, *Israel J. Math* **40**, 65–73.

W. K. Hayman and P. B. Kennedy [1976] *Subharmonic Functions I*, Academic Press, London-New York-San Francisco.

H. Hejcman [1959] Boundedness in uniform spaces and topological groups, *Czehoslovak Math. J.* **9**, 544–563.

W. Hildenbrand [1982] Core of an economy, *Handbook of mathematical economies* Vol. 2 (K. J. Arrow and M. D. Inriligator, eds.), North-Holland.

J. Horvath [1966] *Topological Vector Spaces and Distributions*, Addison-Wesley, Reading, Massaschusets.

U. Höhle [1982] A general theory of fuzzy plausability measures, *J. Math. Anal. Appl.* **127**, 346–364.

U. Höhle [1984] Fuzzy probability measures, *In: Zimmermann, Zadeh and Gaines, eds., Fuzzy Sets and Decision Analysis*, North-Holland.

W. Hua [1988] The properties of some non-additive measures, *Fuzzy Sets and Systems* **27**, 373–377.

P. J. Huber [1973] The use of Choquet Capacities in Statistiqs, *Bull. Internat. Inst. Statistics* **45**, 181–191.

P. J. Huber and V. Strassen [1973] Minimax Tests and the Neyman-Pearson Lemma for Capacities, *Ann. Statist.* **1**, 251–263.

H. Ichihashi, M. Tanaka, K. Asai [1985] An application of the fuzzy integral to multi-attribute decision problem, *Proc. First IFSA Congress*, Palma De Mallorca.

H. Ichihashi, M. Tanaka, K. Asai [1988] Fuzzy Integrals Based on Pseudo-Additions and Multiplications, *J. Math. Anal. Appl.* **130**, 354–364.

Y. Imai and K. Iseki [1966] On axiom systems of propositional calculi XIV, *Proc. Japan Acad.* **42**, 19–22.

K. Ishi and M. Sugeno [1985] A model of human evaluation process using fuzzy measure, *Internat. J. Man-Machine Studies* **22**, 19–38.

K. Iseki and S. Tanaka [1978] An introduction to the theory of BCK-algebras, *Math. Japon.* **23**, 1–26.

J.-Y. Jaffray [1989] Linear Utility Theory for Belief Functions, *Oper. Res. Lett.* **8**, 107–112.

D. Kahnemann and A. Tversky [1979] Prospect Theory: An Analysis of Decision Under Risk, *Econometrica* **47**, 263–291.

G. Kalmbach [1983] *Orthomodular Lattices*, Academic Press, London, New York.

G. Kalmbach and Z. Riečanová [1994] An axiomatization for abelian relative inverses, *Demonstr. Math.* (to appear).

N. J. Kalton [1989] The Maharam problem, *Publ. Math. Univ. Pierre Marie Curie* **94**.

N. J. Kalton and J. W. Roberts [1989] Uniformly exhaustive submeasures and nearly additive set functions, *Trans. Amer. Math. Soc.* **278**, 803–816.

N. J. Kalton and S. J. Montgomery-Smith [1993] Set-functions and factorization, *Arch. Math.* **61**, 183–200.

A. Kandel [1982] *Fuzzy Techniques in Pattern Recognition*, John Wiley, New York.

J. Keller, H. Qiu and H. Tahani [1986] Fuzzy integral and image segmentation, *Proc. North American Fuzzy Information Processing Soc.*, New Orleans, 324–388.

R. Kennes [1992] Computational aspects of the Möbius transformation of graphs, *IEEE Trans. Systems Man Cybernet.* **22**, 201–223.

A. P. Kirman and D. Sondermann [1972] Arrow's Theorem. Many Agents, and Invisible Dictators, *J. Econom. Theory* **5**, 267–277.

E. P. Klement and D. Ralescu [1983] Nonlinearity of the fuzzy integral, *Fuzzy Sets and Systems* **11**, 309–315.

E. P. Klement and S. Weber [1991] Generalized measures, *Fuzzy Sets and Systems* **40**, 375–394.

V. M. Klimkin [1989] Uniform boundedness of a family of nonadditive set functions (Russian), *Mat. Sb.* **180**, no. 3, 385–396.

G. J. Klir and T. A. Folger [1988] *Fuzzy Sets, Uncertainty, and Information*, Prentice Hall, Englewood Ciffs, New Jersey.

C. Klis [1978] An example of noncomplete normed (K)-space, *Bull. Acad. Polon. Sci. Ser. Math. Astronom. Phys.* **26**, 414–420.

J. Kohlas [1988] Conditional belief structures, *Probab. in Engineering Information Science* **2**, 415–433.

F. Kôpka [1992] D-posets of fuzzy sets, *Tatra Mount. Math. Publ.* **1**, 83–87.

F. Kôpka and F. Chovanec [1994] D-posets, *Math. Slovaca* **44**, 21–34.

R. Kruse [1982] On the construction of fuzzy measure, *Fuzzy Sets and Systems* **8**, 323–327.

R. Kruse and K. D. Meyer [1987] *Statistics with Vague data*, Reidel, Dordrecht.

W. Kuich [1986] *Semirings, Automata, Languages*, Springer-Verlag, Berlin.

N. S. Landkopf [1966] *Foundations of modern potential theory* (in Russian), Nauka.

R. Li and C. Swartz [1992] Space for which the Uniform Boundedness Principle Holds, *Studia Sci. Math. Hungar.* **27**, 379–384.

C. M. Ling [1965] Representation of associative functions, *Publ. Math. Debrecen* **12**, 189–212.

X. Liu [1993] Hahn decomposition theorem for infinite signed fuzzy measure, *Fuzzy Sets and Systems* **57**, 377–380.

X. Liu and G. Zhang [1994] Lattice-valued fuzzy measure and lattice-valued fuzzy integral, *Fuzzy Sets and Systems* **62**, 319–332.

A. A. Lyapunov [1940] Sur les fonctions-vecteurs complètement additives, *Bull. Acad. Sci. URSS Math.* **4**, 465–478.

M. Machina [1987] Choice Under Uncertainty: Problems Solved and Unsolved, *Econom. Persp.* **1**, 121–154.

D. Maharam [1947] An algebraic characterization of measure algebras, *Ann. of Math.* **2(48)**, 154–167.

V. P. Maslov [1987] *Asymptotical Methods for Solving Pseudo-Differential Equations*, (in Russian), Moscow.

P. A. Meyer [1966] *Probability and Potentials*, Massachusetts-Toronto-London.

R. Mesiar [1992] Fuzzy sets and probability theory, *Tatra Mount. Math. Publ.* **1**, 105–123.

R. Mesiar [1994a] Fuzzy Difference Posets and MV-algebras, in Proc. IPMU'94 (eds. B. Bouchon-Meunière and R. R. Yager), Paris, 202–206.

R. Mesiar [1994b] Differences on $[0,1]$, *Tatra Mount. Math. Publ.* **6** (to appear).

R. Mesiar [1995] Choquet-like integrals, *J. Math. Anal. Appl.*, (to appear).

R. Mesiar and E. Pap [1992] On additivity and pseudo-additivity of a pseudo-additive measure based integral, preprint.

R. Mesiar and J. Šipoš [1993] Radon-Nikodym-like theorem for fuzzy measures, *Fuzzy Math.* **1**, 873–878.

R. Mesiar and J. Šipoš [1993b] A theory of fuzzy measures: Integration and its additivity, Internat. J. Gen. Systems **23**.

J. Mikusiński [1970] A theorem on vector matrices and its applications in measure theory and functional analysis, *Bull. Acad. Polon. Sci. Ser. Math.* **18**, 193–196.

A. Moltó [1981] On the Vitali-Hahn-Saks theorem, *Proc. Roy. Soc. Edinburg Sect.* **A 90**.

T. Murofushi and M. Sugeno [1989] An interpretation of fuzzy measures and the Choquet integral as an integral with respect to a fuzzy measure, *Fuzzy Sets and Systems* **29**, 201–227.

T. Murofushi and M. Sugeno [1991a] A theory of fuzzy measures. Representation, the Choquet integral and null sets, *J. Math. Anal. Appl.* **159**, 532–549.

T. Murofushi and M. Sugeno [1991b] Fuzzy t-conorm integral with respect to fuzzy measures: Generalization of Sugeno integral and Choquet integral, *Fuzzy Sets and Systems* **42**, 57–71.

T. Murofushi and M. Sugeno [1992] Non-additivity of Fuzzy Measures Representing Preferential Dependence, *Proc. of 2nd International Conference on Fuzzy Logic and Neural Networks (Iizuka, Japan, July 17-22, 1992)*, 617–620.

T. Murofushi and M. Sugeno [1993] Some quantities represented by the Choquet integral, *Fuzzy Sets and Systems* **56**, 229–235.

T. Murofushi and M. Sugeno [1994] Non-monotonic fuzzy measures and the Choquet integral, *Fuzzy Sets and Systems* **64**, 73–86.

J. Musielak [1987] Approximation of elements of a generalized Orlicz sequence space by nonlinear, singular kernels, *J. Approx. Theory* **50**, 366–372.

J. Musielak [1991] Approximation by nonlinear singular integral operators in generalized Orlicz spaces, *Comm. Math.* **31**, 79–88.

J. Musielak [1993] Nonlinear approximation in some modular function spaces I, *Math. Japonica* **38** 83–90.

Navara and P. Pták [1994] Difference posets and orthoalgebras (submitted).

C. V. Negoita and D. A. Ralescu [1987] *Simulation, Knowledge-Based Computing, and Fuzzy Statistics*, Van Nostrand, New York.

G. L. O'Brien and W.Vervaat [1994] How subadditive are subadditive capacities?, *Comment. Math. Univ. Carolinae* **35**, 311–324.

T. Onisawa, M. Sugeno, Y. Nishiwaki, H. Kawai and Y. Harima [1986] Fuzzy measure analysis of public attitude towards the use of nuclear energy, *Fuzzy sets and Systems* **20**, 259–289.

W. Orlicz [1968] On spaces $L^{*\phi}$ based on the notion of a finitely additive integral, *Comm. Math.* **12**, 99–113.

E. Pap [1974] A generalization of the Diagonal theorem on a block-matrix, *Mat. vesnik* **11(26)**, 66–71.

E. Pap [1976] Uniform boundedness of a family of exhaustive set functions, *Mat. vesnik* **13 (28)**, 319–326.

E. Pap [1982] *Functional Analysis (Sequential convergences. Some principles of functional analysis)* (with English Summary), Novi Sad.

E. Pap [1985a] Functional analysis with K-convergence, *Proceedings of the Conference on Convergence, Běchyne, Czechoslovakia 1984*, Akademie-Verlag, Berlin, 245–250.

E. Pap [1985b] The adjoint operator and K-convergence, *Univ. u Novom Sadu Zb. Rad. Prirod. — Mat. Fak. Ser. Mat.* **15, 2**, 51–56.

E. Pap [1986] A generalization of a theorem Dieudonné for k-triangular set functions, *Acta Sci. Math. (Szeged)* **50**, 159–167.

E. Pap [1987] The Vitali-Hahn-Saks theorems for k-triangular set functions, *Atti Sem. Mat. Fis. Univ. Modena* **26**, 21–32.

E. Pap [1990a] Lebesgue and Saks decompositions of $\perp$-decomposable measures, *Fuzzy Sets and Systems* **38**, 345–353.

E. Pap [1990b] Extension of the continuous t-conorm decomposable measure, *Univ. u Novom Sadu Zb. Rad. Prirod. — Mat.Fak. Ser. Mat.* **20, 2**, 121–130.

E. Pap [1990c] Regular Borel t-conorm decomposable measures, *Univ. u Novom Sadu Zb. Rad. Prirod. — Mat. Fak. Ser. Mat.* **20, 2**, 113–120.

E. Pap [1990d] Integral generated by decomposable measure, *Univ. u Novom Sadu Zb. Rad. Prirod. — Mat. Fak. Ser. Mat.* **20, 1**, 135–144.

E. Pap [1991a] On non-additive set functions, *Atti. Sem. Mat. Fis. Univ. Modena* **39**, 345–360.

E. Pap [1991b] Many dimensional Lyapunov's convexity theorem for E_λ-decomposable measure, *Univ. u Novom Sadu Zb. Rad. Prirod. — Mat. Fak. Ser. Mat.* **21, 1**, 65–74.

E. Pap [1991c] The Brooks-Jewett theorem for non-additive set functions, *Univ. u Novom Sadu Zb. Rad. Prirod. — Mat.Fak.Ser.Mat.* **21, 1**, 75–82.

E. Pap [1992] Decomposable measures and applications on nonlinear partial differential equations, *Rend. Circ. Mat. Palermo* **ser II–num 28**, 387–403.

E. Pap [1993a] Regular null-additive monotone set functions, preprint; (*Univ. u Novom Sadu Zb. Rad. Prirod. — Mat. Fak. Ser. Mat.* (to appear).)

E. Pap [1993b] Hewitt-Yosida decomposition for $\Box$-decomposable measures, *Tatra Mount. Math. Publ.* **3**, 147–154.

E. Pap [1993c] g-calculus, *Univ. u N. Sadu, Zb. Rad. Prirod. — Mat.* **23, 1**, 145–156.

E. Pap [1994a] The range of null-additive fuzzy and non-fuzzy measures, *Fuzzy Sets and Systems* **65**, 105–115.

E. Pap [1994b] Decompositions of supermodular functions and □-decomposable measures, *Fuzzy Sets and Systems* **65**, 71–83.

E. Pap and N. Ralević [1992] Applications of g-calculus on nonlinear differential equations, SYM-OP-IS'92, Belgrade, 25–28.

E. Pap and C. Swartz [1990a] The closed graph theorem for locally convex spaces, *Boll. Un. Mat. Ital.* **(7)4-B**, 109–111.

E. Pap and C. Swartz [1990] A locally convex version of Adjoint Theorem, preprint.

E. Pap and C. Swartz, [1990c] On the Closed Graph Theorem, *Proc. of the Conference on Generalised Functions and Convergence, Katowice 1988*, World Sci. Publ, 355–360.

C. J. Preston [1971] A theory of capacities and its applications to some convergence results, *Adv. Math.* **6, 1**, 78–106.

P. Pták and S. Pulmannová [1991] *Orthomodular Structures as Quantum Logic*, Kluwer Academic Publishers, Dordrecht.

M. L. Puri and D. Ralescu [1982] A possibility measure is not a fuzzy measure, *Fuzzy Sets and Systems* **7**, 311–313.

Z. Qiao [1989] On the extension of possibility measures, *Fuzzy Sets and Systems* **32**, 3 15–320.

Z. Qiao [1990] On fuzzy measure and fuzzy integral on fuzzy set, *Fuzzy Sets and Systems* **37**, 77–92.

Z. Qiao [1991] Fuzzy integrals on L-fuzzy sets, *Fuzzy Sets and Systems* **38**, 61–79.

D. Ralescu [1993] Fuzzy Information in Statistical Analysis, *Fifth IFSA World Congress*, 748–752.

D. Ralescu and G. Adams [1980] The fuzzy integral, *J. Math. Anal. Appl.* **75**, 562–570.

C. Randall and D. Foulis [1979] New definitions and theorems, University of Massachusetts Mimeographed Notes, Amherst, Massachusetts.

M. M. Rao [1987] *Measures Theory and Integration*, Academic Press.

K. P. S. B. Rao and M. B. Rao [1983] *Theory of Charges*, Academic Press, London.

Z. Riečanová [1982] About σ-additive and σ-maxitive measures, *Math. Slovaca* **32**, 389–395.

Z. Riečanová and D. Bršel [1994] Contraexamples in difference posets and orthoalgebras, *Internat. J. Theoret. Phys.* **33**, 133–141.

J. Rosenmuller [1971] On Core and Value, *Methods Oper. Res.* **9**, 84–104.

J. Rosenmuller [1972] Some Properties of Convex Set Functions, Part II, *Methods Oper. Res.* **17**, 287–307.

S. Saeki [1992] The Vitali-Hahn-Saks Theorem and measuroids, *Proc. Amer. Math. Soc.* **114**, 775–782.

L. J. Savage [1954] *The Foundations of Statistics*, Wiley, New York.

A. N. Sаženkov [1982] Uniform boundedness principle for topological measures (in Russian), *Mat. Zametki* **31**, 263–267.

D. Schmeidler [1986] Integral Representation without Additivity, *Proc. Amer. Math. Soc.* **97**, 253–261.

D. Schmeidler [1989] Subjective Probability and Expected Utility without Additivity, *Econometrica* **57**, 571–587.

B. Schweizer, A. Sklar [1963] Associative functions and abstract semigroups, *Publ. Math. Debrecen* **10**, 69–81.

G. Shafer [1976] *A Mathematical Theory of Evidence*, Princeton University Press, Princeton, New Jersey.

G. Shafer [1978] Non-additive probabilities in the works of Bernoulli and Lambert, *Arch. Hist. Exact Sci.* **19**, 309–370.

G. Shafer [1990] Perspectives on the theory and practice of belief functions, *Inter. J. Approx. Reas.* **4**, 323–362.

L. S. Shapley [1965] Cores of Convex Games, *Internat. J. Game Theory* **1**, 12–26.

N. Schilkret [1971] Maxitive measure and integration, *Indag. Math.* **33**, 109–116.

J. R. Sims and Z. Wang [1990] Fuzzy measures and fuzzy integrals: An overview, *Internat. J. Gen. Systems* **17**, 157–189.

P. Smets [1981] The Degree of Belief in A Fuzzy Event, *Inform. Sci.* **25**, 1.

P. Smets [1990] The combination of evidence in the transferable belief model, *IEEE Trans. Pattern Anal. Machine Intell.* **12**, 447–458.

M. Squillante and A. G. S. Ventre [1989] Representations of the fuzzy integral, *Fuzzy Sets and Systems* **29**, 165–170.

T. M. Strat [1990] Decision analysis using belief functions, *Internat. J. Approx. Reas.* **4**, 391–417.

M. Sugeno [1974] *Theory of fuzzy integrals and its applications*, Ph. D. Thesis, Tokyo Institute of Technology.

M. Sugeno [1977] Fuzzy measures and fuzzy integrals: A survey, *in Ed. M. M. Gupta, G. N. Saridis and B. R. Gaines, Fuzzy Automata and Decision Processes*, North-Holland, New York, 89–102.

M. Sugeno and T. Murofushi [1987] Pseudo-additive measures and integrals, *J. Math. Anal. Appl.* **122**, 197–222.

Q. Sun [1992] On the pseudo-autocontinuity of fuzzy measures, *Fuzzy Sets and Systems* **45**, 59–68.

Q. Sun [1994] Property (S) of fuzzy measure and Riesz's theorem, *Fuzzy Sets and Systems* **62**, 117–119.

H. Suzuki [1988] On fuzzy measures defined by fuzzy integrals, *J. Math. Anal. Appl.* **132**, 87–101.

H. Suzuki [1991] Atoms of fuzzy measures and fuzzy integrals, *Fuzzy Sets and Systems* **41**, 329–342.

C. Swartz [1987] The closed graph theorem without category, *Bull. Austral. Math. Soc.* **36**, 283–288.

C. Swartz [1989a] The uniform boundedness principle for order bounded operators, Internat. *J. Math. and Math. Sci.* **12, 3**, 487–492.

C. Swartz [1989b] A generalization of Mackey's Theorem, and the Uniform Boundedness Principle, *Bull. Austral. Math. Soc.* **49**, 123–128.

C. Swartz [1992] *Introduction to Functional Analysis*, Marcel Dekker, Inc., New York.

J. Šipoš [1979a] Integral with respect to a pre-measure, *Math. Slovaca* **29**, 141–155.

J. Šipoš [1979b] Non linear integrals, *Math. Slovaca* **29**, 257–270.

J. Šipoš [1979c] Integral representations of non-linear functionals, *Math. Slovaca* **29**, 333–345.

H. Tahani and J. M. Keller [1990] Information fusion in computer vision using the fuzzy integral, *IEEE Trans. Systems Man Cybernet.* **20**, 733–741.

M. Talagrand [1980] A simple example of a pathological submeasure, *Math. Ann.* **252**, 97–102.

M. Takahashi [1966] On topological-additive-group-valued measures, *Proc. Japan Acad.* **42**, 330–334.

H. Tanaka and I. Hayasi [1989] Possibilistic linear regression analysis for fuzzy data, *European J. Oper. Res.* **40**, 389–396.

H. Tanaka and M. Sugeno [1991] A study on subjective evaluations of printed color images, *Internat. J. Approx. Reas.* **5**, 213–222.

T. Terano and M. Sugeno [1977] Macroscopic optimization using conditional fuzzy measures, *in Ed. M. M. Gupta, G. N. Saridis and B. R. Gaines, Fuzzy Automata and Decision Processes*, North-Holland, New York, 197–208.

F. Topsœ [1974] On Construction of Measures, *Kobenhavns Universitet, Mat. Inst. Preprint Series* **27**.

D. A. Vladimirov [1969] *Boolean Algebra* (in Russian), Moskva.

J. von Neumann and O. Morgenstern [1944] *Theory of Games and Economic Behavior*, Princeton University Press, Princeton, New Jersey.

P. Wakker [1989] Continuous Subjective Expected Utility with Non-Additive Probabilities, *J. Math. Econom.* **18**, 1–27.

P. Wakker [1990] A behavioral foundation for fuzzy measures, *Fuzzy Sets and Systems* **37**, 327–350.

P. Walley [1987] Belief Function Representation of Statistical Evidence, *Ann. Statist.* **15**, 1439–1465.

P. Walley [1991] *Statistical Reasoning with Imprecise Probabilities*, Chapman and Hall, London.

P. Walley and T. Fine [1979] Varieties of model (cassificatory) and comparative probability, *Synthese* **41**, 321–374.

P. Walley and T. Fine [1982] Towards a Frequentist Theory of Upper and Lower Probabilities, *Ann. Statist.* **10**, 741–761.

X. Wang and M. Ha [1990] Pan-fuzzy integral, *BUSEFAL* **43**, 37–41.

Z. Wang [1984] The Autocontinuity of Set Function and the Fuzzy Integral, *J. Math. Anal. Appl.* **99**, 195–218.

Z. Wang [1985] Asymptotic structural characteristics of fuzzy measure and their applications, *Fuzzy Sets and Systems* **16**, 277–290.

Z. Wang [1990a] Absolute continuity and extension of fuzzy measures, *Fuzzy Sets and Systems* **36**, 395–399.

Z. Wang [1990b] Structural characteristics of fuzzy measures on S-compact spaces, *Internat. J. Gen. Systems* **17**, 309–316.

Z. Wang [1992] On the null-additivity and the autocontinuity of a fuzzy measure, *Fuzzy Sets and Systems* **45**, 223–226.

Z. Wang and G. Klir [1992] *Fuzzy Measure Theory*, Plenum Press, New York and London.

Z. Wang and L. Qiao [1990] Transformation theorems for fuzzy integrals on fuzzy sets, *Fuzzy Sets and Systems* **34**, 355–364.

L. A. Wasserman [1990] Belief Functions and Statistical Inference, *Canad. J. Statist.* **18**, 183–196.

L. A. Wasserman and J. Kadane [1990] Bayes's Theorem for Choqet Capacities, *Ann. Stat.* (to appear).

H. Weber [1976] Fortsetzung von Massen mit Werten in uniformen Halbgruppen, *Arch. Math.* **27**, 412–423.

H. Weber [1984] Topological Boolean rings. Decomposition of finitely additive set functions, *Pacific J. Math.* **110**, 471–495.

H. Weber [1986] Compactness in spaces of group valued contents, the Vitali-Hahn-Saks theorem and Nikodym's boundedness theorem, *Rocky Mountain J. Math.* **16**, 253–275.

S. Weber [1984] $\perp$-decomposable measures and integrals for Archimedean t-conorm, *J. Math. Anal. Appl.* **101**, 114–138.

S. Weber [1986] Two integrals and some modified version — critical remarks, *Fuzzy Sets and Systems* **20**, 97–105.

S. Weber [1991] Conditional measures and their applications to fuzzy sets, *Fuzzy Sets and Systems* **42**, 73–85.

R. Witula [1990] Continuity and the Darboux property of nonatomic finitely additive measures, *Proc. of the Conference Generalized Functions and Convergence, Katowice 1988*, World Scientific, 377–386.

S. T. Wierzchen [1983] An algorithm for identification of fuzzy measure, *Fuzzy Sets and Systems* **9**, 69–78.

S. K. M. Wong, Y. Y. Yao and P. Bollman [1992] Characterization of comparative belief structures, *Internat. J. Man-Machine Studies* **37**, 123–133.

C. Wu and M. Ma [1989] Some properties of fuzzy integrable function space $L^1(\mu)$, *Fuzzy Sets and Systems* **31**, 397–400.

B. Yan [1993] Semiconormed possibility integrals for application oriented modelling, *Fuzzy Sets and Systems* **57**, 239–248.

B. Yan [1994] Seminormed possibility integrals for application oriented modelling, *Fuzzy Sets and Systems* **61**, 189–198.

B. Yan and J. M. Keller [1991] Conditional fuzzy measures and image segmentation, *Proc. North American Fuzzy Information Processing Society*, Columbia, Missouri, 32–36.

J. Yen [1989] GERTIS: A Dempster-Shafer approach to diagnosing hierarchial hypothesis, *ACM Communications* **32**, 573–585.

J. Yen [1990] Generalising the Dempster-Shafer theory of fuzzy sets, *IEEE Trans. Systems Man Cybernet.* **20**, 559–570.

K. R. Yoo [1991] The Iterative Law of Expectation and Non-Additive Probability Measure, *Econom. Lett.* **37**, 145–149.

L. A. Zadeh [1968] Probability measures of fuzzy events, *J. Math. Anal. Appl.* **23**, 421–427.

L. A. Zadeh [1978] Fuzzy sets as a basis for a theory of possibility, *Fuzzy Sets and Systems* 1, 3–28.

H. J. Zimmermann [1987] *Fuzzy Sets, Decision Making, and Expert Systems*, Kluwer Academic Publishers, Boston.

Index

GPSR Compliance

The European Union's (EU) General Product Safety Regulation (GPSR) is a set of rules that requires consumer products to be safe and our obligations to ensure this.

If you have any concerns about our products, you can contact us on ProductSafety@springernature.com

In case Publisher is established outside the EU, the EU authorized representative is:

Springer Nature Customer Service Center GmbH
Europaplatz 3
69115 Heidelberg, Germany

Batch number: 10583998

Printed by Printforce, the Netherlands